Univalent Functions in Quantum Probability Theory

Mathematical
Surveys
and
Monographs
Volume 294

Univalent Functions in Quantum Probability Theory

Sebastian Schleissinger

AMERICAN MATHEMATICAL SOCIETY
Providence, Rhode Island

2020 *Mathematics Subject Classification*. Primary 30C55, 46L53, 60G51;
Secondary 30C35, 30C45, 60E07, 60E10.

Library of Congress Cataloging-in-Publication Data

Cataloging-in-Publication Data has been applied for by the AMS.
See http://www.loc.gov/publish/cip/.

∞ The paper used in this book is acid-free and falls within the guidelines
established to ensure permanence and durability.
Visit the AMS home page at `https://www.ams.org/`

10 9 8 7 6 5 4 3 2 1   30 29 28 27 26 25

Contents

Notation

2^M	power set of the set M
$\mathbb{N}$	$= \{1, 2, \dots\}$, set of natural numbers
$\mathbb{N}_0$	$= \{0, 1, 2, \dots\}$, set of non-negative integers
$\mathbb{R}$	set of real numbers
$\mathbb{C}$	set of complex numbers
$\hat{\mathbb{C}}$	the Riemann sphere $\mathbb{C} \cup \{\infty\}$
$\mathbb{D}$	$= \{z \in \mathbb{C} \mid \lvert z\rvert < 1\}$, unit disk
$\mathbb{H}$	$= \{z \in \mathbb{C} \mid \mathrm{Im}(z) > 0\}$, upper half-plane
$\mathbb{H}^-$	$= \{z \in \mathbb{C} \mid \mathrm{Im}(z) < 0\}$, lower half-plane
$\mathbb{P}$	probability measure
$\mathbb{E}[X]$	classical expectation
$\varphi(X)$	quantum expectation
$\mathcal{B}(M)$	Borel σ-algebra of a metric space M
$\mathcal{P}(\mathbb{R})$	set of all Borel probability measures on $\mathbb{R}$
$\mathcal{P}_c(\mathbb{R})$	set of all $\mu \in \mathcal{P}(\mathbb{R})$ with compact support
$\mathcal{P}^0(\mathbb{R})$	set of all $\mu \in \mathcal{P}(\mathbb{R})$ with finite variance and mean zero
$\mathcal{P}$	the set of all holomorphic $p : \mathbb{D} \to \mathbb{C}$ with $\mathrm{Re}(p) > 0$ and $p(0) = 1$
φ_μ, φ_X	characteristic function
$\mathcal{N}(\mu, \sigma^2)$	normal distribution with mean μ and variance σ^2
$Arc(\mu, \sigma^2)$	arcsine distribution with mean μ and variance σ^2
$W(\mu, \sigma^2)$	semicircle distribution with mean μ and variance σ^2
G_μ	Cauchy transform
F_μ	F-transform
R_μ	R-transform
B_μ	Boolean transform
$\mathcal{H}_\mu$	Hilbert transform

$*$	classical (tensor) convolution
$\rhd$	monotone convolution / comb product
$\lhd$	anti-monotone convolution
$\boxplus$	free convolution
$\uplus$	Boolean convolution
$a^{\star n}$	the n-th power of a with respect to the operation $\star$, $a^{\star n} = a \star \cdots \star a$
$\mathbf{1}_A(x)$	the function $x \mapsto 1$ if $x \in A$, $x \mapsto 0$ if $x \notin A$
"a.s."	almost surely
"a.e."	almost every
"*iid*"	independent and identically distributed
"variance σ^2"	always implies that σ is the standard deviation, i.e., $\sigma \geq 0$

CHAPTER 1

Introduction

This book treats relations between the following three subjects:

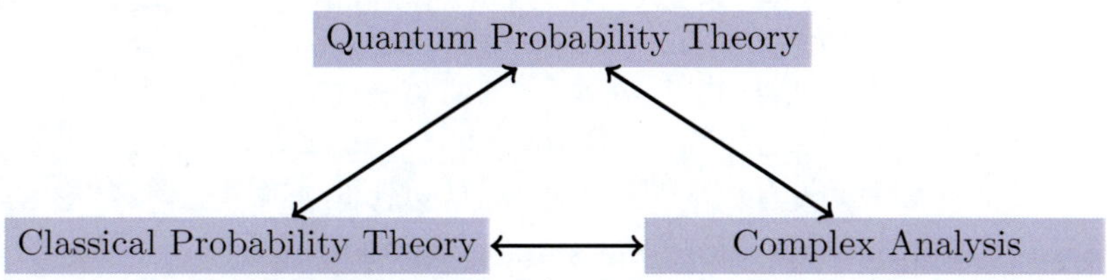

The complex analyst might enjoy how holomorphic functions, in particular conformal mappings, appear as actors on the stage of quantum random variables, while the probabilist might discover that quantum probability offers a powerful framework that enriches the classical theory and is far from being used only in quantum mechanics.

Prerequisites for this book include a solid foundation in real and complex analysis, classical probability theory, and the theory of Hilbert spaces. It is intended for graduate students who have a working knowledge of these areas.

1.1. Quantum probability spaces

We look for a mathematical model that is able to explain a number of observations

$$x_1, \ldots, x_N,$$

which are real-number outputs of a black box function of reality. We might think of repeated observations of a physical experiment. A mathematical model of randomness should give us a probability distribution μ on $\mathbb{R}$, which we can use to get deeper insight into an underlying phenomenon or to make predictions for future observations.

Classical probability theory. In the 1930s, A. Kolmogorov developed the definition of a probability space as a triple

$$(\Omega, \mathcal{F}, \mathbb{P}),$$

where Ω is a sample space, $\mathcal{F}$ is a set of events, which is a σ-algebra consisting of subsets of Ω, and $\mathbb{P}$ is a function from $\mathcal{F}$ to the interval $[0, 1]$, yielding probabilities of events. We often think of Ω as a mysterious, large set of outcomes working in the background.

In order to model the real numbers $x_1, \ldots, x_N$, we would use a random variable on the probability space, which is simply a measurable function $X : \Omega \to \mathbb{R}$. Such a quantity serves as a coordinate that projects Ω into $\mathbb{R}$ and it induces a probability measure μ on $\mathbb{R}$ by

$$\mu(A) = \mathbb{P}[X \in A], \quad \text{for any Borel subset } A \subseteq \mathbb{R}.$$

Quantum mechanics. Assume that $x_1, \ldots, x_N$ are measurements of some property of a particle, e.g., the position or momentum of an electron. In quantum mechanics, we also have a mathematical model for these measurements that yields a probability distribution on $\mathbb{R}$ rather than a single-point prediction. Let H be a Hilbert space with inner product $\langle \cdot, \cdot \rangle$ (we assume it is linear in the second argument) and let $X : H \to H$ be a (possibly unbounded) self-adjoint operator. Furthermore, let $\Psi \in H$ be a unit vector representing the state of the quantum system. The probability distribution μ for the observable X with respect to the quantum state Ψ is given by

$$\mu(A) = \langle \Psi, E_X(A)\Psi \rangle \quad \text{for any Borel subset } A \subset \mathbb{R},$$

where E_X is the projection-valued measure associated with X.

Both models produce a distribution μ for the observed values $x_1, \ldots, x_N$. The differences between these models become clear when we want to incorporate measurements $y_1, \ldots, y_N$ of a further variable, which we model by adding a second random variable $Y : \Omega \to \mathbb{R}$ or a second operator $Y : H \to H$ respectively. Both frameworks also assign meaning to composed variables, such as

$$X + Y, \quad X \cdot Y, \quad X \cdot Y \cdot X + Y^2.$$

While the product in the classical case (pointwise multiplication) is commutative, i.e., $X \cdot Y = Y \cdot X$, the product of operators on H (the composition) is not commutative in general.

Fortunately, *quantum probability theory* (or *non-commutative probability theory*) offers a framework for a more general probability theory which contains both models as special cases. In both cases, we have an algebra $\mathcal{A}$ of (bounded) random variables and an expectation functional $\varphi : \mathcal{A} \to \mathbb{R}$. In the classical case, $\varphi(X) = \mathbb{E}[X]$, and in the quantum case, $\varphi(X) = \langle \Psi, X\Psi \rangle$. The distribution μ of X can be defined via the moments $\varphi(X^n) = \int_{\mathbb{R}} x^n \mu(dx)$ in both cases. In quantum probability theory, one defines an abstract quantum probability space as an algebra $\mathcal{A}$ (with more or less additional structures), together with an expectation functional $\varphi : \mathcal{A} \to \mathbb{R}$.

There are often multiple ways to extend classical concepts into the quantum realm, and the important notion of *independence* of random variables is no exception. The independence of classical random variables can be reformulated as an algebraic relation, i.e., a *dependency*, between them. The non-commutative theory, with its rich historical development, allows for a variety of types of "independence", i.e., constraints of different types among quantum random variables. We refer to the surveys [**2**, **6**].

In this book, we consider five types of independence: tensor (or classical) independence, Boolean independence (von Waldenfels, 1970s), free independence

(Voiculescu, 1980s), monotone and anti-monotone independence (Muraki, De Giosa, Lu, 1990s).

Due to the works of Speicher, Schürmann, Ben Ghorbal, and Muraki, these five types of independence can be described through an elegant algebraic axiomatic framework. This characterization is discussed in [**18**, Section "The Five Universal Independences"]. These five notions of independence lead to five central limit theorems for independent and identically distributed quantum random variables, five Poisson limit theorems, five classes of quantum stochastic processes with independent increments, etc.

The primary emphasis of this book is on monotone independence, noting that anti-monotone independence is merely an order reversal of monotone independence. We prove the monotone central limit theorem and study quantum stochastic processes with monotonically independent increments, which possess a deep connection to complex analysis. We also compare our findings to the other notions of independence, though proofs for those are typically referenced from the existing literature.

1.2. Characteristic function and Cauchy transform

Instead of real algebras we will consider complex algebras $\mathcal{A}$ with a linear mapping $\varphi : \mathcal{A} \to \mathbb{C}$. This is more helpful and more elegant, even if we are only interested in real random variables. In fact, this technique is already used in classical probability theory. Consider the characteristic function, or Fourier transform of a real random variable X, given by

$$\varphi_X(t) = \int_\Omega e^{itX(\omega)} d\mathbb{P}(\omega), \quad t \in \mathbb{R}.$$

This function is simply the expected value of the complex random variable e^{itX}, i.e., $\varphi_X(t) = \mathbb{E}[e^{itX}]$. It plays an important role in classical probability theory, as it encodes not only distributions as functions from $\mathbb{R}$ to $\mathbb{C}$, but also the weak convergence of distributions, which corresponds to pointwise convergence of the characteristic functions. Moreover, the independence of random variables can be expressed simply using higher-dimensional characteristic functions.

Let us consider another example in which the characteristic function elegantly encodes a property of distributions. An *additive process* is a stochastic process $(X_t)_{t\geq 0}$ with $X_0 = 0$, having independent increments and being continuous in probability.

THEOREM 1.1. *Let μ be a probability measure on $\mathbb{R}$. The following statements are equivalent:*

(1) *There exists an additive process $(X_t)_{t\geq 0}$ such that μ is the distribution of X_1.*
(2) *(Lévy–Khintchine representation) There exist $a \in \mathbb{R}$, $\sigma \geq 0$, and a non-negative measure ν with $\nu(\{0\}) = 0$ and $\int_\mathbb{R} \min(1, t^2)\nu(dt) < \infty$ such that*

$$\int_\mathbb{R} e^{ixt}\mu(dx) = \exp\left(iat - \frac{1}{2}\sigma^2 t^2 + \int_\mathbb{R} \left(e^{its} - 1 - its\mathbf{1}_{\{|s|<1\}}(s)\right)\nu(ds)\right), \ t \in \mathbb{R}.$$

A significant amount of research in the field of quantum probability theory concerns the translation of notions and theorems from the classical, commutative theory (Markov processes, Brownian motion, random fields, etc.) to the non-commutative case. In some parts of quantum probability theory, the role of φ_X is replaced by the Cauchy transform

$$G_X(z) = \mathbb{E}\left[(z - X)^{-1}\right], \quad z \in \mathbb{H} = \{w \in \mathbb{C} \mid \operatorname{Im}(w) > 0\}.$$

G_X is a holomorphic function on the upper half-plane $\mathbb{H}$ and maps it into the lower half-plane. A complex analyst could pose the following intriguing question: what distribution μ must X have such that

$$G_\mu(z) := G_X(z) = \int_{\mathbb{R}} \frac{1}{z - x} \mu(dx)$$

is injective on $\mathbb{H}$, and, therefore, maps $\mathbb{H}$ conformally onto a simply connected subdomain of the lower half-plane? By the Riemann mapping theorem, we know that many conformal mappings exist on $\mathbb{H}$, so some of them might indeed be Cauchy transforms.

It turns out that the answer to this question has a deeper meaning, which can be formulated using quantum probability theory. We call a quantum stochastic process $(X_t)_{t\geq 0}$ (with $X_0 = 0$ and weakly continuous distributions) with monotonically independent increments a *monotone additive process*. The main motivation for this book is the following monotone analog to Theorem 1.1.

THEOREM 1.2. *Let μ be a probability measure on $\mathbb{R}$. The following statements are equivalent:*

(1) *There exists a monotone additive process $(X_t)_{t\geq 0}$ such that μ is the distribution of X_1.*
(2) *G_μ is univalent, i.e., G_μ is an injective function on $\mathbb{H}$.*

Furthermore, the evolution $t \mapsto \mu_t$ of the distributions of such processes gives rise to so-called Loewner chains, where $t \mapsto G_{\mu_t}(\mathbb{H})$ describes a decreasing family of simply connected domains in $\mathbb{C}$.

1.3. Outline of the book

Chapter 2: We review the foundations of classical probability theory, covering topics such as classical random variables, independence, the central limit theorem, Markov processes, and Lévy processes.

Chapter 3: This chapter introduces essential concepts from complex analysis, such as integral representations, the Cauchy transform, conformal mappings, and semigroups of holomorphic functions.

Chapter 4: We define a quantum probability space as the Banach space of all linear and bounded operators on a Hilbert space, together with an expectation functional. Instead of referring to the spectral theorem from functional analysis, we use elementary properties of Hilbert spaces to illustrate how self-adjoint operators can be regarded as non-commutative real random variables. Furthermore, we introduce the five notions of independence and take a glimpse at applications to random matrix theory.

Chapters 5 – 9 form the core of this book. While we usually allow for probability measures with unbounded support, we restrict quantum random variables to bounded operators on a Hilbert space in order to avoid becoming overly technical, including when demonstrating that (2) implies (1) in Theorem 1.2. As the construction of the process requires a certain class of classical Markov processes, the proof of Theorem 1.2 effectively connects quantum and classical probability theory to complex analysis.

Chapter 5: This chapter explores the convolutions arising from the different notions of independence, and their respective central limit theorems. Monotone independence was introduced by N. Muraki in [**139**–**141**], based on earlier works on monotone Fock spaces [**59**, **124**, **137**, **138**]. The F-transform of a probability measure μ is defined as the multiplicative inverse of its Cauchy transform:

$$F_\mu : \mathbb{H} \to \mathbb{H}, \quad F_\mu(z) := \frac{1}{G_\mu(z)}.$$

The monotone convolution of two probability measures μ and ν, denoted by $\mu \rhd \nu$, can now be elegantly expressed as

$$F_{\mu \rhd \nu} = F_\mu \circ F_\nu.$$

The monotone central limit theorem takes the following form, where the arcsine distribution $Arc(0,1)$ replaces the normal distribution from classical probability theory.

THEOREM 1.3 (Monotone central limit theorem). *Let $X_1, X_2, \ldots$ be a sequence of monotonically independent identically distributed random variables with mean c and variance $\sigma^2 > 0$. Let μ_n be the distribution of $S_n = (X_1 + \cdots + X_n - nc)/(\sigma\sqrt{n})$. Then μ_n converges weakly to the arcsine distribution $Arc(0,1)$, given by the density*

$$\frac{1}{\pi\sqrt{2-x^2}}, \quad x \in (-\sqrt{2}, \sqrt{2}).$$

Furthermore, we define additive processes and so-called hemigroups as the distributions of such processes. For an additive process $(X_t)_{t\geq 0}$ with monotonically independent increments, let $\mu_{s,t}$ denote the distribution of the quantum random variable $X_t - X_s$, where $0 \leq s \leq t$. The family $(\mu_{s,t})_{0\leq s\leq t}$ forms a monotone hemigroup, i.e., it satisfies:

(1) $\mu_{s,s} = \delta_0$ for all $s \geq 0$,
(2) $(s,t) \mapsto \mu_{s,t}$ is continuous with respect to weak convergence,
(3) $\mu_{s,u} = \mu_{s,t} \rhd \mu_{t,u}$ for all $0 \leq s \leq t \leq u$.

For a given monotone hemigroup, we construct a corresponding monotone additive process. This construction is due to U. Franz and T. Hasebe [**70**, Section 4], with preliminary work in [**69**].

If $\mu_{s,t} = \mu_{0,t-s}$ for all $0 \leq s \leq t$, then $(\mu_{0,t})_{t\geq 0}$ forms a $\rhd$-semigroup, and the corresponding process is the monotone version of a Lévy process. In this case, $(F_{\mu_t})_{t\geq 0}$ is a continuous semigroup of holomorphic functions on $\mathbb{H}$. Its infinitesimal generator $G : \mathbb{H} \to \mathbb{H} \cup \mathbb{R}$, defined by $\frac{d}{dt}F_t = G(F_t)$, has the form

$$G(z) = a + \int_{\mathbb{R}} \frac{1+xz}{x-z}\gamma(dx), \tag{1.1}$$

where $a \in \mathbb{R}$ and γ is a finite, non-negative measure on $\mathbb{R}$.

In the case where $\mu_{s,t} = Arc(0, t-s)$, the associated additive process is called a monotone Brownian motion, which was constructed by Muraki in [**138**] via operators acting on a certain Fock space, see also [**59, 124**].

Chapter 6: We study the F-transforms of monotone hemigroups, which form a certain class of Loewner chains, which we will refer to as *additive Loewner chains*. These Loewner chains satisfy a particular partial differential equation, which is a special case of Loewner's differential equations. If $(f_t)_{t\geq 0}$ is an additive Loewner chain with a sufficiently regular parametrization, then it satisfies the partial differential equation

$$\frac{\partial}{\partial t} f_t(z) = \frac{\partial}{\partial z} f_t(z) \cdot M(z,t) \quad \text{for almost every } t \in [0,\infty) \text{ and all } z \in \mathbb{H}, \tag{1.2}$$

where, for almost every $t \geq 0$, the function $z \mapsto M(z,t)$ is a holomorphic self-mapping of $\mathbb{H}$ of the form (1.1).

Chapter 7: We explore classical Loewner chains on the unit disk $\mathbb{D}$, which first appeared in Loewner's original paper [**123**]. These chains are closely related to monotone multiplicative processes, i.e., processes involving unitary operators whose multiplicative increments are monotonically independent. If U is a unitary operator, i.e., $U^{-1} = U^*$, then its distribution μ within a quantum probability space is supported on the unit circle $\partial\mathbb{D}$. We define the ψ- and η-transform of U by

$$\psi_\mu(z) := \varphi\left(\frac{Uz}{1-Uz}\right) = \int_{\partial\mathbb{D}} \frac{xz}{1-xz}\mu(dx), \quad \eta_\mu(z) := \frac{\psi_\mu(z)}{1+\psi_\mu(z)},$$

where $z \in \mathbb{D}$. The η-transforms arising in this way are exactly the holomorphic self-mappings of $\mathbb{D}$ that fix 0. Furthermore, the distributions of operators from monotone multiplicative processes correspond exactly to those distributions on $\partial\mathbb{D}$ that have a univalent η-transform.

Chapter 8: This chapter is devoted to additive and multiplicative Loewner chains consisting of slit mappings. A slit mapping on $\mathbb{H}$ is a conformal mapping $f : \mathbb{H} \to \mathbb{H} \setminus \gamma[0,1]$, where $\gamma : [0,1] \to \mathbb{H} \cup \mathbb{R}$ is a simple curve, i.e., γ is continuous and injective, and $\gamma(0) \in \mathbb{R}$, $\gamma(0,1] \subset \mathbb{H}$. By considering conformal mappings $(f_t)_{t\in[0,1]}$ of the form $f_t : \mathbb{H} \to \mathbb{H} \setminus \gamma[0,t]$, we describe the growth of a simple curve through a decreasing Loewner chain. Under an appropriate parametrization, $(f_t)_{t\in[0,1]}$ satisfies equation (1.2) for all $t \in [0,1]$, and not just for almost every $t \in [0,1]$, and

$$\frac{\partial}{\partial t} f_t(z) = \frac{\partial}{\partial z} f_t(z) \cdot \frac{1}{U(t)-z} \quad \text{for all } t \in [0,1] \text{ and } z \in \mathbb{H},$$

for a continuous function $U : [0,1] \to \mathbb{R}$.

It is natural to consider additive processes for the remaining three notions of independence. The anti-monotone case corresponds to a time reversal of the monotone case, and Theorem 1.2 applies here as well. In the Boolean case, any probability measure can be embedded in a process with Boolean independent increments. The free case is a mixture of Theorems 1.1 and 1.2.

Chapter 9: In the case of free independence, a representation similar to the Lévy–Khintchine formula exists for the so-called R-transform $R_\mu(z) = G_\mu^{-1}(z) - \frac{1}{z}$, which replaces the characteristic function. The F-transforms of such distributions are again special univalent functions. Additive processes with freely independent increments also produce Loewner chains – specifically, Loewner chains associated with nonlinear resolvents.

Let G be an infinitesimal generator of a continuous semigroup on $\mathbb{H}$. Let $t \geq 0$. If there exists a holomorphic mapping $J_t : \mathbb{H} \to \mathbb{H}$ such that the equation

$$w = z - t \cdot G(z)$$

has a unique solution $z = J_t(w)$ for all $w \in \mathbb{H}$, then J_t is called the *nonlinear resolvent* of G (at "time" t).

THEOREM 1.4. *Let $G : \mathbb{H} \to \mathbb{H} \cup \mathbb{R}$ be of the form* (1.1). *There exists a unique solution $(f_t)_{t\geq 0}$ of holomorphic mappings $f_t : \mathbb{H} \to \mathbb{H}$ to*

$$\frac{\partial}{\partial t} f_t(z) = \frac{\partial}{\partial z} f_t(z) \cdot G(f_t(z)), \quad t \geq 0, \quad f_0(z) = z \in \mathbb{H}.$$

The solution can be expressed as $(f_t)_{t\geq 0} = (F_{\mu_t})_{t\geq 0}$ for a semigroup $(\mu_t)_{t\geq 0}$ with respect to free convolution, and $(f_t)_{t\geq 0}$ is a decreasing Loewner chain. Furthermore, $(f_t)_{t\geq 0}$ is the family of nonlinear resolvents of G. Conversely, every semigroup $(\mu_t)_{t\geq 0}$ with respect to free convolution is of this form.

Chapter 10: The five notions of independence have concrete applications in the spectral analysis of graphs, as discussed in the books [**93**], [**145**]. For each notion, there exists a corresponding graph product, providing a visual representation of the abstract concepts of quantum random variables and non-commutative convolutions. In this chapter, we focus on the monotone case, where special graphs, so-called spidernets, can be used to approximate certain monotone additive processes.

Chapter 11: This chapter provides an outlook on higher-dimensional generalizations, highlighting some open problems and offering extensive references for further study.

Acknowledgments

The author gratefully acknowledges the Institute of Mathematics of the Polish Academy of Sciences (IMPAN) for granting permission to incorporate portions of the paper, "Monotone increment processes, classical Markov processes, and Loewner chains", *Dissertationes Mathematicae*, **552** (2020), by U. Franz, T. Hasebe, and S. Schleissinger, into this textbook. Some of the results, proofs, and expository material presented here are based on that work, with appropriate modifications and adaptations for the purposes of this book.

CHAPTER 2

Classical Probability Theory

In this chapter, we recall several basic notions and theorems from classical probability theory. Standard references for classical probability theory include the books [**40**] and [**103**]. We also refer to [**41**], which comprehensively covers the convergence of probability measures.

2.1. The basic notions

DEFINITION 2.1 (Classical probability space). A probability space is given by a triple $(\Omega, \mathcal{F}, \mathbb{P})$, where Ω is a non-empty set (the sample space), $\mathcal{F} \subset 2^{\Omega}$ is a σ-algebra (the set of events), and $\mathbb{P} : \mathcal{F} \to [0, 1]$ is a probability measure.

We recall that a set $\mathcal{F} \subset 2^{\Omega}$ is called a σ-algebra if it satisfies the following conditions:

- $\Omega \in \mathcal{F}$,
- if $A \in \mathcal{F}$, then $\Omega \setminus A \in \mathcal{F}$,
- if $A_1, A_2, \dots \in \mathcal{F}$, then $\cup_{n=1}^{\infty} A_n \in \mathcal{F}$.

A function $\mathbb{P} : \mathcal{F} \to [0, 1]$ is called a probability measure if

- $\mathbb{P}(\Omega) = 1$, and
- if $A_1, A_2, \dots \in \mathcal{F}$ are pairwise disjoint, then $\mathbb{P}(\cup_{n=1}^{\infty} A_n) = \sum_{n=1}^{\infty} \mathbb{P}(A_n)$.

The σ-algebra $\mathcal{F}$ encodes the observable information within our probabilistic model.

EXAMPLE 2.2. In the trivial case $\Omega = \{\emptyset, \Omega\}$, we only observe one event, namely Ω, and we are completely blind to any further details (the elements of Ω). If $\mathcal{F} = 2^{\Omega}$, then every subset of Ω is an event, and we have the largest possible amount of information. This is the usual choice in case Ω is finite. If $\Omega = \{\omega_1, \dots, \omega_n\}$ and $\mathcal{F} = 2^{\Omega}$, then $\mathbb{P}$ is completely determined by the probabilities $\mathbb{P}(\{\omega_1\}), \dots, \mathbb{P}(\{\omega_n\})$.

EXAMPLE 2.3. Consider a fair dice with six faces. We could choose

$$\Omega = \{1, 2, 3, 4, 5, 6\}, \quad \mathcal{F} = 2^{\Omega}, \quad \text{and} \quad \mathbb{P}(\{n\}) = \frac{1}{6}, \quad n = 1, \dots, 6.$$

Compare this with the probability space

$$(\Omega, \quad \mathcal{F}_2 = \{\emptyset, \{3\}, \{6\}, \{3, 6\}, \Omega \setminus \{3\}, \Omega \setminus \{6\}, \Omega \setminus \{3, 6\}, \Omega\}, \quad \mathbb{P}|_{\mathcal{F}_2}).$$

We can still observe an event like "The outcome can be divided by 3", i.e., $\{3, 6\}$, but we can no longer observe the event "The outcome is a prime number", i.e., $\{2, 3, 5\}$. In this second case, we can imagine that someone removed the numbers from the faces $1, 2, 4, 5$. They are still possible outcomes of a random experiment, but they have become indistinguishable to us.

For any subset $M \subset 2^{\Omega}$, one can construct the σ-algebra $\mathcal{F}$ generated by M, i.e., $\mathcal{F}$ is defined as the smallest σ-algebra that contains M. In other words,

$$\mathcal{F} = \bigcap_{\substack{\sigma\text{-algebra } \mathcal{A} \\ M \subset \mathcal{A}}} \mathcal{A}.$$

EXAMPLE 2.4. If Ω is a metric space and $\mathcal{F}$ is the σ-algebra generated by all open subsets of Ω, then $\mathcal{F} = \mathcal{B}(\Omega)$ is called the *Borel σ-algebra* of Ω. A probability measure $\mathbb{P} : \mathcal{B}(\Omega) \to [0,1]$ is then called a *Borel probability measure.* For example, consider $[0,1]$ with the usual Euclidean metric. There exists a unique probability measure λ on $\mathcal{B}([0,1])$, the Lebesgue measure, such that $\lambda([a,b]) = \lambda((a,b)) = b-a$ for all intervals, see [**40**, Sections 2 and 3].

DEFINITION 2.5 (Random variable). Let M be a metric space endowed with its Borel σ-algebra. Then a measurable function $X : \Omega \to M$ is called an *M-valued random variable.* (Recall that measurable means that $X^{-1}(A)$ belongs to the σ-algebra attached to Ω for every $A \in \mathcal{B}(M)$.)

- The push-forward of $\mathbb{P}$ with respect to X yields the Borel probability measure $\mu(A) = \mathbb{P}(X \in A)$, which is called the *distribution* of X. (We write $\mathbb{P}(X \in A)$ short for $\mathbb{P}(\{\omega \in \Omega \mid X(\omega) \in A\})$.)
- Likewise, the pullback $\sigma(X) = \{X^{-1}(A) \mid A \in \mathcal{B}(M)\}$ of the Borel σ-algebra yields a σ-algebra on Ω, consisting of all the information encoded by X.

We will mostly work with *real-valued random variables*, i.e., with $M = \mathbb{R}$. Unless stated otherwise, we assume all random variables to be real-valued. However, in our study of real-valued random variables, we will also encounter the cases $M = \mathbb{C}$ and $M = \mathbb{R}^n$. In each of these cases, we use the usual Euclidean metric.

DEFINITION 2.6 (Moments, mean, variance). Let X be a real- or complex-valued random variable. Provided the integral exists, the expectation $\mathbb{E}[X]$ is defined by

$$\mathbb{E}[X] = \int_{\Omega} X(\omega) d\mathbb{P}(\omega).$$

If the corresponding integrals exist, then

- $\mathbb{E}[X^n]$ is called the nth moment of X and
- $\mathrm{Var}(X) = \mathbb{E}[(X - \mathbb{E}[X])^2] = \mathbb{E}[X^2] - (\mathbb{E}[X])^2$ is called the variance of X.

If μ is the distribution of X, then $\mathbb{E}[X]$ is also called the *mean*, $\mathbb{E}[(X - \mathbb{E}[X])^2]$ the *variance*, and $\sqrt{\mathbb{E}[(X - \mathbb{E}[X])^2]}$ the *standard deviation of μ.*

DEFINITION 2.7. We denote by $\mathcal{P}(\mathbb{R})$ the set of all Borel probability measures on $\mathbb{R}$. The support $\mathrm{supp}(\mu)$ of μ is defined as $\mathrm{supp}(\mu) = \{x \in \mathbb{R} \mid x \in U \subset \mathbb{R},\ U \text{ open} \implies \mu(U) > 0\}$. The support is always a closed subset of $\mathbb{R}$. We denote by $\mathcal{P}_c(\mathbb{R})$ the set of all $\mu \in \mathcal{P}(\mathbb{R})$ with compact support.

EXAMPLE 2.8. The most important element of $\mathcal{P}(\mathbb{R})$, and the most important of all distributions, is the Gaussian normal distribution $\mathcal{N}(c, \sigma^2)$, with mean c and variance σ^2, given by

$$\mathcal{N}(\mu, \sigma^2)(A) = \frac{1}{\sigma\sqrt{2\pi}} \int_A e^{-\frac{1}{2}(\frac{x-c}{\sigma})^2} dx,$$

where $A \subset \mathbb{R}$ is a Borel subset; see Figure 2.1.

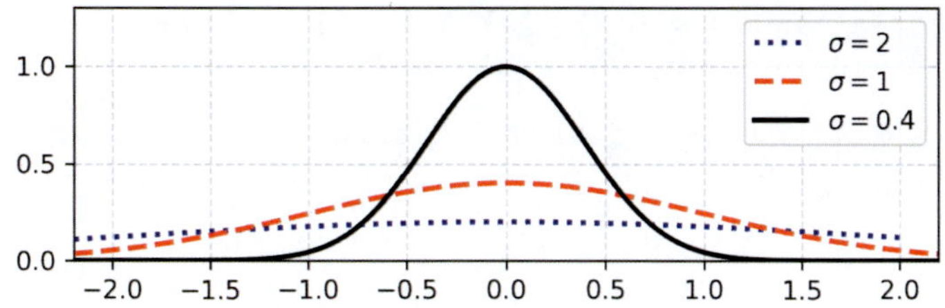

FIGURE 2.1. Densities of centered Gaussian normal distributions.

REMARK 2.9. Let $(\Omega, \mathcal{F}, \mathbb{P})$ be a probability space and let us call two complex-valued random variables equivalent if they differ only on sets of measure 0. Fix some $p \geq 1$. The space $L^p(\Omega, \mathcal{F}, \mathbb{P})$ consists of all equivalence classes $[X]$ of random variables X with $\mathbb{E}[|X|^p] < \infty$. Note that this definition does not depend on the choice of representative. This space becomes a complex Banach space with the norm $\|[X]\|_p = \mathbb{E}[|X|^p]$, see [**103**, Chapter 1].

For $p = 2$, the norm is induced by an inner product, and $L^2(\Omega, \mathcal{F}, \mathbb{P})$ becomes a complex Hilbert space via

$$\langle X, Y \rangle := \mathbb{E}[\overline{X}Y] = \int_\Omega \overline{X}(\omega)Y(\omega)d\mathbb{P}(\omega).$$

REMARK 2.10. On $\Omega = [0, 1]$, the Lebesgue measure λ also yields the probability space $([0, 1], \mathcal{F}_\lambda, \lambda)$, where $\mathcal{F}_\lambda$ consists of all Lebesgue measurable subsets of $[0, 1]$ (the completion of the Borel σ-algebra $\mathcal{B}([0, 1])$). We have

$$\mathcal{B}([0, 1]) \subsetneq \mathcal{F}_\lambda \subsetneq 2^{[0,1]}.$$

In probability theory, the term "uniform distribution on $[0, 1]$" usually refers to the probability space $([0, 1], \mathcal{B}([0, 1]), \lambda)$, and we typically use the Borel σ-algebra to define random variables rather than $\mathcal{F}_\lambda$. The reason is that this approach yields a broader class of random variables. This is similar to how measurable functions are defined in analysis: a function $f : [0, 1] \to [0, 1]$ is called Lebesgue measurable if it is measurable as a function from $([0, 1], \mathcal{F}_\lambda)$ to $([0, 1], \mathcal{B}([0, 1]))$.

Furthermore, it should be noted that the proof of $\mathcal{F}_\lambda \subsetneq 2^{[0,1]}$, i.e., the existence of a non-Lebesgue measurable subset of $[0, 1]$, due to Vitali, requires the axiom of choice, see [**40**, p.45]. One can study a different kind of mathematics by excluding the axiom of choice (and keeping the Zermelo-Fraenkel set theory) and adding the axiom that all subsets of $\mathbb{R}$ are Lebesgue measurable (the Solovay model).

2.2. Independence

Let us fix a probability space $(\Omega, \mathcal{F}, \mathbb{P})$. If we are interested in only a single random variable X on Ω, this entire setup might seem overly complex, as we could simply consider the push-forward measure of X on $\mathbb{R}$. However, this changes when we examine at least two random variables on Ω, as their behavior may be interrelated in more or less complex ways. A simple measure to analyze their interaction is the covariance.

DEFINITION 2.11 (Covariance). Let X and Y be square-integrable random variables. The *covariance* of X and Y, denoted by $cov(X,Y)$, is defined by

$$cov(X,Y) = \mathbb{E}[(X - \mathbb{E}[X])(Y - \mathbb{E}[Y])] = \mathbb{E}[XY] - \mathbb{E}[X]\mathbb{E}[Y].$$

If $cov(X,Y) = 0$, X and Y are called *uncorrelated.* Under the assumption of positive variances, the *correlation coefficient* $\rho(X,Y)$ is defined as

$$\rho(X,Y) = \frac{cov(X,Y)}{\sigma(X)\sigma(Y)} \in [-1,1].$$

The fact that $\rho(X,Y) \in [-1,1]$ follows from the Cauchy–Schwarz inequality

$$|cov(X,Y)|^2 \leq \sigma^2(X)\sigma^2(Y)$$

(recall the Hilbert space from Remark 2.9).[1]

The notion of independence of random variables is of paramount importance in probability theory.

DEFINITION 2.12 (Independence of events and σ-algebras). Let $(A_j)_{j\in J} \subset \mathcal{F}$ be a family of events with a non-empty index set J. Then $(A_j)_{j\in J}$ is called *independent* if for any distinct $j_1, \dots, j_n \in J$, we have

$$\mathbb{P}\left(\bigcap_{k=1}^{n} A_{j_k}\right) = \prod_{k=1}^{n} \mathbb{P}(A_{j_k}).$$

A family $(\mathcal{A}_j)_{j\in J}$ of subsets $\mathcal{A}_j \subset \mathcal{F}$ is called *independent* if any family $(A_j)_{j\in J}$ with $A_j \in \mathcal{A}_j$ is independent.

DEFINITION 2.13 (Independence of random variables). Random variables X_1, $\dots$, X_n are called *independent* if the σ-algebras $\sigma(X_1), \dots, \sigma(X_n)$ are independent. Equivalently, the joint distribution of $X_1, \dots, X_n$ is the product distribution, i.e.,

$$\mathbb{P}[X_1 \leq x_1 \wedge \cdots \wedge X_n \leq x_n] = \mathbb{P}[X_1 \leq x_1] \cdot \cdots \cdot \mathbb{P}[X_n \leq x_n]$$

for all $x_1, \dots, x_n \in \mathbb{R}$. An infinite sequence $X_1, X_2, \dots$ of random variables is called independent if every finite collection $X_{k_1}, \dots, X_{k_n}$, for n distinct indices $k_1, \dots, k_n$, is independent.

If $X_1, X_2, \dots$ are independent and identically distributed random variables, we will simply call them "*iid*" random variables.

2.3. The characteristic function

DEFINITION 2.14 (Characteristic function). Let X be a random variable. The function

$$\varphi_X(t) : \mathbb{R} \to \mathbb{C}, \quad \varphi_X(t) = \mathbb{E}[e^{itX}],$$

is called the *characteristic function* of X. Similarly, if μ is a probability measure on $\mathbb{R}$, then φ_μ is defined as the function

$$\varphi_\mu(t) : \mathbb{R} \to \mathbb{C}, \quad \varphi_\mu(t) = \int_{\mathbb{R}} e^{itx} \mu(dx).$$

[1] In the case of two complex-valued random variables, we would define the covariance as $cov(X,Y) = \mathbb{E}[\overline{(X - \mathbb{E}[X])}(Y - \mathbb{E}[Y])]$.

The characteristic function is also called the *Fourier transform*. In contrast to other transforms, such as the Laplace transform

$$\mathbb{E}[e^{-tX}]$$

or the Hilbert transform

$$\mathcal{H}_\mu(t) := \lim_{\varepsilon \searrow 0} \frac{1}{\pi} \int_{|t-x|>\varepsilon} \frac{1}{x-t}\, \mu(dx),$$

the Fourier transform always exists for all $t \in \mathbb{R}$. We have the following simple properties.

THEOREM 2.15. *Let X be a random variable. Then*

- $|\varphi_X(t)| \leq 1$ *for all* $t \in \mathbb{R}$,
- $t \mapsto \varphi_X(t)$ *is continuous,*
- *and* $\varphi_X(-t) = \overline{\varphi_X(t)}$ *for all* $t \in \mathbb{R}$.

Furthermore, for any $n \in \mathbb{N}$, complex numbers $c_1, \ldots, c_n$, and real numbers $t_1, \ldots, t_n$,

$$\sum_{1 \leq k,l \leq n} c_k \overline{c_l} \varphi_X(t_k - t_l) \geq 0. \tag{2.1}$$

PROOF. Let μ be the distribution of X. Clearly, $|\varphi_X(t)| \leq \int_{\mathbb{R}} |e^{itx}| \mu(dx) = 1$ and

$$\varphi_X(-t) = \int_{\mathbb{R}} e^{-itx} \mu(dx) = \int_{\mathbb{R}} \overline{e^{itx}} \mu(dx) = \overline{\int_{\mathbb{R}} e^{itx} \mu(dx)} = \overline{\varphi_X(t)}$$

for all $t \in \mathbb{R}$. Furthermore

$$|\varphi_X(t) - \varphi_X(s)| = \left| \int_{\mathbb{R}} e^{itx} - e^{isx} \mu(dx) \right| \leq \int_{\mathbb{R}} |e^{i(t-s)x} - 1| \mu(dx).$$

As $|e^{i(t-s)x} - 1| \leq 2$, the dominated convergence theorem implies that the last integral converges to 0 as $s \to t$ for fixed t. (In fact, we even have uniform continuity of $\varphi_X(t)$.)

Finally,

$$\sum_{1 \leq k,l \leq n} c_k \overline{c_l} \varphi_X(t_k - t_l) = \sum_{1 \leq k,l \leq n} c_k \overline{c_l} \mathbb{E}[e^{i(t_k - t_l)X}]$$

$$= \mathbb{E}\left[\sum_{k=1}^{n} c_k e^{it_k X} \cdot \overline{\sum_{l=1}^{n} c_l e^{it_l X}} \right] = \mathbb{E}\left[\left| \sum_{k=1}^{n} c_k e^{it_k X} \right|^2 \right] \geq 0. \qquad \square$$

REMARK 2.16. Bochner's theorem states that any continuous function $\varphi : \mathbb{R} \to \mathbb{C}$ with $\varphi(0) = 1$ and property (2.1) is, in fact, a characteristic function; see [**164**, Section 1.4.3].

EXAMPLE 2.17. We calculate φ_μ for some distributions μ.

(a) If $\mu = \delta_{x_0}$ is a point measure, then $\varphi_\mu(t) = e^{itx_0}$. The two-point measure $\mu = p\delta_1 + (1-p)\delta_{-1}$, for some $p \in [0,1]$, yields $\varphi_\mu(t) = \cos(t) + i(2p-1)\sin(t)$. The image curve is a (possibly degenerate) ellipse.

(b) Let μ be the Poisson distribution, that is, $\mu(\{k\}) = \frac{\lambda^k e^{-\lambda}}{k!}$, $k \in \mathbb{N}_0$, where $\lambda > 0$ is the first moment as well as the variance of μ. We have

$$\varphi_\mu(t) = e^{-\lambda} \sum_{k=0}^{\infty} \frac{\lambda^k}{k!} e^{itk} = e^{-\lambda} \sum_{k=0}^{\infty} \frac{(\lambda e^{it})^k}{k!} = e^{\lambda(e^{it}-1)}.$$

(c) Let μ be the uniform distribution on $[a,b]$, $a<b$. Then $\varphi_\mu(0)=1$ and $\varphi_\mu(t)=\frac{e^{itb}-e^{ita}}{it(b-a)}$ for $t\neq 0$.

(d) If $\mu=\mathcal{N}(0,\sigma^2)$, then

$$\varphi_\mu(t)=e^{-\sigma^2t^2/2}.$$

This can be shown by considering $\frac{d}{dt}\varphi_\mu(t)$. Via exchanging differentiation and integration, we obtain

$$\frac{d}{dt}\varphi_\mu(t)=\frac{1}{\sigma\sqrt{2\pi}}\int_{\mathbb{R}}\frac{d}{dt}e^{itx}e^{-\frac{1}{2}(\frac{x}{\sigma})^2}dx=\frac{i}{\sigma\sqrt{2\pi}}\int_{\mathbb{R}}xe^{itx}e^{-\frac{1}{2}(\frac{x}{\sigma})^2}dx.$$

As $e^{itx}=\cos(tx)+i\sin(tx)$ and $\int_{\mathbb{R}}x\cos(tx)e^{-\frac{1}{2}(\frac{x}{\sigma})^2}dx=0$, we obtain

$$\begin{aligned}\frac{d}{dt}\varphi_\mu(t)&=\frac{-1}{\sigma\sqrt{2\pi}}\int_{\mathbb{R}}x\sin(tx)e^{-\frac{1}{2}(\frac{x}{\sigma})^2}dx\\
&=\frac{\sigma^2}{\sigma\sqrt{2\pi}}\int_{\mathbb{R}}\sin(tx)\frac{d}{dx}\left[e^{-\frac{1}{2}(\frac{x}{\sigma})^2}\right]dx=\lim_{R\to\infty}\frac{\sigma^2}{\sigma\sqrt{2\pi}}\sin(tR)e^{-\frac{1}{2}(\frac{R}{\sigma})^2}\\
&-\lim_{R\to-\infty}\frac{\sigma^2}{\sigma\sqrt{2\pi}}\sin(tR)e^{-\frac{1}{2}(\frac{R}{\sigma})^2}-\frac{\sigma^2}{\sigma\sqrt{2\pi}}\int_{\mathbb{R}}t\cos(tx)e^{-\frac{1}{2}(\frac{x}{\sigma})^2}dx\\
&=-\sigma^2t\int_{\mathbb{R}}\cos(tx)\frac{1}{\sigma\sqrt{2\pi}}e^{-\frac{1}{2}(\frac{x}{\sigma})^2}dx=-\sigma^2t\int_{\mathbb{R}}e^{itx}\frac{1}{\sigma\sqrt{2\pi}}e^{-\frac{1}{2}(\frac{x}{\sigma})^2}dx\\
&=-\sigma^2t\varphi_\mu(t);\end{aligned}$$

where we used that $\int_{\mathbb{R}}\sin(tx)e^{-\frac{1}{2}(\frac{x}{\sigma})^2}dx=0$ in the second to last line. So $\frac{d}{dt}\varphi_\mu(t)=-\sigma^2t\varphi_\mu(t)$ and together with $\varphi_\mu(0)=1$, we obtain $\varphi_\mu(t)=e^{-\sigma^2t^2/2}$ as the unique solution to this initial value problem. More generally, if $\mu=\mathcal{N}(c,\sigma^2)$, then

$$\varphi_\mu(t)=e^{ict-\sigma^2t^2/2}.$$

The characteristic function encodes the distribution μ completely and we can recover μ from φ_μ by an inversion formula.

THEOREM 2.18. *Let $\mu\in\mathcal{P}(\mathbb{R})$.*

(a) *We have the inversion formula*

$$\mu((a,b))+\frac{1}{2}\mu(\{a\})+\frac{1}{2}\mu(\{b\})=\frac{1}{2\pi}\lim_{T\to\infty}\int_{-T}^{T}\frac{e^{-ita}-e^{-itb}}{it}\varphi_\mu(t)dt$$

for all $a<b$. In particular, if $\varphi_\mu=\varphi_\nu$ for two probability measures μ and ν, then $\mu=\nu$.

(b) *(Inverse Fourier transform) Assume that $\int_{\mathbb{R}}|\varphi_\mu(t)|dt<\infty$. Then μ is absolutely continuous with respect to the Lebesgue measure λ and its density f is given by*

$$f(x)=\frac{1}{2\pi}\int_{\mathbb{R}}e^{-itx}\varphi_\mu(t)dt,\quad x\in\mathbb{R}.$$

PROOF.

(a) By Fubini's theorem, and since $\int_a^b e^{-ity}dy=\frac{e^{-ita}-e^{-itb}}{it}$, the right integral can be expressed as

$$F(a,b,T)=\int_{[-T,T]\times[a,b]}e^{-ity}\varphi_\mu(t)dydt.$$

Using Fubini's theorem again, we obtain:

$$
\begin{aligned}
F(a,b,T) &= \int_{[-T,T]\times[a,b]\times\mathbb{R}} e^{-ity}e^{itx}dydt\mu(dx)\\
&= \int_{\mathbb{R}}\left(\int_{[-T,T]}\int_{[a,b]} e^{-it(y-x)}dydt\right)\mu(dx)\\
&= \int_{\mathbb{R}}\left(\int_{[-T,T]}\frac{1}{it}(e^{-it(a-x)}-e^{-it(b-x)})dt\right)\mu(dx).
\end{aligned}
$$

Put $f(a,b,T,x) = \int_{-T}^{T}\frac{1}{it}(e^{-it(a-x)}-e^{-it(b-x)})dt$. As $\sin(x)$ is an odd and $\cos(x)$ an even function, we have

$$
\begin{aligned}
f(a,b,T,x) &= \int_{-T}^{T}\frac{\sin(-t(a-x))}{t}-\frac{\sin(-t(b-x))}{t}dt\\
&= 2\int_{0}^{T}\frac{\sin(t(x-a))}{t}-\frac{\sin(t(x-b))}{t}dt.
\end{aligned}
$$

We have

$$
\lim_{T\to\infty}\int_0^T\frac{\sin(ct)}{t} = \begin{cases}\frac{\pi}{2} & \text{if } c>0,\\ 0 & \text{if } c=0,\\ -\frac{\pi}{2} & \text{if } c<0.\end{cases}
$$

Consequently,

$$
\lim_{T\to\infty} f(a,b,T,x) = \begin{cases}\pi & \text{if } x=a \text{ or } x=b,\\ 0 & \text{if } x<a \text{ or } x>b,\\ 2\pi & \text{if } x\in(a,b).\end{cases}
$$

As $\sup\{|f(a,b,T,x)|\,|\,x\in\mathbb{R}, T\geq 0\}<\infty$, we can use the dominated convergence theorem to obtain

$$
\lim_{T\to\infty}\frac{1}{2\pi}F(a,b,T) = \lim_{T\to\infty}\frac{1}{2\pi}\int_{\mathbb{R}} f(a,b,T,x)\mu(dx)
$$
$$
= \frac{1}{2\pi}\int_{\mathbb{R}}\lim_{T\to\infty} f(a,b,T,x)\mu(dx) = \mu((a,b))+\frac{1}{2}\mu(\{a\})+\frac{1}{2}\mu(\{b\}).
$$

Assume that $\varphi_\mu=\varphi_\nu$. The set S of all x with $\mu(\{x\})>0$ or $\nu(\{x\})>0$ is at most countably infinite, see Exercise 2.8.1. For all $x_0, x\in\mathbb{R}\setminus S$, $x_0<x$, we have $\mu((x_0,x))=\nu((x_0,x))$ by the inversion formula. With $x_0\to-\infty$, we get $\mu((-\infty,x))=\nu((-\infty,x))$. If $x\in S$, then we can write $(-\infty,x)=\cup_{n\in\mathbb{N}}(-\infty,x_n)$ for an increasing sequence $(x_n)\subset\mathbb{R}\setminus S$, and we obtain

$$
\begin{aligned}
\mu((-\infty,x)) &= \mu(\cup_{n\in\mathbb{N}}(-\infty,x_n)) = \lim_{n\to\infty}\mu((-\infty,x_n))\\
&= \lim_{n\to\infty}\nu((-\infty,x_n)) = \nu(\cup_{n\in\mathbb{N}}(-\infty,x_n))\\
&= \nu((-\infty,x)).
\end{aligned}
$$

Now we can show that $\mu(I)=\nu(I)$ for all open and closed intervals I, and thus, $\mu=\nu$.

(b) The assumption allows us to define the function $f(x) = \frac{1}{2\pi}\int_{\mathbb{R}} e^{-itx}\varphi_\mu(t)dt$. For $a < b$ we have

$$\begin{aligned}\int_a^b f(x)dx &= \frac{1}{2\pi}\int_a^b\int_{\mathbb{R}} e^{-itx}\varphi_\mu(t)dtdx\\ &= \frac{1}{2\pi}\int_{\mathbb{R}}\varphi_\mu(t)\int_a^b e^{-itx}dxdt = \frac{1}{2\pi}\int_{\mathbb{R}}\varphi_\mu(t)\frac{e^{-ita}-e^{-itb}}{it}dt\\ &= \frac{1}{2\pi}\lim_{T\to\infty}\int_{-T}^{T}\varphi_\mu(t)\frac{e^{-ita}-e^{-itb}}{it}dt = \mu((a,b)) + \frac{1}{2}\mu(\{a,b\}).\end{aligned}$$

As $\int_a^b f(x)dx$ varies continuously with respect to a and b, we have $\mu(\{a\}) = \mu(\{b\}) = 0$. Thus, $\mu((a,b)) = \int_a^b f(x)dx$, and we conclude that $\mu(A) = \int_A f(x)dx$ for any Borel subset $A \subset \mathbb{R}$.

□

THEOREM 2.19. *Let X be a random variable with $\mathbb{E}[|X|^n] < \infty$. Then φ_X is n times differentiable at every $t \in \mathbb{R}$ with*

$$\frac{d^n}{dt^n}\varphi_X(t) = \mathbb{E}[e^{itX}(iX)^n]. \quad \textit{In particular,} \quad \mathbb{E}[X^n] = (-i)^n\left.\frac{d^n}{dt^n}\varphi_X(t)\right|_{t=0}.$$

PROOF. The statement holds trivially for $n = 0$. So we prove the statement by induction starting at $n = 0$. Let μ be the distribution of X. Assume that $\mathbb{E}[|X|^{n+1}] < \infty$ and that $\frac{d^n}{dt^n}\varphi_X(t) = \mathbb{E}[e^{itX}(iX)^n]$ holds for some $n \in \mathbb{N}$ and all $t \in \mathbb{R}$. Then

$$\begin{aligned}\lim_{h\to 0}\frac{\frac{d^n}{dt^n}\varphi_X(t+h) - \frac{d^n}{dt^n}\varphi_X(t)}{h} &= \lim_{h\to 0}(\mathbb{E}[(e^{i(t+h)X}(iX)^n] - \mathbb{E}[e^{itX}(iX)^n])/h\\ = \lim_{h\to 0}\int_{\mathbb{R}}\frac{e^{i(t+h)x}(ix)^n - e^{itx}(ix)^n}{h}\mu(dx) &= \int_{\mathbb{R}}(ix)^n\lim_{h\to 0}\frac{e^{i(t+h)x}-e^{itx}}{h}\mu(dx)\\ = \int_{\mathbb{R}}(ix)^{n+1}e^{itx}\mu(dx) = \mathbb{E}[e^{itX}(iX)^{n+1}].\end{aligned}$$

The exchange of the limit and the integral is justified by the estimate

$$\left|(ix)^n\frac{e^{i(t+h)x}-e^{itx}}{h}\right| = |x|^n\left|\frac{e^{ihx}-1}{h}\right| \le 2|x|^{n+1},$$

the fact that $\int_{\mathbb{R}}|x|^{n+1}\mu(dx) < \infty$, and by the dominated convergence theorem.

The above estimate follows from $|\frac{\sin(h)}{h}| \le 1$ and $|\frac{\cos(h)-1}{h}| \le 1$ for all $h \in \mathbb{R}\setminus\{0\}$, as $|\frac{\sin(hx)}{h}| \le |x|$ and $|\frac{\cos(hx)-1}{h}| \le |x|$, and thus, $\left|\frac{e^{ihx}-1}{h}\right| \le \left|\frac{\cos(hx)-1}{h}\right| + \left|\frac{\sin(hx)}{h}\right| \le 2|x|$ for all $h \in \mathbb{R}\setminus\{0\}$ and $x \in \mathbb{R}$. □

EXAMPLE 2.20. Consider $\mu = \mathcal{N}(0,1)$. Then we have $\varphi_\mu(t) = e^{-t^2/2}$, see Example 2.17 (d). Clearly, all odd moments of μ are equal to 0. We use Theorem 2.19 to compute the even moments. From

$$\varphi_\mu(t) = \sum_{n=0}^{\infty}\frac{(-1)^n t^{2n}}{2^n n!},$$

we obtain

$$\begin{aligned}\mathbb{E}[X^{2n}] &= (-i)^{2n} \frac{d^{2n}}{dt^{2n}}\varphi_\mu(t)\Big|_{t=0} = (-1)^n \cdot (2n)! \cdot \frac{(-1)^n}{2^n n!} \\ &= \frac{(2n)!}{2^n n!} = 1 \cdot 3 \cdot 5 \cdot \dots \cdot (2n-1).\end{aligned}$$

EXAMPLE 2.21. Let μ be the distribution defined by

$$\mu(\{n\}) = \frac{C}{2n^2 \log(|n|)}, \qquad n = \pm 2, \pm 3, \pm 4, \dots,$$

where C is chosen such that all probabilities add up to 1. The first moment of μ does not exist:

$$\int_{\mathbb{R}} |x| \mu(dx) = \sum_{n=2}^{\infty} 2n \frac{C}{2n^2 \log(n)} = \sum_{n=2}^{\infty} \frac{C}{n \log(n)} = \infty.$$

The characteristic function φ_μ is given by

$$\varphi_\mu(t) = \sum_{n=2}^{\infty} \frac{Ce^{itn}}{2n^2 \log(n)} + \frac{Ce^{-itn}}{2n^2 \log(n)} = C \sum_{n=2}^{\infty} \frac{\cos(nt)}{n^2 \log(n)}.$$

For $x \in (1, \infty)$, the function $f(x) = 1/\left(x\sqrt{\log(x)}\right)$ is strictly decreasing and $f(x) \to 0$ as $x \to \infty$. For $|h|$ small enough, we let $N(h)$ be the smallest natural number with $f(N(h)) \le |h|$. Then $f(N(h)-1) > |h|$ and $1 \le |h| N(h) \sqrt{\log(N(h))} < K$ for some constant $K > 0$. So

$$|h| N(h) \to 0 \quad \text{and} \quad |h| N(h) \log(N(h)) \to \infty \quad \text{as } h \to 0. \tag{2.2}$$

We have

$$\frac{\varphi_\mu(h) - 1}{h} = C \sum_{n=2}^{\infty} \frac{\cos(nh) - 1}{hn^2 \log(n)} = C \sum_{n=2}^{N(h)} \frac{\cos(nh) - 1}{hn^2 \log(n)} + C \sum_{n=N(h)+1}^{\infty} \frac{\cos(nh) - 1}{hn^2 \log(n)}.$$

Now we use that $|\cos(x) - 1| \le x^2/2$ for all $x \in \mathbb{R}$ for the first term to get

$$\begin{aligned}\left|\frac{\varphi_\mu(h) - 1}{h}\right| &\le C \sum_{n=2}^{N(h)} \frac{|\cos(nh) - 1|}{|h| n^2 \log(n)} + C \sum_{n=N(h)+1}^{\infty} \frac{|\cos(nh) - 1|}{|h| n^2 \log(n)} \\ &\le C \sum_{n=2}^{N(h)} \frac{|h|}{2\log(2)} + C \sum_{n=N(h)+1}^{\infty} \frac{2}{|h| n^2 \log(n)} \\ &\le C|h|N(h) + C \sum_{n=N(h)+1}^{\infty} \frac{2}{|h| n^2 \log(N(h))} \\ &= C|h|N(h) + \frac{2C}{|h| \log(N(h))} \sum_{n=N(h)+1}^{\infty} \frac{1}{n^2} \\ &\le C|h|N(h) + \frac{2C}{|h| \log(N(h))} \int_{N(h)}^{\infty} \frac{1}{x^2} dx \\ &= C|h|N(h) + \frac{2C}{|h| N(h) \log(N(h))} \to 0\end{aligned}$$

as $h \to 0$, due to (2.2). So $\varphi_\mu(t)$ is differentiable at $t = 0$. This example is taken from [**197**] and shows that the converse of Theorem 2.19 is not true.

In a certain sense, the converse of Theorem 2.19 is true for even moments, see [**126**, Lemma 1]. In particular, if all the derivatives of φ_μ exist at $t = 0$, then all moments of μ are finite [**126**, Corollary 1]. There has also been research on the analytic extension of φ_μ and on entire characteristic functions. For further relations between φ_μ and μ, see [**42, 125**].

THEOREM 2.22. *Let $X_1, \dots, X_n$ be random variables on a common probability space. Then $X_1, \dots, X_n$ are independent if and only if for any bounded Borel measurable functions $f_1, \dots, f_n : \mathbb{R} \to \mathbb{C}$ we have*

$$\mathbb{E}[f_1(X_1) \cdots f_n(X_n)] = \mathbb{E}[f_1(X_1)] \cdots \mathbb{E}[f_n(X_n)]. \tag{2.3}$$

If all $X_1, \dots, X_n$ are bounded, then they are independent if and only if

$$\mathbb{E}[X_1^{k_1} \cdots X_n^{k_n}] = \mathbb{E}[X_1^{k_1}] \cdots \mathbb{E}[X_n^{k_n}] \tag{2.4}$$

for all $k_1, \dots, k_n \in \mathbb{N}_0$.

PROOF. Assume that $X_1, \dots, X_n$ are independent. Then the joint distribution of $(X_1, \dots, X_n)$ is the product distribution. Let $\mu_1, \dots, \mu_n$ be the corresponding distributions. Then Fubini's theorem yields

$$\begin{aligned} \mathbb{E}[f_1(X_1) \cdots f_n(X_n)] &= \int_{\mathbb{R}} f_1(x_1)\mu_1(dx_1) \cdots \int_{\mathbb{R}} f_n(x_n)\mu_n(dx_n) \\ &= \mathbb{E}[f_1(X_1)] \cdots \mathbb{E}[f_n(X_n)]. \end{aligned}$$

If $X_1, \dots, X_n$ are bounded, then this holds also for $f_j(x) = x^{k_j}$, $k_j \in \mathbb{N}_0$.

Now consider arbitrary random variables $X_1, \dots, X_n$ on a common probability space. The multivariate characteristic function $\varphi_{(X_1,\dots,X_n)}$ is defined by

$$\varphi_{(X_1,\dots,X_n)} : \mathbb{R}^n \to \mathbb{C}, \quad \varphi_{(X_1,\dots,X_n)}(t_1, \dots, t_n) = \mathbb{E}[e^{i(t_1 X_1 + \cdots + t_n X_n)}].$$

If $Y_1, \dots, Y_N$ are random variables with $\varphi_{(X_1,\dots,X_n)} = \varphi_{(Y_1,\dots,Y_n)}$, then the distributions of $(X_1, \dots, X_n)$ and $(Y_1, \dots, Y_n)$ are identical. This can be shown as in the univariate case. The inversion formula for the multivariate characteristic function can be found in [**136**, Theorem 10.6.1]. If $X_1, \dots, X_n$ are independent, then

$$\varphi_{(X_1,\dots,X_n)}(t_1, \dots, t_n) = \varphi_{X_1}(t_1) \cdots \varphi_{X_n}(t_n)$$

for all $t_1, \dots, t_n \in \mathbb{R}$.

Now assume that the random variables $X_1, \dots, X_n$ satisfy (2.3). For $f_k(x) = e^{it_k x}$, we obtain that $\varphi_{(X_1,\dots,X_n)}(t_1, \dots, t_n) = \varphi_{X_1}(t_1) \cdots \varphi_{X_n}(t_n)$ for all $t_1, \dots, t_n \in \mathbb{R}$. Hence, the joint distribution of $(X_1, \dots, X_n)$ is the product distribution, i.e., $X_1, \dots, X_n$ are independent.

Finally, assume that $X_1, \dots, X_n$ are bounded and satisfy (2.4). We find $M > 0$ such that the support of the distribution of X_k is contained in $[-M, M]$ for all $k = 1, \dots, n$. Then $X_1, \dots, X_n$ satisfy (2.3) for any polynomials $f_1, \dots, f_n : \mathbb{R} \to \mathbb{C}$. By approximating arbitrary bounded measurable functions on $[-M, M]$ with polynomials, we obtain that (2.3) also holds for bounded measurable functions. □

Theorem 2.22 is crucial for defining a notion of independence in quantum probability theory. It shows that the independence of bounded random variables can be expressed by an algebraic property involving only the expectation $\mathbb{E}$ and products of these random variables.

THEOREM 2.23. *Let $X_1, \ldots, X_n$ be random variables on a common probability space. If $X_1, \ldots, X_n$ are independent, then*

$$\mathbb{E}[e^{it(X_1+\cdots+X_n)}] = \mathbb{E}[e^{itX_1}]\cdots\mathbb{E}[e^{itX_n}] \quad \text{for all} \quad t \in \mathbb{R}. \tag{2.5}$$

Furthermore, if two square-integrable random variables X_1, X_2 satisfy (2.5), *then they are uncorrelated.*

PROOF. The first statement follows directly from Theorem 2.22, equation (2.3). Let X_1, X_2 be square-integrable random variables which satisfy (2.5). Then Theorem 2.19 implies

$$\begin{aligned}&\mathbb{E}[X_1^2] + 2\mathbb{E}[X_1X_2] + \mathbb{E}[X_2^2] = \mathbb{E}[(X_1+X_2)^2] = -\frac{d^2}{dt^2}\varphi_{X_1+X_2}(0)\\ &= -\frac{d^2}{dt^2}(\varphi_{X_1}\cdot\varphi_{X_2})(0) = -\left(\frac{d^2}{dt^2}\varphi_{X_1}(0) + 2\frac{d}{dt}\varphi_{X_1}(0)\frac{d}{dt}\varphi_{X_2}(0) + \frac{d^2}{dt^2}\varphi_{X_2}(0)\right)\\ &= \mathbb{E}[X_1^2] + 2\mathbb{E}[X_1]\mathbb{E}[X_2] + \mathbb{E}[X_2^2],\end{aligned}$$

and it follows that $\mathbb{E}[X_1X_2] = \mathbb{E}[X_1]\mathbb{E}[X_2]$. □

Two random variables satisfying (2.5) are also called *subindependent*. Hence, we have

$$\begin{aligned}&X, Y \text{ independent} \Longrightarrow X, Y \text{ subindependent}\\ &\Longrightarrow X, Y \text{ uncorrelated (if } X, Y \text{ are square-integrable).}\end{aligned}$$

In Exercise 2.8.7, we see that these implications cannot be reversed.

DEFINITION 2.24. If X and Y are subindependent, the distribution α of $X+Y$ only depends on the distributions μ and ν of X and Y. The probability measure $\mu * \nu := \alpha$ is called the (classical) additive *convolution* of μ and ν.

REMARK 2.25. For a random variable X, the function $M_X(t) = \mathbb{E}[e^{tX}]$ is called the *moment generating function of* X, provided the expectation exists for all t in a neighborhood of 0. For the sake of simplicity, let us assume that the distribution μ of X has compact support and denote the n-th moment of μ by m_n. Then

$$M_X(t) = \int_{\mathbb{R}} e^{tx}\mu(dx) = \int_{\mathbb{R}}\sum_{n=0}^{\infty}\frac{t^nx^n}{n!}\mu(dx) = \sum_{n=0}^{\infty}\int_{\mathbb{R}}x^n\mu(dx)\frac{t^n}{n!} = \sum_{n=0}^{\infty}m_n\frac{t^n}{n!}.$$

The function $K_X(t) = \log(M_X(t)) = \sum_{n=1}^{\infty}\kappa_n\frac{t^n}{n!}$, for $|t|$ small enough, is called the *cumulant generating function* of X, and the coefficients κ_n are the *cumulants* of X (see [**40**, Section 9]). So we have

$$m_n = \frac{d^n}{dt^n}M_X(t)\bigg|_{t=0}, \qquad \kappa_n = \frac{d^n}{dt^n}\log(M_X(t))\bigg|_{t=0}.$$

From the power series expansions, one can derive polynomials $p_n(x_1, \ldots, x_n)$ that are independent of X such that $\kappa_n = p_n(m_1, \ldots, m_n)$. For example,

$$\kappa_1 = m_1, \quad \kappa_2 = m_2 - m_1^2, \quad \kappa_3 = m_3 - 3m_1m_2 + 2m_1^3.$$

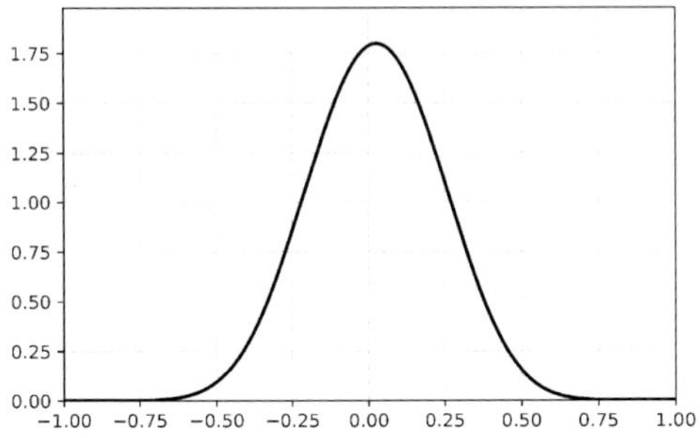

FIGURE 2.2. The density of μ.

If X, Y are two independent random variables with compactly supported distributions, then $M_{X+Y}(t) = M_X(t) \cdot M_Y(t)$ by (2.3) and $K_{X+Y}(t) = K_X(t) + K_Y(t)$, and thus,

$$\kappa_n(X+Y) = \kappa_n(X) + \kappa_n(Y) \quad \text{for all } n \in \mathbb{N}.$$

EXAMPLE 2.26. The Riemann zeta function is defined by

$$\zeta(s) = \sum_{n=1}^{\infty} \frac{1}{n^s}$$

for $s \in \mathbb{C}$ with $\mathrm{Re}(s) > 1$. It can be shown that the function

$$\xi(s) = \frac{1}{2} s(s-1)\pi^{-s/2}\Gamma\left(\frac{s}{2}\right)\zeta(s)$$

extends analytically to the whole complex plane, where $\Gamma(z) = \int_0^\infty t^{z-1}e^{-t}dt$. ξ possesses the symmetry $\xi(1-s) = \xi(s)$. The Riemann hypothesis (**RH**) states that all nontrivial zeros of ζ lie on the line $\mathrm{Re}\, s = \frac{1}{2}$, in other words:

(**RH**) $\qquad$ if $\xi(s_0) = 0$, then $\mathrm{Re}(s_0) = \frac{1}{2}$.

For the following we refer to [**38**]. For $y > 0$, let

$$f(y) = 4y \cdot \sum_{n=1}^{\infty} (2\pi^2 n^4 y^2 - 3\pi n^2) e^{-\pi n^2 y^2}.$$

Then $f(y)dy$ defines a probability distribution on $(0, \infty)$. Let Y be a random variable with that distribution. Then, for any $s \in \mathbb{C}$,

$$\mathbb{E}[Y^s] = \int_0^\infty y^s f(y)dy = 2\xi(s),$$

see [**38**, Proposition 2.1]. Define $X = -\log(Y)$ and let μ be the distribution of X. (μ has the density $e^{-x}f(e^{-x})$, see Figure 2.2.) Then the moment generating function $M_X(t) = \mathbb{E}[e^{tX}]$ satisfies

$$M_X(t) = \int_{\mathbb{R}} e^{tx}\mu(dx) = \int_0^\infty y^{-t} f(y)dy = 2\xi(-t),$$

and we see that $2\xi(-t)$ extends $M_X(t)$ analytically to $\mathbb{C}$.

Denote by $\kappa_1, \kappa_2, \ldots$ the cumulants of X, i.e.,

$$\kappa_n = \frac{d^n}{dt^n} \log(M_X(t))|_{t=0} = \frac{d^n}{dt^n} \log(2\xi(-t))|_{t=0}.$$

Define

$$\lambda_n = \frac{1}{(n-1)!}\frac{d^n}{ds^n}(s^{n-1}\log\xi(s))|_{s=1}.$$

λ_n can also be expressed as

$$\lambda_n = n\sum_{j=0}^{n-1}\binom{n-1}{j}\frac{\kappa_{n-j}}{(n-j)!}$$

or via the power series

$$\frac{d}{dz}\log M_X\left(\frac{z}{1-z}\right) = \sum_{n=0}^{\infty}\lambda_{n+1}z^n.$$

The numbers λ_n are also called the Keiper–Li coefficients. In [**119**], Li has proven that the Riemann hypothesis is true if and only if these numbers are non-negative:

$$(\mathbf{RH}) \quad \iff \quad \lambda_n \geq 0 \text{ for all } n \in \mathbb{N}.$$

Keiper has proven that (**RH**) implies positivity of all λ_n, see [**104**, p.769]. So (**RH**) $\iff \lambda_n > 0$ for all $n \in \mathbb{N}$.

Finally, we turn to the weak convergence of probability measures.

DEFINITION 2.27. Let μ and $\mu_1, \mu_2, \cdots \in \mathcal{P}(\mathbb{R})$. We say that μ_n *converges weakly* to μ if

$$\int_{\mathbb{R}} f(x)\mu_n(dx) \to \int_{\mathbb{R}} f(x)\mu(dx)$$

for every bounded and continuous $f : \mathbb{R} \to \mathbb{C}$. For random variables $X, X_1, X_2, \ldots,$ we say that the sequence $(X_n)_{n\in\mathbb{N}}$ *converges in distribution* to X if the distribution of X_n converges weakly to the distribution of X as $n \to \infty$.

The random variables do not need to be defined on the same probability space. Again, we can use the characteristic function to translate the notion of weak convergence into a very simple condition. First, we need the following estimate.

LEMMA 2.28. *Let $\mu \in \mathcal{P}(\mathbb{R})$ and $r > 0$. Then*

$$\mu(\{x \in \mathbb{R} \,|\, |x| \geq r\}) \leq \frac{r}{2}\int_{-2/r}^{2/r} 1 - \varphi_\mu(t)dt.$$

PROOF. Let $c > 0$. By Fubini's theorem, we have

$$\int_{-c}^{c} 1 - \varphi_\mu(t)dt = \int_{\mathbb{R}}\int_{-c}^{c} 1 - e^{itx}dt\mu(dx) = \int_{\mathbb{R}}\int_{-c}^{c} 1 - \cos(tx)dt\mu(dx)$$
$$= 2c\int_{\mathbb{R}} 1 - \frac{\sin(cx)}{cx}\mu(dx) \geq 2c\int_{|cx|\geq 2} 1 - \frac{\sin(cx)}{cx}\mu(dx).$$

For $x \geq 2$, we have $\sin(x) \leq x/2$. Hence,

$$\int_{-c}^{c} 1 - \varphi_\mu(t)dt \geq 2c\int_{|cx|\geq 2}\frac{1}{2}\mu(dx) = c\mu(\{x \in \mathbb{R} \,|\, |cx| \geq 2\}).$$

The statement now follows by taking $c = 2/r$. □

THEOREM 2.29 (Lévy's continuity theorem). *Let $\mu, \mu_1, \mu_2, \cdots \in \mathcal{P}(\mathbb{R})$. Then $\mu_n \to \mu$ weakly if and only if $\varphi_{\mu_n}(t) \to \varphi_\mu(t)$ for all $t \in \mathbb{R}$.*

PROOF. If $\mu_n \to \mu$, then $\varphi_{\mu_n}(t) \to \varphi_\mu(t)$ for all $t \in \mathbb{R}$ by the definition of weak convergence.

Conversely, assume that $\varphi_{\mu_n}(t) \to \varphi_\mu(t)$ for all $t \in \mathbb{R}$. Let $r > 0$. By Lemma 2.28, we get

$$\limsup_{n\to\infty} \mu_n(\{x \in \mathbb{R} \,|\, |x| > r\}) \le \lim_{n\to\infty} \frac{r}{2} \int_{-2/r}^{2/r} 1 - \varphi_{\mu_n}(t)dt = \frac{r}{2} \int_{-2/r}^{2/r} 1 - \varphi_\mu(t)dt.$$

The right-hand side converges to 0 as $r \to \infty$. Thus, given $\varepsilon > 0$, we find $r > 0$ large enough such that $\mu_n(\{x \in \mathbb{R} \,|\, |x| > r\}) \le \varepsilon$ for all n and also $\mu(\{x \in \mathbb{R} \,|\, |x| > r\}) \le \varepsilon$.

Now let $f : \mathbb{R} \to \mathbb{C}$ be a bounded continuous function, bounded by $M > 0$. Denote by f_r the restriction of f to all $|x| \le r$, and extend f_r to a continuous function $\tilde{f} : \mathbb{R} \to \mathbb{C}$, which is again bounded by M and $2\pi r$-periodic. Let g be a linear combination of functions $e^{ikx/r}$, $k \in \mathbb{Z}$, such that $|\tilde{f}(x) - g(x)| \le \varepsilon$ for all $x \in \mathbb{R}$ (Weierstraß approximation theorem for trigonometric polynomials). We obtain

$$\begin{aligned} &\left| \int_{\mathbb{R}} f(x)\mu_n(dx) - \int_{\mathbb{R}} g(x)\mu_n(dx) \right| \\ &\le \mu_n(\{x \in \mathbb{R} \,|\, |x| > r\}) \sup_{x\in\mathbb{R}} |f(x) - g(x)| + \sup_{x\in\mathbb{R}} |\tilde{f}(x) - g(x)|. \end{aligned}$$

The function $g = \tilde{f} + g - \tilde{f}$ is bounded by $M + \varepsilon$, so $\sup_{x\in\mathbb{R}} |f(x) - g(x)| \le 2M + \varepsilon$, and hence,

$$\left| \int_{\mathbb{R}} f(x)\mu_n(dx) - \int_{\mathbb{R}} g(x)\mu_n(dx) \right| \le \varepsilon(2M + \varepsilon) + \varepsilon = (2M + \varepsilon + 1)\varepsilon.$$

This is true for any $n \in \mathbb{N}$ and also for μ instead of μ_n. Thus,

$$\begin{aligned} \left| \int_{\mathbb{R}} f(x)\mu_n(dx) - \int_{\mathbb{R}} f(x)\mu(dx) \right| &\le \left| \int_{\mathbb{R}} g(x)\mu_n(dx) - \int_{\mathbb{R}} g(x)\mu(dx) \right| \\ + \left| \int_{\mathbb{R}} f(x)\mu_n(dx) - \int_{\mathbb{R}} g(x)\mu_n(dx) \right| &+ \left| \int_{\mathbb{R}} g(x)\mu(dx) - \int_{\mathbb{R}} f(x)\mu(dx) \right| \\ \le \left| \int_{\mathbb{R}} g(x)\mu_n(dx) - \int_{\mathbb{R}} g(x)\mu(dx) \right| &+ 2(2M + \varepsilon + 1)\varepsilon. \end{aligned}$$

As $\varphi_{\mu_n}(t) \to \varphi_\mu(t)$ for all $t \in \mathbb{R}$, we can choose $n \in \mathbb{N}$ large enough to have

$$\left| \int_{\mathbb{R}} g(x)\mu_n(dx) - \int_{\mathbb{R}} g(x)\mu(dx) \right| \le \varepsilon.$$

As f was chosen arbitrarily, we conclude that μ_n converges weakly to μ as $n \to \infty$. □

REMARK 2.30. The weak convergence turns $\mathcal{P}(\mathbb{R})$ into a topological space, and its topology is induced by a metric. Define the Lévy distance between $\mu, \nu \in \mathcal{P}(\mathbb{R})$ as

$$d_{\text{Lévy}}(\mu, \nu) = \inf\{\delta > 0 \,|\, \mu((-\infty, x - \delta]) - \delta \le \nu((-\infty, x]) \le \mu((-\infty, x + \delta]) + \delta \text{ for all } x \in \mathbb{R}\}.$$

Then $\mu_n \to \mu$ if and only if $d_{\text{Lévy}}(\mu_n, \mu) \to 0$, see [**41**, Section 7].

2.4. Central limit theorem

Assume that the expectation $c = \mathbb{E}[X] \in \mathbb{R}$ of a random variable X exists. The law of large numbers states that the arithmetic average of independent samples of X, the sample mean, converges to c as the number of the samples tends to ∞. This clarifies in which sense we should *expect* c.

THEOREM 2.31 (Strong law of large numbers). *If $X_1, X_2, \dots$ are iid random variables with finite mean $c = \mathbb{E}[X_k]$, then $(X_1 + \cdots + X_n)/n$ converges with probability 1 to c.*

PROOF. A proof can be found in [**103**, Theorem 3.23]. □

In particular, the distribution μ_n of $(X_1 + \cdots + X_n)/n$ converges weakly to the distribution δ_c. This convergence is further refined by the famous central limit theorem. We first need a small auxiliary lemma.

LEMMA 2.32. *Let $z_1, \dots, z_n, w_1, \dots, w_n$ be complex numbers with $|z_k| \leq 1$, $|w_k| \leq 1$ for all k. Then*

$$\left| \prod_{k=1}^{n} z_k - \prod_{k=1}^{n} w_k \right| \leq \sum_{k=1}^{n} |z_k - w_k|.$$

PROOF. This is proved by induction. For $n = 1$, we have equality. So assume the statement holds for some $n \in \mathbb{N}$. We have

$$\left| \prod_{k=1}^{n+1} z_k - \prod_{k=1}^{n+1} w_k \right| = \left| (z_{n+1} - w_{n+1}) \left(\prod_{k=1}^{n} z_k \right) + w_{n+1} \left(\prod_{k=1}^{n} z_k - \prod_{k=1}^{n} w_k \right) \right|$$

$$\leq |z_{n+1} - w_{n+1}| + \left| \prod_{k=1}^{n} z_k - \prod_{k=1}^{n} w_k \right| \leq \sum_{k=1}^{n+1} |z_k - w_k|.$$ □

THEOREM 2.33 (Central limit theorem). *If $X_1, X_2, \dots$ are iid random variables with finite mean c and finite positive variance σ^2, then*

$$S_n := (X_1 + \cdots + X_n - nc)/(\sigma\sqrt{n})$$

converges in distribution to the normal distribution $\mathcal{N}(0,1)$.

PROOF. Put $Y_k = (X_k - c)/\sigma$ and let φ be the characteristic function of Y_k, which does not depend on k as all Y_k are *iid*. By Theorem 2.23 we have

$$\varphi_{S_n}(t) = \varphi(t/\sqrt{n})^n.$$

Fix $t \in \mathbb{R} \setminus \{0\}$. Then $\varphi(t/\sqrt{n}) = 1 - \frac{t^2}{2n} + o(1/n)$ by Theorem 2.19. Assume that n is so large that $|\varphi(t/\sqrt{n})| \leq 1$ and $|1 - \frac{t^2}{2n}| \leq 1$. Lemma 2.32 implies

$$\left| \left(1 - \frac{t^2}{2n} + o(1/n)\right)^n - \left(1 - \frac{t^2}{2n}\right)^n \right| \leq n o(1/n) \to 0$$

as $n \to \infty$. As $\left(1 - \frac{t^2}{2n}\right)^n \to e^{-t^2/2}$, we have $\varphi_{S_n}(t) \to e^{-t^2/2}$ as $n \to \infty$, which is the characteristic function of the normal distribution, see Example 2.17 (d). Lévy's continuity theorem implies that S_n converges in distribution to $\mathcal{N}(0,1)$. □

EXAMPLE 2.34. Imagine there are N lectures being held at a university in N different classrooms. Also, assume that the numbers of students attending these lectures are all positive. (Rumor has it that there have been math professors teaching in front of zero students, talking to the blackboard as they would with an audience of positive size.) We may assume that the heights of the students are *iid* random variables. Therefore, if we go to each classroom and calculate the average height, we obtain N numbers, and the histogram of these averages will have the shape of a normal distribution by the central limit theorem. Figure 2.3 shows histograms derived from simulations with $N = 100$ and $N = 100{,}000$ classrooms, respectively. We put 100 students in each classroom and model their heights by independent random variables with a uniform distribution μ between 1.5 and 1.7 meters.

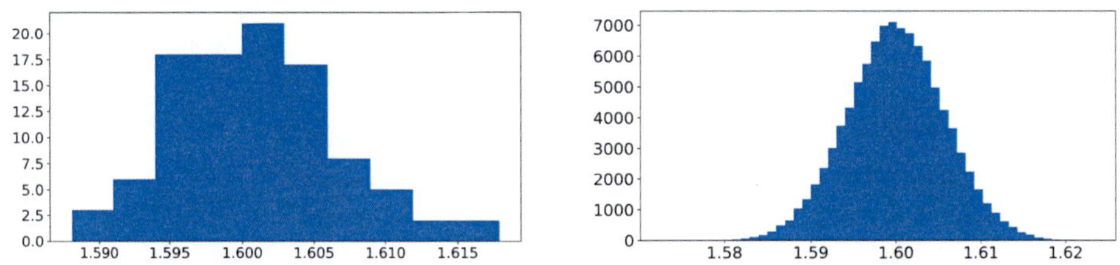

FIGURE 2.3. Histograms for the average heights.

The mean of μ is 1.6 and the variance is given by $\sigma^2 = (1.7 - 1.5)^2/12$. If we model the $n = 100$ heights by *iid* random variables $X_1, \ldots, X_n$, each with distribution μ, then the distribution of $\frac{X_1+\cdots+X_n-n\cdot 1.6}{\sigma\sqrt{n}}$ can be approximated by the normal distribution $\mathcal{N}(0,1)$, and thus:

$$\text{distribution of } \frac{X_1 + \cdots + X_n}{n} \approx \mathcal{N}(1.6, \sigma^2/n).$$

The normal distribution everywhere:

The normal distribution appears widely in statistical models and real-world data. The central limit theorem explains why: when random variables arise as the sum of many independent noise terms, their distribution tends toward a normal shape.

The assumption in the previous example that adult human height follows a uniform distribution was simplified. In reality, height (for a group like adult males or females) is closer to a normal distribution. This can be understood by considering that height depends on numerous genetic and environmental factors. Say an individual's final height, X, represents the sum of monthly growth increments $G_1, \ldots, G_{216}$ over 18 years. By treating the G_k as independent and identically distributed variables, the distribution of the sum X approximates a normal distribution.

The normal distribution not everywhere:

Not all random variables around us follow a normal distribution. Let us return to the example of measuring student heights in classrooms. Instead of collecting the N arithmetic averages of classroom heights, we could record all N maximum heights, medians, 0.75-quantiles, and so on. How are these values distributed? It is not always possible to provide nice analytic characterizations. In case of the maximum (or minimum) values, there exist well-known limit theorems (*extreme value theory*). For practical purposes, simulations are often sufficient to gain reasonable insight. Figure 2.4 shows the result of the simulation from Example 2.34 when we replace the average by the maximum height.

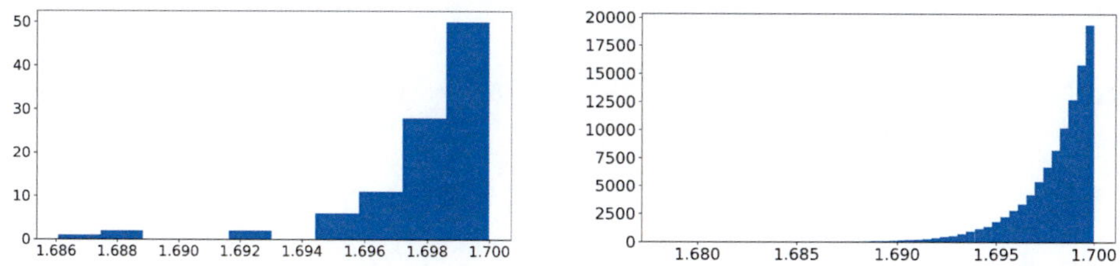

FIGURE 2.4. Histograms for the largest heights.

Apparently, there should be a limiting distribution for the maximum that has a simple exponential shape. In case of our uniform distribution, this limit can be calculated quite easily. For $a \in \mathbb{R}$ and $b > 0$, let $\mu_{Wei}(a, b)$ be the distribution given by the density $be^{b(x-a)}\mathbf{1}_{(-\infty,a]}(x)$ (a shifted Weibull distribution).

If $X_1, \ldots, X_n$, with $n = 100$, are *iid* random variables with the uniform distribution μ from Example 2.34, then the distribution of $n \cdot \frac{\max\{X_1,\ldots,X_n\}-1.7}{0.2}$ can be approximated by $\mu_{Wei}(0, 1)$, see Exercise 2.8.10, and thus:

$$\text{distribution of } \max\{X_1, \ldots, X_n\} \approx \mu_{Wei}(1.7, n/0.2).$$

REMARK 2.35. We conclude this section by noting that the Poisson distribution, given by $\mathbb{P}[X = k] = \frac{\lambda^k e^{-\lambda}}{k!}$ for $k \in \mathbb{N}_0$, is also an important limiting distribution. It appears as the limit of a binomial distribution in the following way (see [**103**, Theorem 4.7] for a more general Poisson limit theorem).

For $n \in \mathbb{N}$, let $(X_{k,n})_{1 \le k \le n}$ be *iid* random variables with

$$\mathbb{P}[X_{k,n} = 1] = 1 - \mathbb{P}[X_{k,n} = 0] = c_n,$$

and assume that $nc_n \to c > 0$ as $n \to \infty$. We determine the limiting distribution of $X_{1,n} + \cdots + X_{n,n}$ similarly to the proof of the central limit theorem. We have $\varphi_{X_{k,n}}(t) = 1 - c_n + c_n e^{it}$. Let $\psi_n(t)$ be the characteristic function of $X_{1,n} + \cdots + X_{n,n}$. Then

$$\psi_n(t) = (1 - c_n + c_n e^{it})^n = \left(1 + \frac{nc_n e^{it} - nc_n}{n}\right)^n = \left(1 + \frac{ce^{it} - c}{n} + o(1/n)\right)^n.$$

By Lemma 2.32, $\lim_{n\to\infty} \psi_n(t) = e^{c(e^{it}-1)}$, which is the characteristic function of the Poisson distribution with mean c, see Example 2.17 (b). Lévy's continuity theorem implies that $X_{1,n} + \cdots + X_{n,n}$ converges in distribution to the Poisson distribution with mean c.

2.5. Conditional expectation

Let X be a random variable on $P_1 = (\Omega, \mathcal{F}, \mathbb{P})$. If $\mathcal{G} \subset \mathcal{F}$ is a sub-σ-algebra, we can pass to the probability space $P_2 = (\Omega, \mathcal{G}, \mathbb{P}|_{\mathcal{G}})$. This process can be viewed as simplifying the model P_1, since the information encoded by $\mathcal{F}$ is now reduced to that of $\mathcal{G}$. Can we also "simplify" X to a $\mathcal{G}$-measurable random variable $\hat{X}$? If X is integrable, we would like to have

$$\int_A \hat{X}(\omega) d\mathbb{P}(\omega) = \int_A X(\omega) d\mathbb{P}(\omega)$$

for all $A \in \mathcal{G}$. Indeed, this is possible, and $\hat{X} = \mathbb{E}[X|\mathcal{G}]$ is called the *conditional expectation* with respect to the σ-algebra $\mathcal{G}$.

THEOREM 2.36. *There exists an almost surely unique linear operator* $\mathbb{E}[\cdot|\mathcal{G}] : L^1(P_1) \to L^1(P_2)$ *such that*

$$\mathbb{E}[\mathbb{E}[X|\mathcal{G}] \cdot \mathbf{1}_A] = \mathbb{E}[X \cdot \mathbf{1}_A] \tag{2.6}$$

for all $X \in L^1(P_1)$ *and* $A \in \mathcal{G}$*. Furthermore, the following properties hold:*

(a) *(L^1-contractivity)* $\mathbb{E}[|\mathbb{E}[X|\mathcal{G}]|] \leq \mathbb{E}[|X|]$ *for all* $X \in L^1(P_1)$.
(b) *(Positivity) If* $X \in L^1(P_1)$ *and* $X \geq 0$*, then* $\mathbb{E}[X|\mathcal{G}] \geq 0$ *a.s.*
(c) *($L^1(P_2)$-linearity) If* $X \in L^1(P_1)$ *and* $Y \in L^1(P_2)$*, then* $\mathbb{E}[XY|\mathcal{G}] = Y \cdot \mathbb{E}[X|\mathcal{G}]$ *a.s.*

PROOF. First, assume that $X \in L^2(P_1)$. We regard $L^2(P_1)$ as a Hilbert space with $\langle X, Y\rangle = \mathbb{E}[\overline{X}Y]$ as in Remark 2.9. (Formally, it contains equivalence classes of random variables whose elements differ only on sets of measure 0.) Then $L^2(P_2)$ is a closed linear subspace of $L^2(P_1)$ and we can define $\mathbb{E}[X|\mathcal{G}]$ as the projection of X to $L^2(P_2)$. Then $\mathbb{E}[X|\mathcal{G}]$ is a $\mathcal{G}$-measurable random variable on Ω, uniquely defined almost everywhere. We have $\langle f, \mathbb{E}[X|\mathcal{G}]\rangle = \langle f, X\rangle$ for all $f \in L^2(P_2)$, which implies (2.6).

For $A = \{\omega \in \Omega \,|\, \mathbb{E}[X|\mathcal{G}](\omega) \geq 0\}$, we get

$$\mathbb{E}[|\mathbb{E}[X|\mathcal{G}]|] = \mathbb{E}[\mathbb{E}[X|\mathcal{G}]\mathbf{1}_A] - \mathbb{E}[\mathbb{E}[X|\mathcal{G}]\mathbf{1}_{\Omega\setminus A}] = \mathbb{E}[X\mathbf{1}_A] - \mathbb{E}[X\mathbf{1}_{\Omega\setminus A}] \leq \mathbb{E}[|X|],$$

which is property (a). It follows that the mapping $\mathbb{E}[\cdot|\mathcal{G}]$ is uniformly $L^1(P_1)$-continuous on $L^2(P_1)$. As $L^2(P_1)$ is dense in $L^1(P_1)$, we can extend $\mathbb{E}[\cdot|\mathcal{G}]$ uniquely to a linear and continuous mapping on $L^1(P_1)$, and (a) holds for all $X \in L^1(P_1)$.

Assume that $X \in L^1(P_1)$ with $X \geq 0$. Let $A = \{\omega \in \Omega \,|\, \mathbb{E}[X|\mathcal{G}](\omega) < 0\}$. Then

$$\mathbb{E}[\mathbb{E}[X|\mathcal{G}] \cdot \mathbf{1}_A] = \mathbb{E}[X \cdot \mathbf{1}_A] \geq 0.$$

This implies $\mathbb{P}(A) = 0$, and thus, $\mathbb{E}[X|\mathcal{G}] \geq 0$ a.s., which proves (b).

Finally, let $X \in L^2(P_1)$ and $Y \in L^2(P_2)$. We have $\langle \mathbb{E}[XY|\mathcal{G}], f\rangle = \langle XY, f\rangle$, and also

$$\langle Y\mathbb{E}[X|\mathcal{G}], f\rangle = \langle \mathbb{E}[X|\mathcal{G}], \overline{Y}f\rangle = \langle X, \overline{Y}f\rangle = \langle XY, f\rangle$$

for all $f \in L^2(P_2)$. This implies that $\mathbb{E}[XY|\mathcal{G}] = Y\mathbb{E}[X|\mathcal{G}]$ almost surely. The general case $X \in L^1(P_1)$, $Y \in L^1(P_2)$ follows by approximation. □

EXAMPLE 2.37. The extreme cases are $\mathcal{G} = \mathcal{F}$ and $\mathcal{G} = \{\emptyset, \Omega\}$, for which we obtain

$$\mathbb{E}[X|\mathcal{F}] = X \text{ a.s.} \quad \text{and} \quad \mathbb{E}[X|\{\emptyset, \Omega\}] = \mathbb{E}[X] \text{ a.s.}$$

EXAMPLE 2.38. Let $B \in \mathcal{F}$ with $p := \mathbb{P}(B) \in (0,1)$. Another simple case is $\mathcal{G} = \sigma(B) = \{\emptyset, B, \Omega \setminus B, \Omega\}$. We have

$$\mathbb{E}[X|\mathcal{G}] = p^{-1}\mathbb{E}[X\mathbf{1}_B]\mathbf{1}_B + (1-p)^{-1}\mathbb{E}[X\mathbf{1}_{\Omega\setminus B}]\mathbf{1}_{\Omega\setminus B} \quad \text{a.s.}$$

Note the similarity to the projection in vector spaces: $\mathbf{1}_B/\sqrt{p} = \mathbf{1}_B/\|\mathbf{1}_B\| =: v$ and $\mathbf{1}_{\Omega\setminus B}/\sqrt{1-p} = \mathbf{1}_{\Omega\setminus B}/\|\mathbf{1}_{\Omega\setminus B}\| =: w$ are vectors of norm 1 and

$$\mathbb{E}[X|\mathcal{G}] = \mathbb{E}[Xv]v + \mathbb{E}[Xw]w = \langle v, X\rangle\, v + \langle w, X\rangle\, w \quad \text{a.s.}$$

LEMMA 2.39. *Let $X, Y \in L^1(P_1)$. Then*

$$\mathbb{E}[X \cdot \mathbb{E}[Y|\mathcal{G}]] = \mathbb{E}[Y \cdot \mathbb{E}[X|\mathcal{G}]].$$

PROOF. We use Theorem 2.36 (c) to obtain:

$$\begin{aligned}\mathbb{E}[X \cdot \mathbb{E}[Y|\mathcal{G}]] &= \mathbb{E}[\mathbb{E}[X \cdot \mathbb{E}[Y|\mathcal{G}]|\mathcal{G}]] = \mathbb{E}[\mathbb{E}[X|\mathcal{G}] \cdot \mathbb{E}[Y|\mathcal{G}]],\\ \mathbb{E}[Y \cdot \mathbb{E}[X|\mathcal{G}]] &= \mathbb{E}[\mathbb{E}[Y \cdot \mathbb{E}[X|\mathcal{G}]|\mathcal{G}]] = \mathbb{E}[\mathbb{E}[X|\mathcal{G}] \cdot \mathbb{E}[Y|\mathcal{G}]].\end{aligned}$$ □

The conditional probability of an event $A \in \mathcal{F}$, given a sub-σ-algebra $\mathcal{G} \subset \mathcal{F}$, is defined as

$$\mathbb{P}[A|\mathcal{G}] = \mathbb{E}[\mathbf{1}_A|\mathcal{G}].$$

Then $\mathbb{P}[A|\mathcal{G}]$ is a random variable with $0 \le \mathbb{P}[A|\mathcal{G}] \le 1$ a.s., and for all $B \in \mathcal{G}$,

$$\mathbb{E}[\mathbb{P}[A|\mathcal{G}] \cdot \mathbf{1}_B] = \mathbb{E}[\mathbf{1}_A \cdot \mathbf{1}_B] = \mathbb{P}(A \cap B).$$

EXAMPLE 2.40. For $X = \mathbf{1}_A$ in Example 2.38, we obtain

$$\begin{aligned}\mathbb{P}[A|\mathcal{G}] &= \frac{\mathbb{E}[\mathbf{1}_A\mathbf{1}_B]\mathbf{1}_B}{\mathbb{P}(B)} + \frac{\mathbb{E}[\mathbf{1}_A\mathbf{1}_{\Omega\setminus B}]\mathbf{1}_{\Omega\setminus B}}{\mathbb{P}(\Omega\setminus B)}\\ &= \frac{\mathbb{P}(A\cap B)\mathbf{1}_B}{\mathbb{P}(B)} + \frac{\mathbb{P}(A\cap(\Omega\setminus B))\mathbf{1}_{\Omega\setminus B}}{\mathbb{P}(\Omega\setminus B)} \quad \text{a.s.}\end{aligned}$$

For two events A, B with $\mathbb{P}(B) > 0$,

$$\mathbb{P}[A|B] := \frac{\mathbb{P}(A\cap B)}{\mathbb{P}(B)}$$

is called the *conditional probability* of A given B.

DEFINITION 2.41. Let $\mathcal{F}_1, \mathcal{F}_2, \mathcal{G}$ be sub-σ-algebras of $\mathcal{F}$. Then $\mathcal{F}_1$ and $\mathcal{F}_2$ are called *conditionally independent given* $\mathcal{G}$ if

$$\mathbb{P}[A_1 \cap A_2|\mathcal{G}] = \mathbb{P}[A_1|\mathcal{G}] \cdot \mathbb{P}[A_2|\mathcal{G}] \quad \text{a.s. for all } A_1 \in \mathcal{F}_1,\ A_2 \in \mathcal{F}_2.$$

EXAMPLE 2.42. Let $A_1, A_2 \in \mathcal{F}$ and $\mathcal{F}_1 = \sigma(A_1), \mathcal{F}_2 = \sigma(A_2)$. We choose $\mathcal{G}$ as in Example 2.38. Then

$$\begin{aligned}\mathbb{P}[A_1 \cap A_2|\mathcal{G}] &= \frac{\mathbb{P}(A_1\cap A_2\cap B)\mathbf{1}_B}{\mathbb{P}(B)} + \frac{\mathbb{P}(A_1\cap A_2\cap(\Omega\setminus B))\mathbf{1}_{\Omega\setminus B}}{\mathbb{P}(\Omega\setminus B)},\\ \mathbb{P}[A_1|\mathcal{G}] \cdot \mathbb{P}[A_2|\mathcal{G}] &= \frac{\mathbb{P}(A_1\cap B)\mathbb{P}(A_2\cap B)\mathbf{1}_B}{\mathbb{P}(B)^2}\\ &\quad + \frac{\mathbb{P}(A_1\cap(\Omega\setminus B))\mathbb{P}(A_2\cap(\Omega\setminus B))\mathbf{1}_{\Omega\setminus B}}{\mathbb{P}(\Omega\setminus B)^2},\end{aligned}$$

and $\mathcal{F}_1$ and $\mathcal{F}_2$ are conditionally independent given $\mathcal{G}$ if and only if

$$\mathbb{P}[A_1 \cap A_2|B] = \mathbb{P}[A_1|B] \cdot \mathbb{P}[A_2|B].$$

THEOREM 2.43. *Assume that $\mathcal{G} \subset \mathcal{F}_2$. Then $\mathcal{F}_1$ and $\mathcal{F}_2$ are conditionally independent given $\mathcal{G}$ if and only if*

$$\mathbb{P}[A|\mathcal{F}_2] = \mathbb{P}[A|\mathcal{G}] \quad \textit{a.s. for all } A \in \mathcal{F}_1. \tag{2.7}$$

PROOF. Assume that $\mathcal{F}_1$ and $\mathcal{F}_2$ are conditionally independent given $\mathcal{G}$, and let $A_1 \in \mathcal{F}_1$. Since $\mathcal{G} \subset \mathcal{F}_2$, $\mathbb{E}[\mathbf{1}_{A_1}|\mathcal{G}]$ is $\mathcal{F}_2$-measurable. If we can show that $\mathbb{E}[\mathbb{E}[\mathbf{1}_{A_1}|\mathcal{G}]\mathbf{1}_{A_2}] = \mathbb{E}[\mathbf{1}_{A_1}\mathbf{1}_{A_2}]$ for all $A_2 \in \mathcal{F}_2$, it follows that $\mathbb{E}[\mathbf{1}_{A_1}|\mathcal{G}] = \mathbb{E}[\mathbf{1}_{A_1}|\mathcal{F}_2]$ a.s. by the definition of conditional independence (2.6). We have

$$\begin{aligned}\mathbb{E}[\mathbf{1}_{A_1}\mathbf{1}_{A_2}] &= \mathbb{E}[\mathbf{1}_{A_1}\mathbf{1}_{A_2}\mathbf{1}_\Omega] = \mathbb{E}[\mathbb{E}[\mathbf{1}_{A_1}\mathbf{1}_{A_2}|\mathcal{G}]\mathbf{1}_\Omega] = \mathbb{E}[\mathbb{E}[\mathbf{1}_{A_1}\mathbf{1}_{A_2}|\mathcal{G}]] \\ &= \mathbb{E}[\mathbb{E}[\mathbf{1}_{A_1\cap A_2}|\mathcal{G}]] = \mathbb{E}[\mathbb{E}[\mathbf{1}_{A_1}|\mathcal{G}]\mathbb{E}[\mathbf{1}_{A_2}|\mathcal{G}]] = \mathbb{E}[\mathbb{E}[\mathbb{E}[\mathbf{1}_{A_1}|\mathcal{G}]\mathbf{1}_{A_2}|\mathcal{G}]] \\ &= \mathbb{E}[\mathbb{E}[\mathbf{1}_{A_1}|\mathcal{G}]\mathbf{1}_{A_2}].\end{aligned}$$

Conversely, suppose that (2.7) holds. Let $A_1 \in \mathcal{F}_1$, $A_2 \in \mathcal{F}_2$. Then

$$\begin{aligned}\mathbb{E}[\mathbf{1}_{A_1\cap A_2}|\mathcal{G}] &= \mathbb{E}[\mathbf{1}_{A_1}\mathbf{1}_{A_2}|\mathcal{G}] = \mathbb{E}[\mathbb{E}[\mathbf{1}_{A_1}\mathbf{1}_{A_2}|\mathcal{F}_2]|\mathcal{G}] = \mathbb{E}[\mathbf{1}_{A_2}\mathbb{E}[\mathbf{1}_{A_1}|\mathcal{F}_2]|\mathcal{G}] \\ &= \mathbb{E}[\mathbf{1}_{A_2}\mathbb{E}[\mathbf{1}_{A_1}|\mathcal{G}]|\mathcal{G}] = \mathbb{E}[\mathbf{1}_{A_1}|\mathcal{G}]\mathbb{E}[\mathbf{1}_{A_2}|\mathcal{G}] \quad \text{a.s.}\end{aligned}$$

□

2.6. Markov processes

A stochastic process is a family $(X_t)_{t\in T}$ of random variables defined on a common probability space $(\Omega, \mathcal{F}, \mathbb{P})$, where $T \subset \mathbb{R}$ is a non-empty index set. It can also be viewed as a random function via the *sample paths*

$$t \mapsto X_t(\omega).$$

For $t \in T$, the σ-algebra generated by the set $\cup_{s\in T, s\le t}\sigma(X_s)$ is denoted by $\sigma((X_s)_{s\le t})$. It encodes all the information described by the stochastic process up to time t.

REMARK 2.44. A stochastic process $(X_t)_{t\in T}$ is often associated with a filtration $(\mathcal{F}_t)_{t\in T}$, which is a family of σ-subalgebras of $\mathcal{F}$ such that $\mathcal{F}_s \subset \mathcal{F}_t$ whenever $s \le t$. This filtration describes an increasing amount of information over time. The process (X_t) is called *adapted* to $(\mathcal{F}_t)$ if X_t is $\mathcal{F}_t$-measurable for every $t \in T$, meaning that X_t cannot "see into the future" of our available information. Every stochastic process is adapted to its natural filtration $\mathcal{F}_t = \sigma((X_s)_{s\le t})$. For our purposes, it will be enough to consider this filtration only.

The distributions of all X_t may be as interdependent as one desires. A stochastic process is said to have the *Markov property* if, given the present state, the future is independent of the past. On the one hand, this property is natural for many applications, on the other hand, it defines a class of processes having a rich mathematical theory.

DEFINITION 2.45. Let $T = [0, \infty)$ or $T = \mathbb{N}_0$, and let $S \subset \mathbb{C}$ be a Borel subset, the state space, equipped with its Borel σ-algebra $\mathcal{B}(S)$. An S-valued stochastic process $(X_t)_{t\in T}$ on a probability space $(\Omega, \mathcal{F}, \mathbb{P})$ is called a *Markov process* if, for all $s, t \in T$ with $s \le t$,

$$\sigma((X_\tau)_{\tau\le s}) \text{ and } \sigma(X_t) \text{ are conditionally independent given } \sigma(X_s).$$

In the case $T = \mathbb{N}_0$, a Markov process is also called a *Markov chain*.

With $\mathcal{F}_s = \sigma((X_\tau)_{\tau \le s})$, we have $\sigma(X_s) \subset \mathcal{F}_s$. We can rephrase the Markov property by using Theorem 2.43: $(X_t)_{t \in T}$ is a Markov process if and only if

$$\mathbb{P}[A|\mathcal{F}_s] = \mathbb{P}[A|\sigma(X_s)] \quad \text{a.s.}$$

for all $A \in \sigma(X_t)$ and all $s, t \in T$ with $s \le t$.

REMARK 2.46. This definition implies that $\sigma((X_\tau)_{\tau \le s})$ and $\sigma((X_\tau)_{\tau \ge s})$ are conditionally independent given $\sigma(X_s)$, see [**103**, Lemma 7.1].

DEFINITION 2.47. A *probability kernel* k on $(S, \mathcal{B}(S))$ is a map $k : S \times \mathcal{B}(S) \to [0, 1]$ such that

(i) $B \mapsto k(x, B)$ is a probability measure for each $x \in S$,
(ii) $x \mapsto k(x, B)$ is a measurable function for each $B \in \mathcal{B}(S)$.

For two probability kernels k_1 and k_2, we define its composition

$$(k_1 \star k_2)(x, B) = \int_S k_2(y, B) k_1(x, dy) \qquad \text{for } x \in S, B \in \mathcal{B}(S).$$

Two S-valued random variables X, Y produce a kernel k such that $\mathbb{P}[Y \in B|\sigma(X)] = k(X, B)$ almost surely, see [**103**, Theorem 5.3]. A Markov process thus produces a family $k_{s,t}$ of probability kernels, called *transition kernels*, where $s, t \in T$ with $s \le t$, such that

$$k_{s,t}(X_s, B) = \mathbb{P}[X_t \in B|\sigma(X_s)] = \mathbb{P}[X_t \in B|\mathcal{F}_s]$$

almost surely, for every $B \in \mathcal{B}(S)$. We can rewrite this as

$$\int_S \mathbf{1}_B(x) k_{s,t}(X_s, dx) = k_{s,t}(X_s, B) = \mathbb{P}[X_t \in B|\mathcal{F}_s] = \mathbb{E}[\mathbf{1}_{X_t \in B}|\mathcal{F}_s],$$

and by linear extension and approximation, this implies

$$\mathbb{E}[f(X_t)|\mathcal{F}_s] = \int_S f(x) k_{s,t}(X_s, dx) \tag{2.8}$$

almost surely for all bounded measurable functions $f : S \to \mathbb{C}$.

THEOREM 2.48. *Let $(k_{s,t})_{s,t \in T, s \le t}$ be the transition kernels of a Markov process, and let $s, t, u \in T$ with $s \le t \le u$. Then*

$$k_{s,u} = k_{s,t} \star k_{t,u}. \qquad \textbf{\textit{(Chapman–Kolmogorov relation)}} \tag{2.9}$$

(The relation holds almost surely with respect to the distribution of X_s.)

Conversely, let ν be a probability measure on S, and let $(k_{s,t})_{s,t \in T, s \le t}$ be a family of transition kernels on S that satisfies (2.9). *Then there exists a Markov process $(X_t)_{t \in T}$ on S with transition kernels $k_{s,t}$ and initial distribution ν, i.e., ν is the distribution of X_0.*

PROOF. The proof, in particular the construction of the process, relies on additional foundational results from probability theory. We refer to [**103**, Corollary 7.3] and [**103**, Theorem 7.4]. □

EXAMPLE 2.49. The simplest example of a Markov process is a Markov chain with a finite state space $S = \{x_1, \dots, x_N\}$. The transition kernel can then be represented by a transition matrix. For $s, t \in \mathbb{N}_0$, $s \le t$, we define the $N \times N$-matrix $P_{s,t}$ by

$$P_{s,t} = (p_{j,k,s,t})_{1 \le j,k \le N} \quad \text{with} \quad p_{j,k,s,t} = \mathbb{P}[X_t = x_j|X_s = x_k] \quad \text{a.s.}$$

(If $\mathbb{P}[X_s = x_k] = 0$, we let $\mathbb{P}[X_t = x_j | X_s = x_k]$ be arbitrary probabilities that sum up to 1, such that $P_{s,t}$ is a kernel.) The product $\star$ now simply becomes the matrix product, and the Chapman–Kolmogorov relation takes the form

$$P_{s,u} = P_{t,u} \cdot P_{s,t}.$$

If the Markov chain is time-homogeneous, i.e., $P_{s,t} = P_{0,t-s}$ for all $s \leq t$, then the Markov chain is uniquely determined by an initial distribution on S and the matrix $P_{0,1}$, since $P_{s,t} = P_{0,t-s} = P_{0,1}^{t-s}$.

Markov chains are of paramount importance for stochastic modeling, and the interested reader will find an abundance of applications, such as modeling processes in physics, chemistry, biology ([**147**], [**185**]), Markov chain Monte Carlo algorithms ([**50**]), models in queuing theory, and models for reinforcement learning ([**158**], [**183**]).

More on the theory of Markov processes and Markov chains can be found in the books [**40**], [**103**], and [**118**]. We will be particularly interested in the subclass of space-homogeneous Markov processes.

2.7. Space-homogeneous Markov processes

In this section, we let $T = [0, \infty)$ and $S = \mathbb{R}$.

DEFINITION 2.50. A Markov process $(X_t)_{t \in T}$ on $\mathbb{R}$ with transition kernels $k_{s,t}$ is called *space-homogeneous* if $k_{s,t}(x, B) = k_{s,t}(0, B - x)$ for all $s, t \in T$, with $s \leq t$, and all $x \in \mathbb{R}$, $B \in \mathcal{B}(\mathbb{R})$.

DEFINITION 2.51. A stochastic process $(X_t)_{t \in T}$ on $\mathbb{R}$ is said to have *independent increments* if

$$X_0, X_{t_1} - X_0, \ldots, X_{t_n} - X_{t_{n-1}}$$

are independent for any choice of $n \geq 1$ and $0 \leq t_1 \leq \cdots \leq t_n$.

If $X_0 = 0$ a.s., then $(X_t)_{t \in T}$ has independent increments if and only if the n increments $X_{t_1} - X_0, \ldots, X_{t_n} - X_{t_{n-1}} = X_{t_1}, \ldots, X_{t_n} - X_{t_{n-1}}$ are independent.

THEOREM 2.52 (Proposition 7.5 in [**103**]). *A stochastic process $(X_t)_{t \in T}$ is a space-homogeneous Markov process if and only if it has independent increments. In this case, the transition kernels are given by*

$$k_{s,t}(x, B) = \mathbb{P}[X_t - X_s \in B - x], \quad x \in \mathbb{R}, B \in \mathcal{B}(\mathbb{R}), s, t \in T, s \leq t.$$

The most important space-homogeneous Markov process, and the maybe most important stochastic process in all of probability theory, is the Brownian motion; see Figure 2.5.

DEFINITION 2.53 (Brownian motion). A stochastic process $(B_t)_{t \geq 0}$ is called a *Brownian motion* if

(1) $B_0 = 0$ almost surely,
(2) the distribution of B_t is $\mathcal{N}(0, t)$ for all $t > 0$,
(3) the increments $B_{t_1}, B_{t_2} - B_{t_1}, \ldots, B_{t_n} - B_{t_{n-1}}$ are independent for any choice of $n \geq 1$ and $0 \leq t_1 \leq \cdots \leq t_n$,
(4) the sample paths $t \mapsto B_t$ are continuous almost surely.

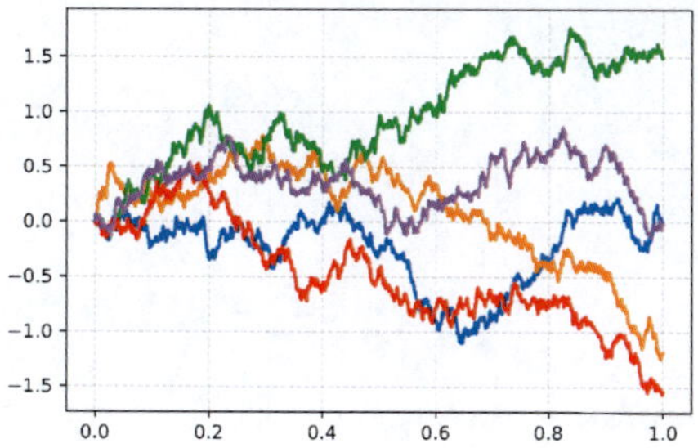

FIGURE 2.5. Sample paths of a Brownian motion.

A Brownian motion can be constructed in several ways. The first proof of its existence was given by N. Wiener in 1923. We follow Lévy's construction of the Brownian motion (from [**135**, Theorem 1.3]).

THEOREM 2.54. *The Brownian motion exists.*

PROOF. We first construct a Brownian motion on the interval $[0, 1]$. For $n \in \mathbb{N}_0$, let

$$D_n = \{k2^{-n} \mid k \in \mathbb{N}_0, 0 \le k \le 2^n\} \quad \text{and} \quad D = \bigcup_{n \in \mathbb{N}_0} D_n.$$

Then D is countable and there exists a probability space $(\Omega, \mathcal{F}, \mathbb{P})$ and a collection $(Z_t)_{t \in D}$ of *iid* random variables with distribution $\mathcal{N}(0, 1)$. First, we define random variables $(B_d)_{d \in D}$ by induction with respect to D_n. For $n = 0$, we let $B_0 = 0$ and $B_1 = Z_1$. Now let $n \ge 1$, and $d \in D_n \setminus D_{n-1}$. We let

$$B_d = \frac{B_{d-2^{-n}} + B_{d+2^{-n}}}{2} + \frac{Z_d}{2^{(n+1)/2}}.$$

It is easy to verify that the differences $B_d - B_{d-2^{-n}}$, $d \in D_n \setminus \{0\}$, are independent and have normal distribution $\mathcal{N}(0, 2^{-n})$.

Next, we define random functions $F_n : [0, 1] \to \mathbb{R}$ for each $n \in \mathbb{N}_0$. For $n = 0$, let $F_0(0) = 0, F_0(1) = B_1 = Z_1$, and define $F_0(t)$ on $(0, 1)$ by linear interpolation. For $n \ge 1$, let $F_n(t) = 0$ for $t \in D_{n-1}$, $F_n(t) = 2^{-(n+1)/2} Z_t$ for $t \in D_n \setminus D_{n-1}$, and define $F_n(t)$ for $t \notin D_n$ by linear interpolation. We have

$$B_d = \sum_{k=0}^{n} F_k(d) = \sum_{k=0}^{\infty} F_k(d), \quad d \in D_n.$$

Fix $c > \sqrt{2 \log 2} > 1$. As each Z_d is a $\mathcal{N}(0, 1)$ random variable, we have

$$\mathbb{P}[|Z_d| \ge c\sqrt{n}] \le 2e^{-c^2 n/2}$$

for $n \in \mathbb{N}_0$, see [**44**, Section 2.2, The Cramér-Chernoff method]. We conclude

$$\sum_{n=0}^{\infty} \mathbb{P}[\exists d \in D_n : |Z_d| \ge c\sqrt{n}] \le \sum_{n=0}^{\infty} \sum_{d \in D_n} \mathbb{P}[|Z_d| \ge c\sqrt{n}] \le \sum_{n=0}^{\infty} (2^n + 1) 2e^{-c^2 n/2} < \infty$$

as $c > \sqrt{2 \log 2}$. The Borel-Cantelli lemma ([**40**, Theorem 4.3]) implies that there exists a random $N \in \mathbb{N}$ that is almost surely finite such that $|Z_d| < c\sqrt{n}$ for all

$d \in D_n$ and $n > N$. In particular,

$$\sup_{t\in[0,1]} |F_n(t)| < c\sqrt{n}2^{-(n+1)/2}$$

for all $n > N$, and we conclude that, almost surely, the limit

$$B_t := \lim_{n\to\infty} \sum_{k=0}^{n} F_k(t)$$

exists with respect to uniform convergence on $[0,1]$. Furthermore, $t \mapsto B_t$ is continuous almost surely. The distribution of B_d, $d \in D$, is $\mathcal{N}(0,d)$, and thus, by approximation, the distribution of B_t is $\mathcal{N}(0,t)$ for all $t \in [0,1]$. Similarly, one can show that B has independent increments.

Finally, we construct a Brownian motion on $[0,\infty)$ by gluing together independent copies of $B|_{[0,1]}$. □

REMARK 2.55. Endow $C([0,1],\mathbb{R})$ with the topology induced by the maximum norm. The *Wiener measure* is the probability measure on

$$(C([0,1],\mathbb{R}),\mathcal{B}(C([0,1],\mathbb{R})))$$

that is induced by a Brownian motion, and it can be seen as a normal distribution for functions. Let $X_1, X_2, \ldots$ be a sequence of *iid* random variables with mean 0 and finite, positive variance σ^2. Define the random function $f_n \in C([0,1],\mathbb{R})$ by

$$f_n(k/n) = (X_1 + \cdots + X_k)/(\sigma\sqrt{n}), \text{ for } k = 0,\ldots,n,$$

and by linear interpolation between the points k/n. *Donsker's Theorem* states that, as $n \to \infty$, the distribution of f_n converges weakly to the Wiener measure, see [**41**, Theorem 8.2].

REMARK 2.56. The Brownian motion has many interesting properties, for which we refer to the lovely book [**135**]. For example, for any $t_0 \geq 0$, B_t is not differentiable at t_0 almost surely. More precisely, the modulus of continuity of a Brownian motion is given by

$$\limsup_{h\searrow 0} \sup_{0\leq t\leq 1-h} \frac{|B_{t+h} - B_t|}{\sqrt{2h\log(1/h)}} = 1,$$

see [**135**, Theorem 1.14]. The Brownian motion is the basis for constructing many further, more complicated stochastic processes, e.g., via stochastic differential equations, see [**146**].

Space-homogeneous Markov processes on $\mathbb{R}$ with a continuity property lead to additive processes and Lévy processes.

DEFINITION 2.57. A stochastic process $(X_t)_{t\geq 0}$ on $\mathbb{R}$ is called an *additive process* if the following three conditions are satisfied:

(1) $X_0 = 0$ almost surely.
(2) The increments $X_{t_0}, X_{t_1} - X_{t_0}, \ldots, X_{t_n} - X_{t_{n-1}}$ are independent for any choice of $n \geq 1$ and $0 \leq t_0 \leq t_1 \leq \cdots \leq t_n$.
(3) Continuity in probability: for any $\varepsilon > 0$ and $s \geq 0$, $\mathbb{P}[|X_{s+t} - X_s| > \varepsilon] \to 0$ as $t \to 0$.

Such a process is called a *Lévy process* if, in addition,

(4) the distribution of $X_{t+s} - X_s$ does not depend on s.

Lévy processes are not only space-homogeneous, but also time-homogeneous, i.e., the transition kernels satisfy

$$k_{s,t}(x, B) = k_{0,t-s}(x, B) = k_{0,t-s}(0, B - x).$$

Let us consider some examples of Lévy processes.

EXAMPLE 2.58. The deterministic process $X_t = a \cdot t$, with $a \in \mathbb{R}$, is clearly a Lévy process.

EXAMPLE 2.59. A Brownian motion is a Lévy process as sample path continuity implies continuity in probability. For the transition kernels $k_{s,t}$ we have $k_{0,t}(0, B) = \frac{1}{\sqrt{2\pi t}} \int_B e^{-y^2/(2t)} dy$ for $t > 0$, and thus,

$$k_{s,t}(x, B) = \frac{1}{\sqrt{2\pi(t-s)}} \int_B e^{-(x-y)^2/(2(t-s))} dy, \quad 0 \leq s < t.$$

EXAMPLE 2.60. Fix $\lambda > 0$ and let $Y_1, Y_2, \ldots$ be *iid* random variables such that the distribution of Y_k is the exponential distribution with parameter λ. The random variable $T_n = Y_1 + \cdots + Y_n$ has a $\Gamma(n, \lambda)$ distribution, given by the density

$$\frac{\lambda^n e^{-\lambda x} x^{n-1}}{(n-1)!}, \quad x \geq 0.$$

Define the stochastic process $(X_t)_{t\geq 0}$ by $X_t = \sum_{n=1}^{\infty} \mathbf{1}_{[0,t]}(T_n)$. $(X_t)_{t\geq 0}$ is called a *Poisson process* and it can be seen as a counting process for events happening at arrival times T_n, see Figure 2.6. If T_n is the time of the nth occurrence of an event, then X_t counts the number of such events occurring within the time interval $[0, t]$. With $T_0 := 0$, we can also write

$$X_t = \sup\{n \geq 0 \,|\, T_n \leq t\}.$$

The Poisson process is a Lévy process with

$$\mathbb{P}[X_t - X_s = n] = \frac{(\lambda(t-s))^n e^{-\lambda(t-s)}}{n!}, \quad n \in \mathbb{N}_0,$$

i.e., the increments are stationary with a Poisson distribution.

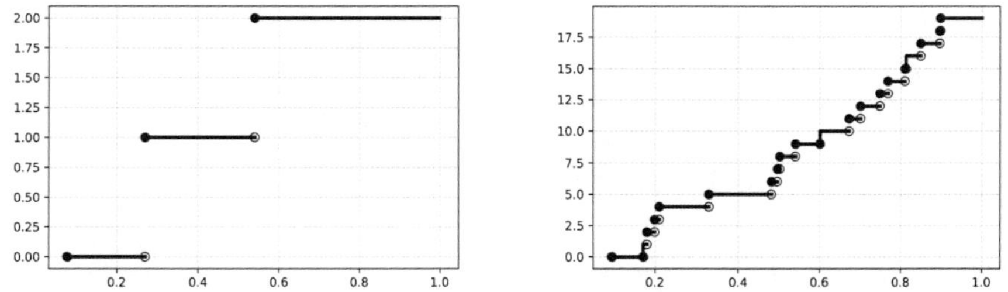

FIGURE 2.6. Sample paths of Poisson processes for $t \in [0, 1]$, with $\lambda = 2$ (left) and $\lambda = 20$ (right).

We can modify the Poisson process to obtain further examples of Lévy processes. Let $D_1, D_2, \ldots$ be *iid* random variables that are independent of $(X_t)_{t\geq 0}$. The Lévy process

$$Z_t := \sum_{n=1}^{X_t} D_n$$

is called a *compound Poisson process.*

The distributions associated with additive processes can be characterized quite elegantly. First, we need the following definition.

DEFINITION 2.61. A probability measure $\mu \in \mathcal{P}(\mathbb{R})$ is called $*$*-infinitely divisible* if, for each $n \in \mathbb{N}$, there exists $\mu_n \in \mathcal{P}(\mathbb{R})$ with

$$\mu = (\mu_n)^{*n} = \mu_n * \cdots * \mu_n.$$

EXAMPLE 2.62. The normal distribution $\mathcal{N}(c, \sigma^2)$ is $*$-infinitely divisible, since

$$\mathcal{N}(c/n, \sigma^2/n)^{*n} = \mathcal{N}(c, \sigma^2)$$

for each $n \in \mathbb{N}$, which follows by computing the characteristic functions, see Example 2.17 (d).

THEOREM 2.63. *Let μ be a probability measure on $\mathbb{R}$. The following statements are equivalent:*

(a) *There exists an additive process $(X_t)_{t\geq 0}$ such that μ is the distribution of X_1.*
(b) *There exists a Lévy process $(X_t)_{t\geq 0}$ such that μ is the distribution of X_1.*
(c) *μ is infinitely divisible.*
(d) *μ is the limit of an infinitesimal array: there exist random variables $(X_{n,j})_{n\in\mathbb{N}, 1\leq j\leq k_n}$ with $k_n \to \infty$ as $n \to \infty$, $X_{n,1}, \ldots, X_{n,k_n}$ are independent,*

$$\lim_{n\to\infty} \max_{j=1,\ldots,k_n} \mathbb{P}[|X_{n,j}| \geq \varepsilon] = 0$$

for every $\varepsilon > 0$, such that μ is the limit of the distribution of $X_{n,1} + \cdots + X_{n,k_n}$ as $n \to \infty$.
(e) *(Lévy–Khintchine representation) There exist $a \in \mathbb{R}$, $\sigma \geq 0$, and a non-negative measure ν on $\mathbb{R}$ with $\nu(\{0\}) = 0$ and $\int_{\mathbb{R}} \min(1, t^2)\nu(dt) < \infty$ such that*

(2.10)
$$\varphi_\mu(t) = \exp\left(iat - \frac{1}{2}\sigma^2 t^2 + \int_{\mathbb{R}} \left(e^{its} - 1 - its\mathbf{1}_{\{|s|<1\}}(s)\right)\nu(ds)\right), \qquad t \in \mathbb{R}.$$

The triple (a, σ, ν) is called the Lévy triple *of μ and it is uniquely determined.*

Conversely, any triple (a, σ, ν) as in (e) corresponds to an infinitely divisible distribution.

PROOF. We refer to [**19**, Theorems 1.1-1.3], [**74**, §24, Theorem 2] for (d), and [**168**, Theorem 8.1] for (e). Note that slightly different conventions for the Lévy–Khintchine formula can be found throughout the literature. □

It is easy to see that (b) implies (c). Let $(X_t)_{t\geq 0}$ be a Lévy process with distributions $(\mu_t)_{t\geq 0}$. For each $n \in \mathbb{N}$,

$$X_1 = X_1 - X_{1-1/n} + \cdots + X_{1/n} - X_0$$

is a sum of n *iid* random variables, and we conclude that $\mu_1 = \mu_{1/n}^{*n}$. So μ_1 is infinitely divisible. Furthermore, $\varphi_{\mu_T} = \varphi_{\mu_1}^T$ in the sense that, with (2.10) for μ_1, we have

$$\varphi_{\mu_T}(t) = \exp\left(T\left(iat - \frac{1}{2}\sigma^2 t^2 + \int_{\mathbb{R}} \left(e^{its} - 1 - its\mathbf{1}_{\{|s|<1\}}(s)\right)\nu(ds)\right)\right)$$

for all $T \geq 0$ and $t \in \mathbb{R}$.

EXAMPLE 2.64. The Poisson distribution has the characteristic function

$$\varphi_\mu(t) = e^{\lambda(e^{it}-1)}, \quad \lambda > 0,$$

see Example 2.17 (b). The representation (2.10) yields this function for $a = \sigma = 0$, and $\nu = \lambda\delta_1$, while the Lévy triple $(0, \sigma, 0)$ leads to $\varphi_\mu(t) = e^{-\sigma^2 t^2/2}$, i.e., $\mu = \mathcal{N}(0, \sigma^2)$.

The Lévy–Khintchine representation comes along with the Lévy-Itô decomposition theorem, a classification of Lévy processes and additive processes, see [**168**, Section 19].

2.8. Exercises

E 2.8.1. Let μ be a probability measure on $\mathbb{R}$. Show that $S = \{x \in \mathbb{R} \mid \mu(\{x\}) > 0\}$ is at most countably infinite.

E 2.8.2. Let X be a random variable with values in $\mathbb{N}_0$ and $\mathbb{E}[X] < \infty$. Show that

$$\mathbb{E}[X] = \sum_{n=0}^{\infty} \mathbb{P}(X > n).$$

E 2.8.3. Let $\mu \in \mathcal{P}(\mathbb{R})$ and $a \in \mathbb{R}$. Show:

$$\mu(\{a\}) = \lim_{T\to\infty} \frac{1}{2T} \int_{-T}^{T} e^{-iat} \varphi_\mu(t) dt.$$

E 2.8.4.

(a) Show: If μ is a symmetric probability distribution, i.e., $\mu(A) = \mu(-A)$ for every Borel subset $A \subset \mathbb{R}$, then $t \mapsto \varphi_\mu(t)$ is not injective.

(b) Let $\lambda > 0$, and let μ be the exponential distribution defined by the density $\lambda e^{-\lambda x}$, $x \geq 0$. Show that φ_μ is an injective function. How does the image $\varphi_\mu(\mathbb{R})$ look like?

E 2.8.5. Compute the characteristic function of the random variable X.

(a) X has a (centered) Cauchy distribution, given by the density $\frac{1}{\pi}\frac{\gamma}{x^2+\gamma^2}$ with scale $\gamma > 0$. (Hint: consider a complex integral along the path $\Gamma_R = [-R, R] \cup [R, R + iR] \cup [R + iR, -R + iR] \cup [-R + iR, -R]$.)

(b) Let $X_1, X_2, \ldots$ be *iid* random variables with $\mathbb{P}[X_1 = -1] = \mathbb{P}[X_1 = 1] = \frac{1}{2}$, and let $X = \frac{1}{2} + \sum_{k=1}^{\infty} \frac{X_k}{3^k}$.

E 2.8.6. Let X, Y be independent, square-integrable random variables. Show that $\mathrm{Var}(X + Y) = \mathrm{Var}(X) + \mathrm{Var}(Y)$.

E 2.8.7. Recall Theorem 2.23.

(a) Construct random variables X, Y, which are subindependent but not independent.

(b) Construct random variables X, Y, which are uncorrelated but not subindependent.

E 2.8.8. Let μ and ν be probability measures. Show that Parseval's identity holds:

$$\int_{\mathbb{R}} e^{-its} \varphi_\mu(t) \nu(dt) = \int_{\mathbb{R}} \varphi_\nu(t - s) \mu(dt) \quad \text{for all } s \in \mathbb{R}.$$

E 2.8.9. Consider the metric space $M = (\mathcal{P}(\mathbb{R}), d_{\text{Lévy}})$.

(a) Is M sequentially compact? (Does every sequence $(\mu_n)_{n\in\mathbb{N}} \subset \mathcal{P}(\mathbb{R})$ have a convergent subsequence?)

(b) Is M connected?

E 2.8.10. Consider *iid* random variables $X_1, X_2, \ldots$ with a uniform distribution on $[-1, 0]$. Calculate the limiting distribution of $Y_n = \max\{X_1, \ldots, X_n\} \cdot n$ as $n \to \infty$.

E 2.8.11. Let B_t be a Brownian motion. Show that $cov(B_t, B_s) = \min(s, t)$ for all $s, t \geq 0$.

E 2.8.12 (Ornstein–Uhlenbeck process). Let B_t be a Brownian motion, and consider the process $X_t = e^{-t}B_{e^{2t}}$. Show that the process $(X_t)_{t\geq 0}$ is a Markov process, where all X_t, $t \geq 0$, have the same distribution. Is X_t space-homogeneous?

E 2.8.13.

(a) Show that δ_c, $c \in \mathbb{R}$, is infinitely divisible.
(b) Show that if μ is infinitely divisible with compact support, then $\mu = \delta_c$, $c \in \mathbb{R}$.
(c) Show that a centered Cauchy distribution with scale $\gamma > 0$ is infinitely divisible.

2.9. Solutions

S 2.9.1. For $n \in \mathbb{N}$, define $S_n = \{x \in \mathbb{R} \,|\, \mu(\{x\}) \geq 1/n\}$. Then S_n is finite and $S \subset \cup_{n=1}^{\infty} S_n$.

S 2.9.2. We have

$$\begin{aligned}\mathbb{E}[X] &= \sum_{n=0}^{\infty} n\mathbb{P}(X=n) = \mathbb{P}(X=1) + 2\cdot\mathbb{P}(X=2) + 3\cdot\mathbb{P}(X=3) + \dots \\ &= (\mathbb{P}(X=1) + \mathbb{P}(X=2) + \dots) + (\mathbb{P}(X=2) + \mathbb{P}(X=3) + \dots) + \dots \\ &= \mathbb{P}(X>0) + \mathbb{P}(X>1) + \dots = \sum_{n=0}^{\infty} \mathbb{P}(X>n).\end{aligned}$$

S 2.9.3. For $x \in \mathbb{R}$, let $G(a,x) := \int_{-T}^{T} e^{it(x-a)}dt$. Then

$$G(a,a) = 2T \text{ and } G(a,x) = \tfrac{e^{iT(x-a)}}{i(x-a)} - \tfrac{e^{-iT(x-a)}}{i(x-a)} \text{ if } x \neq a.$$

Thus,

$$\begin{aligned}&\lim_{T\to\infty} \frac{1}{2T}\int_{-T}^{T} e^{-iat}\varphi_\mu(t)dt = \lim_{T\to\infty}\frac{1}{2T}\int_{-T}^{T}\int_{\mathbb{R}} e^{it(x-a)}\mu(dx)dt \\ &= \lim_{T\to\infty}\frac{1}{2T}\int_{\mathbb{R}}\int_{-T}^{T} e^{it(x-a)}dt\mu(dx) = \lim_{T\to\infty}\int_{\mathbb{R}} G(a,x)/(2T)\mu(dx) \\ &= \int_{\mathbb{R}} \lim_{T\to\infty} G(a,x)/(2T)\mu(dx) = \int_{\mathbb{R}} \mathbf{1}_{\{a\}}(x)\mu(dx) = \mu(\{a\}).\end{aligned}$$

S 2.9.4.

(a) If $\mu(A) = \mu(-A)$, then

$$\varphi_\mu(t) = \int_{\mathbb{R}} e^{itx}\mu(dx) = \int_{\mathbb{R}} e^{it(-x)}\mu(dx) = \int_{\mathbb{R}} e^{i(-t)x}\mu(dx) = \varphi_\mu(-t),$$

which shows that φ_μ is not injective.

(b) We have

$$\varphi_\mu(t) = \int_0^{\infty} \lambda e^{(it-\lambda)x}dx = \left[\lambda\frac{e^{(it-\lambda)x}}{(it-\lambda)}\right]_{x=0}^{\infty} = \frac{\lambda}{\lambda - it}.$$

The function $t \mapsto \frac{\lambda}{\lambda-it} = \frac{1}{1-i(t/\lambda)}$ is injective. It describes a circle (as the image of a straight line under a Möbius transform) passing through 1 and 0. As it is symmetric with respect to conjugation, the image is the circle with center $\frac{1}{2}$ and radius $\frac{1}{2}$, where the point 0 corresponds to the limits $\lim_{t\to\pm\infty}\varphi_\mu(t)$, i.e., $\varphi_\mu(\mathbb{R}) = \{1/2 + e^{ix}/2 \,|\, x \in (-\pi,\pi)\}$.

S 2.9.5.

(a) We have to calculate the integral

$$\varphi_X(t) = \int_{\mathbb{R}} \frac{\gamma}{\pi}\frac{e^{itx}}{x^2+\gamma^2}dx.$$

Clearly, $\varphi_X(-t) = \varphi_X(t)$. So assume that $t > 0$. Consider the complex integral along the path

$$\Gamma_R = [-R,R] \cup [R, R+iR] \cup [R+iR, -R+iR] \cup [-R+iR, -R]$$

(in counter-clockwise direction). Then the residue theorem gives

$$\int_{\Gamma_R} \frac{\gamma}{\pi} \frac{e^{itz}}{z^2+\gamma^2} dz = 2\gamma i \lim_{z \to i\gamma} (z - i\gamma) \frac{e^{itz}}{z^2+\gamma^2} = 2\gamma i \frac{e^{-\gamma t}}{2i\gamma} = e^{-\gamma t},$$

provided that $R > \gamma$. For $z \in [R, R+iR]$, we have $|\frac{e^{itz}}{z^2+\gamma^2}| \leq \frac{1}{R^2}$, and thus,

$$\left| \int_{[R,R+iR]} \frac{e^{itz}}{z^2+\gamma^2} dz \right| \leq \int_{[R,R+iR]} \frac{1}{R^2} |dz| = \frac{1}{R}.$$

Similarly,

$$\left| \int_{[-R+iR,-R]} \frac{e^{itz}}{z^2+\gamma^2} dz \right| \leq \frac{1}{R} \quad \text{and} \quad \left| \int_{[R+iR,-R+iR]} \frac{e^{itz}}{z^2+\gamma^2} dz \right| \leq \frac{2}{R}.$$

We conclude that

$$\int_{\mathbb{R}} \frac{\gamma}{\pi} \frac{e^{itx}}{x^2+\gamma^2} dx = \lim_{R \to \infty} \int_{\Gamma_R} \frac{\gamma}{\pi} \frac{e^{itz}}{z^2+\gamma^2} dz = e^{-\gamma t},$$

and thus, $\varphi_X(t) = e^{-\gamma|t|}$ for all $t \in \mathbb{R}$.

(b) For each $k \in \mathbb{N}$, we have $\varphi_{X_k/3^k}(t) = \cos(t/3^k)$. By Theorem 2.23, we obtain

$$\varphi_X(t) = e^{it/2} \prod_{k=1}^{\infty} \cos\left(\frac{t}{3^k}\right).$$

S 2.9.6. We have

$$\begin{aligned}
\operatorname{Var}(X+Y) &= \mathbb{E}[(X+Y-\mathbb{E}[X]-\mathbb{E}[Y])^2] \\
&= \mathbb{E}[(X-\mathbb{E}[X])^2] + \mathbb{E}[(Y-\mathbb{E}[Y])^2] + 2\mathbb{E}[(X-\mathbb{E}[X])(Y-\mathbb{E}[Y])] \\
&= \operatorname{Var}(X) + \operatorname{Var}(Y) + 2\mathbb{E}[X-\mathbb{E}[X]]\mathbb{E}[Y-\mathbb{E}[Y]] \\
&= \operatorname{Var}(X) + \operatorname{Var}(Y).
\end{aligned}$$

S 2.9.7.

(a) Let X be a random variable with Cauchy distribution with scale $\gamma > 0$, and let $Y = X$. Then X and Y are not independent. The distribution of $X + Y = 2X$ is again a Cauchy distribution with scale 2γ. We have

$$\varphi_{X+Y}(t) = e^{-2\gamma|t|} = e^{-\gamma|t|} \cdot e^{-\gamma|t|} = \varphi_X(t) \cdot \varphi_Y(t).$$

(b) Let X be a random variable with $\mathbb{P}[X = -1] = \mathbb{P}[X = 1] = \mathbb{P}[X = -2] = \mathbb{P}[X = 2] = \frac{1}{4}$, and let $Y = |X|$. Then $\mathbb{E}[X] = 0$ and also $\mathbb{E}[XY] = \mathbb{E}[X|X|] = 0$, thus $\mathbb{E}[XY] = \mathbb{E}[X]\mathbb{E}[Y]$, and X and Y are uncorrelated. Furthermore, we have

$$\begin{aligned}
\varphi_X(t) \cdot \varphi_Y(t) &= \frac{1}{4}(e^{it} + e^{-it} + e^{i2t} + e^{-i2t}) \cdot \frac{1}{2}(e^{it} + e^{i2t}) \\
&= \frac{1}{8}(2 + e^{-it} + e^{i2t} + 2e^{i3t} + e^{it} + e^{i4t}),
\end{aligned}$$

and

$$\varphi_{X+|X|}(t) = \frac{1}{4}(2 + e^{i2t} + e^{i4t}) \not\equiv \varphi_X(t) \cdot \varphi_Y(t).$$

S 2.9.8.

$$\int_{\mathbb{R}} e^{-its}\varphi_\mu(t)\nu(dt) = \int_{\mathbb{R}}\int_{\mathbb{R}} e^{it(x-s)}\mu(dx)\nu(dt)$$
$$= \int_{\mathbb{R}}\int_{\mathbb{R}} e^{it(x-s)}\nu(dt)\mu(dx) = \int_{\mathbb{R}} \varphi_\nu(x-s)\mu(dx).$$

S 2.9.9.

(a) M is not compact: the sequence $\mu_n = \delta_n$ satisfies $d_{\text{Lévy}}(\mu_n, \mu_m) = 1$ whenever $m \neq n$.

(b) Let $\mu, \nu \in \mathcal{P}(\mathbb{R})$. Then $[0,1] \ni t \mapsto t\nu + (1-t)\mu$ is a continuous path connecting μ to ν. The continuity follows from $\int_{\mathbb{R}} f(x)(t\nu+(1-t)\mu)(dx) = t\int_{\mathbb{R}} f(x)\nu(dx) + (1-t)\int_{\mathbb{R}} f(x)\mu(dx)$ for every bounded and continuous $f : \mathbb{R} \to \mathbb{C}$. Hence, M is pathwise connected, and thus, connected.

S 2.9.10. Clearly, $\mathbb{P}[Y_n \leq -n] = 0$ and $\mathbb{P}[Y_n \leq 0] = 1$ for all n. Let $x \in [-n, 0]$. Then we have

$$\mathbb{P}\left[Y_n \leq x\right] = \mathbb{P}\left[X_1 \leq \frac{x}{n} \wedge \cdots \wedge X_n \leq \frac{x}{n}\right] = \mathbb{P}\left[X_1 \leq \frac{x}{n}\right] \cdots \mathbb{P}\left[X_n \leq \frac{x}{n}\right]$$
$$= \left(1 + \frac{x}{n}\right)^n \to e^x.$$

Hence, the limiting distribution is given by the density e^x on $(-\infty, 0]$ and 0 on $(0, \infty)$.

S 2.9.11. Let $0 \leq s \leq t$. Then

$$\begin{aligned} cov(B_t, B_s) &= \mathbb{E}[B_t B_s] = \mathbb{E}[(B_s + B_t - B_s)B_s] \\ &= \mathbb{E}[B_s^2] + \mathbb{E}[(B_t - B_s)(B_s - B_0)] \\ &= s + \mathbb{E}[B_t - B_s]\mathbb{E}[B_s - B_0] = s + 0 = s. \end{aligned}$$

S 2.9.12. Let $X_t = e^{-t}B_{e^{2t}}$. Then each X_t is $\mathcal{N}(0,1)$ distributed. As B_t is a Markov process, the rescaled process X_t is also a Markov process. We have

$$\begin{aligned} \operatorname{Var}(X_{t+s} - X_s) &= \mathbb{E}[(X_{t+s} - X_s)^2] = \mathbb{E}[X_{t+s}^2 - 2X_{t+s}X_s + X_s^2] \\ &= 1 - 2e^{-t-2s}\mathbb{E}[B_{e^{2(t+s)}}B_{e^{2s}}] + 1 = 1 - 2e^{-t-2s}(e^{2s}) + 1 \\ &= 2(1 - e^{-t}), \end{aligned}$$

where we used Exercise 2.8.11, which shows that $\mathbb{E}[B_{e^{2(t+s)}}B_{e^{2s}}] = e^{2s}$.

Assume that (X_t) is space-homogeneous. Theorem 2.52 implies that X_t has independent increments, and $\operatorname{Var}(X_2 - X_0) = \operatorname{Var}(X_2 - X_1) + \operatorname{Var}(X_1 - X_0)$ (due to Exercise 2.8.6), i.e.,

$$2(1 - e^{-2}) = 2(1 - e^{-1}) + 2(1 - e^{-1}),$$

a contradiction.

S 2.9.13.

(a) We have $\delta_c = (\delta_{c/n})^{*n}$ for all $n \in \mathbb{N}$.

(b) Let X be a random variable with infinitely divisible distribution μ, whose support is contained in $[-M, M]$, $M > 0$. For $n \in \mathbb{N}$, we can write $X = Y_1 + \cdots + Y_n$ with *iid* random variables $Y_1, \ldots, Y_n$. Assume that

$\mathbb{P}(|Y_k| > M/n) \neq 0$. Then there exists $\varepsilon > 0$ such that $\mathbb{P}(|Y_k| \geq M/n + \varepsilon) \neq 0$. Say, we have $\mathbb{P}(Y_k \geq M/n + \varepsilon) = c > 0$. Then

$$\mathbb{P}(X \geq M + n\varepsilon) \geq \mathbb{P}(\cap_{k=1}^n \{Y_k \geq M/n + \varepsilon\}) = c^n > 0,$$

a contradiction. Thus $\mathbb{P}(|Y_k| > M/n) = 0$. So we have $\mathrm{Var}(Y_k) \leq \mathbb{E}[Y_k^2] \leq M^2/n^2$, and thus (with Exercise 2.8.6),

$$\mathrm{Var}(X) = \sum_{k=1}^n \mathrm{Var}(Y_k) \leq nM^2/n^2 = M^2/n \to 0$$

as $n \to \infty$. Thus, $\mathrm{Var}(X) = 0$ and X must be constant a.s., i.e., $\mu = \delta_c$ for some $c \in \mathbb{R}$.

(c) The sum of n independent random variables with Cauchy distribution, each with scale γ, has again a Cauchy distribution, with scale $n\gamma$. This follows from Theorem 2.23 and Exercise 2.8.5 (a). Thus, if μ_n is the Cauchy distribution with scale γ/n, then $\mu = (\mu_n)^{*n}$, and we see that μ is infinitely divisible.

CHAPTER 3

The Complex Toolbox

When we study quantum random variables in the next chapter, the characteristic function will be replaced by the Cauchy transform as well as other transforms derived from it. These transforms are holomorphic functions on the upper half-plane. Complex analysis offers us several powerful theorems that will help us deal with these functions.

3.1. Unit disk and upper half-plane

DEFINITION 3.1. A bijective holomorphic function $f : D \to E$ between two domains $D, E \subset \mathbb{C}$ is called *conformal*.

The most interesting case is when D, and thus also E, is a simply connected, proper subdomain of $\mathbb{C}$, due to the Riemann mapping theorem.

THEOREM 3.2 (Riemann mapping theorem). *Let $D \subsetneq \mathbb{C}$ be a simply connected domain. Then D is conformally equivalent to $\mathbb{D}$. For any $z_0 \in D$ there is exactly one conformal mapping $f : D \to \mathbb{D}$ with $f(z_0) = 0$ and $f'(z_0) > 0$.*

EXAMPLE 3.3. The Cayley map

$$C(z) = \frac{z - i}{z + i}$$

is a Möbius transformation which maps the upper half-plane $\mathbb{H} = \{z \in \mathbb{C} \mid \operatorname{Im}(z) > 0\}$ conformally onto the unit disk $\mathbb{D} = \{z \in \mathbb{C} \mid |z| < 1\}$.

Let us recall the Schwarz lemma, which will also be quite useful in Chapter 6.

THEOREM 3.4 (Schwarz lemma). *Let $f : \mathbb{D} \to \mathbb{D}$ be holomorphic with $f(0) = 0$. Then*

$$|f(z)| \leq |z| \text{ for all } z \in \mathbb{D} \text{ and } |f'(0)| \leq 1.$$

If $|f(z)| = |z|$ for a $z \in \mathbb{D} \setminus \{0\}$, then there is an $a \in \partial\mathbb{D}$ such that $f(z) = az$ for all $z \in \mathbb{D}$.

For a domain $D \subset \mathbb{C}$, we denote by $\operatorname{Aut}(D)$ the set of all conformal self-mappings of D.

THEOREM 3.5.

$$\operatorname{Aut}(\mathbb{D}) = \left\{ z \mapsto a\frac{z - w}{\overline{w}z - 1} \,\middle|\, w \in \mathbb{D}, a \in \partial\mathbb{D} \right\}.$$

PROOF. Let $f : \mathbb{D} \to \mathbb{D}$ be conformal. First, assume that $f(0) = 0$. Then $|f(z)| \leq |z|$ and $|f^{-1}(z)| \leq |z|$, thus $|z| \leq |f(z)|$ for all $z \in \mathbb{D}$. It follows that $|f(z)| = |z|$, which implies $f(z) = az$ for all $z \in \mathbb{D}$ for some $a \in \partial\mathbb{D}$.

Next, assume $f(w) = 0$ for some $w \in \mathbb{D}$. The Möbius transformation $M(z) = \frac{z-w}{\overline{w}z-1}$ belongs to $\mathrm{Aut}(\mathbb{D})$: for $z \in \partial\mathbb{D}$ we have

$$|w-z|^2 = (w-z)(\overline{w}-\overline{z}) = |w|^2 - \overline{w}z - w\overline{z} + 1, \quad |1-\overline{w}z|^2 = 1 - \overline{w}z - w\overline{z} + |w|^2,$$

thus $|M(z)| = 1$. Furthermore, $M(0) = w$, so $M \in \mathrm{Aut}(\mathbb{D})$. For $F := f \circ M$ we have $F(0) = 0$. Hence, $F(z) = az$ for all $z \in \mathbb{D}$ and some $a \in \partial\mathbb{D}$. We conclude (with $M^{-1} = M$)

$$f(z) = aM^{-1}(z) = a\frac{z-w}{\overline{w}z-1}. \qquad \square$$

$\mathrm{Aut}(\mathbb{D})$ has three free real parameters. For any pair $p, q \in \mathbb{D}$, we find an $f \in \mathrm{Aut}(\mathbb{D})$ with $f(p) = q$. This automorphism is unique if we fix, e.g., the argument of $f'(p)$, such as $f'(p) > 0$ as in the Riemann mapping theorem.

COROLLARY 3.6.

$$\mathrm{Aut}(\mathbb{H}) = \left\{ z \mapsto \frac{az+b}{cz+d} \,|\, a,b,c,d \in \mathbb{R}, ad-bc > 0 \right\}.$$

PROOF. $\mathrm{Aut}(\mathbb{H})$ only consists of Möbius transformations: If $f \in \mathrm{Aut}(\mathbb{H})$, then $f = C^{-1} \circ g \circ C$ for some $g \in \mathrm{Aut}(\mathbb{D})$, where C is the Cayley map. The composition of Möbius transformations is again a Möbius transformation.

Let $f \in \mathrm{Aut}(\mathbb{H})$ with $f(\infty) = \infty$. Then $f(z) = az + b$ with $a > 0$ and $b \in \mathbb{R}$. Now let $f(\infty) = p \in \mathbb{R}$. Then $-1/(f(z) - p) \in \mathrm{Aut}(\mathbb{H})$ and it fixes ∞, i.e., $-1/(f(z) - p) = az + b$ or

$$f(z) = p - \frac{1}{az+b} = \frac{paz + pb - 1}{az+b}.$$

We have $pa \cdot b - (pb-1) \cdot a = pab - pab + a = a > 0$.

Conversely, each mapping of the form $z \mapsto \frac{az+b}{cz+d}$, $a, b, c, d \in \mathbb{R}$, $ad - bc > 0$, is an automorphism of $\mathbb{H}$. $\square$

Since $\mathbb{D}$ and $\mathbb{H}$ are conformally equivalent, one might wonder why both domains appear consistently in complex analysis. The reason is that problems related to one domain might become simpler if translated to the other domain. For example, suppose we need to work with the set of all automorphisms of $\mathbb{H}$ that fix some $p \in \mathbb{H}$. By switching to $\mathbb{D}$ and taking $p = 0$, we obtain the much friendlier setting of simple rotations $z \mapsto az$, $a \in \partial\mathbb{D}$. On the other hand, all automorphisms that fix a boundary point take a simpler form in $\mathbb{H}$, with boundary point ∞, namely $z \mapsto az + b$, where $b \in \mathbb{R}$ and $a > 0$.

3.2. The Cauchy transform

DEFINITION 3.7. Let $\mu \in \mathcal{P}(\mathbb{R})$. The *Cauchy transform* (also called *Stieltjes transform*) G_μ of μ is defined as the holomorphic function

$$G_\mu(z) := \int_{\mathbb{R}} \frac{1}{z-x}\mu(dx), \quad z \in \mathbb{H}. \tag{3.1}$$

As $1/(z-x) \in \mathbb{H}^- := \{w \in \mathbb{C} \mid \mathrm{Im}(w) < 0\}$ for all $z \in \mathbb{H}$ and $x \in \mathbb{R}$, G_μ maps $\mathbb{H}$ into $\mathbb{H}^-$. We could also regard G_μ on the lower half-plane $\mathbb{H}^-$, but here we simply get the conjugation as $G_\mu(\overline{z}) = \overline{G_\mu(z)}$.

If μ has compact support, G_μ can in fact be extended analytically to $\hat{\mathbb{C}}\backslash\mathrm{supp}(\mu)$, where $\hat{\mathbb{C}} = \mathbb{C} \cup \{\infty\}$ denotes the Riemann sphere. Then $G_\mu(\infty) = 0$, and the power series expansion of G_μ at $z = \infty$ yields the moments of μ:

$$
\begin{aligned}
G_\mu(z) &= \frac{1}{z}\int_{\mathbb{R}} \frac{1}{1-\frac{x}{z}}\mu(dx) = \frac{1}{z}\int_{\mathbb{R}} \sum_{k=0}^{\infty} \frac{x^k}{z^k}\mu(dx) \\
&= \sum_{k=0}^{\infty} \frac{1}{z^{k+1}} \int_{\mathbb{R}} x^k \mu(dx) = \frac{1}{z} + \frac{\int_{\mathbb{R}} x\mu(dx)}{z^2} + \frac{\int_{\mathbb{R}} x^2\mu(dx)}{z^3} + \dots
\end{aligned} \tag{3.2}
$$

We will primarily regard G_μ as a function on the upper half-plane, but we keep the analytic extension (3.2) for compactly supported measures in mind.

The Cauchy transform completely encodes the measure μ, which can be recovered from G_μ by the Stieltjes–Perron inversion formula.

THEOREM 3.8 (Stieltjes–Perron inversion formula). *Let* $\mu \in \mathcal{P}(\mathbb{R})$.

$$
\frac{1}{2}\mu(\{a\}) + \frac{1}{2}\mu(\{b\}) + \mu((a,b)) = -\frac{1}{\pi}\lim_{y \searrow 0}\int_a^b \mathrm{Im}(G_\mu(x+iy))\,dx, \quad a,b \in \mathbb{R}, a < b, \tag{3.3}
$$

$$
\mu(\{a\}) = \lim_{y\searrow 0} iyG_\mu(a+iy), \quad a \in \mathbb{R}. \tag{3.4}
$$

In particular, if $G_\mu = G_\nu$ *for* $\mu, \nu \in \mathcal{P}(\mathbb{R})$, *then* $\mu = \nu$.

PROOF. Let $y > 0$. Then

$$
\begin{aligned}
&\int_a^b \mathrm{Im}(G_\mu(x+iy))dx = -\int_a^b \int_{\mathbb{R}} \frac{y}{(x-u)^2+y^2}\mu(du)dx \\
&= -y\int_{\mathbb{R}}\int_a^b \frac{1}{(x-u)^2+y^2}dx\mu(du) \\
&= -y\int_{\mathbb{R}} \frac{1}{y}\left(\arctan\left(\frac{b-u}{y}\right) - \arctan\left(\frac{a-u}{y}\right)\right)\mu(du) \\
&= -\int_{\mathbb{R}} \arctan\left(\frac{b-u}{y}\right) - \arctan\left(\frac{a-u}{y}\right)\mu(du).
\end{aligned}
$$

Denote the integrand of the last integral by $s(u)$. Then

$$
\lim_{y\searrow 0} s(u) = \begin{cases} \frac{\pi}{2} & \text{if } u = a \text{ or } u = b, \\ 0 & \text{if } u < a \text{ or } u > b, \\ \pi & \text{if } u \in (a,b). \end{cases}
$$

By the dominated convergence theorem,

$$
\begin{aligned}
&-\frac{1}{\pi}\lim_{y\to 0}\int_a^b \mathrm{Im}(G_\mu(x+iy))dx = \frac{1}{\pi}\lim_{y\to 0}\int_{\mathbb{R}} s(u)\mu(du) \\
&= \frac{1}{\pi}\int_{\mathbb{R}}\lim_{y\to 0} s(u)\mu(du) = \frac{1}{2}\mu(\{a\}) + \frac{1}{2}\mu(\{b\}) + \mu((a,b)).
\end{aligned}
$$

Furthermore,

$$iyG_\mu(a+iy) = \int_{\mathbb{R}} \frac{iy}{a-u+iy}\mu(du) = \int_{\mathbb{R}} \frac{1}{(a-u)/(iy)+1}\mu(du).$$

As $y \searrow 0$, the integrand converges to 0 whenever $u \neq a$ and to 1 for $u = a$. We obtain $\lim_{y\searrow 0} iyG_\mu(a+iy) = \mu(\{a\})$. □

3.3. Integral representations

We can think of $\mathbb{R}$ being embedded within the circle $\hat{\mathbb{R}} = \mathbb{R} \cup \{\infty\}$ in the Riemann sphere, which corresponds to the unit circle $\partial\mathbb{D}$ via the Cayley transform. So $\hat{\mathbb{R}}$ is a metric space and we can regard Borel probability measures on $\hat{\mathbb{R}}$. Let $\mu, \mu_1, \mu_2, \ldots$ be probability measures on $\hat{\mathbb{R}}$. We say that the sequence $(\mu_n)_{n\in\mathbb{N}}$ *converges weakly* to μ if

$$\int_{\hat{\mathbb{R}}} f(x)\mu_n(dx) \to \int_{\hat{\mathbb{R}}} f(x)\mu(dx)$$

for every continuous (and thus bounded) $f : \hat{\mathbb{R}} \to \mathbb{C}$.

In particular, every probability measure on $\mathbb{R}$ extends to a probability measure on $\hat{\mathbb{R}}$. This point of view can be very helpful, as $\hat{\mathbb{R}}$ is compact and we have Helly's selection theorem available, see [**40**, Theorem 25.9].

THEOREM 3.9 (Helly's selection theorem). *Let $(\mu_n)_{n\in\mathbb{N}}$ be a sequence of probability measures on $\hat{\mathbb{R}}$ (or on $\partial\mathbb{D}$). Then there exists a subsequence $(\mu_{n_k})_{k\in\mathbb{N}}$ that converges weakly to a probability measure μ on $\hat{\mathbb{R}}$.*

More generally, if $(\mu_n)_{n\in\mathbb{N}}$ is a uniformly bounded sequence of non-negative measures on $\hat{\mathbb{R}}$, then there exists a weakly converging subsequence $(\mu_{n_k})_{k\in\mathbb{N}}$.

We would like to represent holomorphic functions $f : \mathbb{H} \to \mathbb{H} \cup \mathbb{R}$ by their boundary values on $\mathbb{R}$. Even though f does not extend to $\mathbb{R}$ in general, f induces a non-negative measure on $\mathbb{R}$, which can be used for an integral representation of f. We also look at the corresponding case for $\mathbb{D}$, which concerns holomorphic functions $f : \mathbb{D} \to \mathbb{C}$ with $\operatorname{Re} f \geq 0$.

First, we need the following statement, which gives the solution of the Dirichlet problem on the unit disk.[1]

THEOREM 3.10. *Let $g : \partial\mathbb{D} \to \mathbb{R}$ be continuous, and for $z \in \mathbb{D}$, let*

$$F(z) = \frac{1}{2\pi}\int_{-\pi}^{\pi} \frac{e^{it}+z}{e^{it}-z} g(e^{it})dt.$$

Then F is holomorphic and $\operatorname{Re} F$ extends continuously to $\overline{\mathbb{D}}$ with $\operatorname{Re} F(z) = g(z)$ on $\partial\mathbb{D}$.

[1] Let $D \subset \mathbb{C}$ be a domain and let $g : \partial D \to \mathbb{R}$ be continuous. The *Dirichlet problem* asks for a function u which is continuous on $D \cup \partial D$, harmonic on D, and satisfies $u|_{\partial D} = g$. Theorem 3.10 shows that the Dirichlet problem always has a solution for $D = \mathbb{D}$, given by $u(z) = \operatorname{Re} F(z) = \frac{1}{2\pi}\int_{-\pi}^{\pi} \operatorname{Re}\left(\frac{e^{it}+z}{e^{it}-z}\right) g(e^{it})dt$ for $z \in \mathbb{D}$.

PROOF. F is holomorphic as $z \mapsto \frac{e^{it}+z}{e^{it}-z} g(e^{it})$ is holomorphic. Define

$$u(z) = \operatorname{Re} F(z) \text{ for } z \in \mathbb{D} \text{ and } u(z) = g(z) \text{ for } z \in \partial\mathbb{D}.$$

It suffices to show that u is continuous at $z = 1$ and we may assume that $u(1) = 0$. As g is continuous, there exists an open interval $C \subset [-\pi, \pi]$ with $0 \in C$ and $|u(e^{i\varphi})| < \varepsilon/2$ for all $\varphi \in C$. Now we consider the decomposition of $[-\pi, \pi]$ into $C \cup ([-\pi, \pi] \setminus C)$, and we obtain for every $z \in \mathbb{D}$:

$$u(z) = \underbrace{\frac{1}{2\pi} \int_C \operatorname{Re}\left(\frac{e^{it}+z}{e^{it}-z}\right) g(e^{it})dt}_{:=I_1(z)} + \underbrace{\frac{1}{2\pi} \int_{[-\pi,\pi]\setminus C} \operatorname{Re}\left(\frac{e^{it}+z}{e^{it}-z}\right) g(e^{it})dt}_{:=I_2(z)}.$$

As $z \to 1$, the function $t \mapsto \operatorname{Re}\left(\frac{e^{it}+z}{e^{it}-z}\right)$ converges on $[-\pi, \pi] \setminus C$ uniformly to 0. Thus there exists a neighborhood U of $z = 1$ (in $\overline{\mathbb{D}}$) such that $|I_2(z)| < \varepsilon/2$ for all $z \in \mathbb{D} \cap U$ and $|u(z)| < \varepsilon$ for $z \in \partial\mathbb{D} \cap U$. We assume that $0 \notin U$ and obtain for $I_1(z)$, $z \in \mathbb{D} \cap U$:

$$\begin{aligned}
|I_1(z)| &\leq \frac{1}{2\pi} \int_C \operatorname{Re}\left(\frac{e^{it}+z}{e^{it}-z}\right) \frac{\varepsilon}{2}\, dt \leq \frac{1}{2\pi} \int_{-\pi}^{\pi} \operatorname{Re}\left(\frac{e^{it}+z}{e^{it}-z}\right) \frac{\varepsilon}{2}\, dt \\
&= \frac{\varepsilon}{2} \operatorname{Re}\left(\frac{1}{2\pi i} \int_{-\pi}^{\pi} \frac{e^{it}+z}{(e^{it}-z)e^{it}} ie^{it}\, dt\right) = \frac{\varepsilon}{2} \operatorname{Re}\left(\frac{1}{2\pi i} \int_{\partial\mathbb{D}} \frac{w+z}{(w-z)w} dw\right) \\
&\underset{\text{Residue theorem}}{=} \frac{\varepsilon}{2} \operatorname{Re}(2-1) = \frac{\varepsilon}{2}.
\end{aligned}$$

Consequently, $|u(z)| = |u(z) - u(1)| < \varepsilon$ for all $z \in U$. □

THEOREM 3.11 (Herglotz representation). *Let $f : \mathbb{D} \to \mathbb{C}$ be holomorphic with $\operatorname{Re}(f(z)) > 0$ for all $z \in \mathbb{D}$ and $f(0) = 1$. Then there exists a unique probability measure ρ on $\partial\mathbb{D}$ such that*

$$f(z) = \int_{\partial\mathbb{D}} \frac{w+z}{w-z} \rho(dw).$$

PROOF. Let $(r_n)_{n\in\mathbb{N}}$ be an increasing sequence of numbers in $(0, 1)$ with $r_n \to 1$ as $n \to \infty$. Then $f_n(z) := f(r_n z)$ extends analytically to a neighborhood of $\overline{\mathbb{D}}$ and Theorem 3.10 yields

$$f_n(z) = \frac{1}{2\pi} \int_{-\pi}^{\pi} \frac{e^{it}+z}{e^{it}-z} \operatorname{Re} f_n(e^{it})dt.$$

The measure $\rho_n := \frac{1}{2\pi} \operatorname{Re} f_n(e^{it})dt$ is non-negative and $\int_{-\pi}^{\pi} 1\rho_n(dt) = f_n(0) = 1$. So ρ_n is a probability measure. By regarding it as a probability measure on $\partial\mathbb{D}$, Helly's selection theorem gives us a weakly converging subsequence $(\rho_{n_k})_k$ with limit ρ and as $f_n \to f$ locally uniformly, we obtain $f(z) = \int_{\partial\mathbb{D}} \frac{w+z}{w-z} \rho(dw)$. Uniqueness follows from the proof of Theorem 3.12 below, which only uses the existence of ρ. □

Now we consider holomorphic mappings of the form $f : \mathbb{H} \to \mathbb{H} \cup \mathbb{R}$. Note that either $f(\mathbb{H}) \subset \mathbb{H}$ or f is constant $C \in \mathbb{R}$ due to the open mapping theorem.

THEOREM 3.12 (Nevanlinna representation). *Every holomorphic function $f : \mathbb{H} \to \mathbb{H} \cup \mathbb{R}$ can be written as*

$$f(z) = a + bz + \int_{\mathbb{R}} \frac{1+zx}{x-z} \gamma(dx), \qquad z \in \mathbb{H}, \tag{3.5}$$

with $a \in \mathbb{R}, b \geq 0$, *and a finite, non-negative Borel measure* γ *on* $\mathbb{R}$. *The triple* (a, b, γ) *is uniquely determined and will be called the* Nevanlinna triple *of* f.

Conversely, every such triple (a, b, γ) *produces a holomorphic function from* $\mathbb{H}$ *into* $\mathbb{H} \cup \mathbb{R}$.

PROOF. Let $C : \mathbb{H} \to \mathbb{D}$ be the Cayley transform $C(z) = \frac{z-i}{z+i}$. Let $g : \mathbb{H} \to \mathbb{H}$ be holomorphic with $g(i) = i$. Then $-ig(C^{-1}(w))$ maps $\mathbb{D}$ holomorphically into the right half-plane and 0 onto 1. We apply the Herglotz representation to this function and get

$$-ig(C^{-1}(w)) = \int_{\partial\mathbb{D}} \frac{u+w}{u-w} \rho(du)$$

for a probability measure ρ on $\partial\mathbb{D}$. Now we form the push-forward $\hat{\gamma}$ on $\hat{\mathbb{R}}$ of ρ via C^{-1}. We let $z = C^{-1}(w)$ and $x = C^{-1}(u)$, and we obtain

$$\begin{aligned} -ig(z) &= \int_{\hat{\mathbb{R}}} \frac{C(x) + C(z)}{C(x) - C(z)} \hat{\gamma}(dx) = \int_{\hat{\mathbb{R}}} \frac{\frac{x-i}{x+i} + \frac{z-i}{z+i}}{\frac{x-i}{x+i} - \frac{z-i}{z+i}} \hat{\gamma}(dx) = \\ &= \int_{\hat{\mathbb{R}}} \frac{\frac{2(xz+1)}{(x+i)(z+i)}}{\frac{2(ix-iz)}{(x+i)(z+i)}} \hat{\gamma}(dx) = \int_{\hat{\mathbb{R}}} \frac{1+xz}{ix-iz} \hat{\gamma}(dx). \end{aligned}$$

For a general holomorphic $f : \mathbb{H} \to \mathbb{H} \cup \mathbb{R}$, we can write $f = a + cg$ for $a \in \mathbb{R}$, $c \geq 0$, and $g : \mathbb{H} \to \mathbb{H}$ with $g(i) = i$. Hence

$$f(z) = a + c \int_{\hat{\mathbb{R}}} \frac{1+xz}{x-z} \hat{\gamma}(dx).$$

Write $c\hat{\gamma} = c\hat{\gamma}(\{\infty\})\delta_\infty + \gamma$, for a finite, non-negative measure γ on $\mathbb{R}$. Let $b = c\hat{\gamma}(\{\infty\})$. Then $b \geq 0$ and

$$f(z) = a + bz + \int_{\mathbb{R}} \frac{1+xz}{x-z} \gamma(dx).$$

Furthermore, we have

$$\frac{1+zx}{x-z} = \left(\frac{1}{x-z} - \frac{x}{1+x^2} \right) (1+x^2).$$

Let $\tau(dx) = (1+x^2)\gamma(dx)$. Then τ is a non-negative, locally finite measure on $\mathbb{R}$. (In general, this measure is not finite on $\mathbb{R}$ anymore.) So we have

$$f(z) = a + bz + \int_{\mathbb{R}} \left(\frac{1}{x-z} - \frac{x}{1+x^2} \right) \tau(dx).$$

The Stieltjes–Perron inversion can now be shown for τ as well, i.e.,

$$\begin{aligned} \frac{1}{2}\tau(\{\alpha\}) + \frac{1}{2}\tau(\{\beta\}) + \tau((\alpha,\beta)) &= \frac{1}{\pi} \lim_{y \searrow 0} \int_\alpha^\beta \operatorname{Im}(f(x+iy))\, dx, \quad \alpha, \beta \in \mathbb{R}, \alpha < \beta, \\ \tau(\{\alpha\}) &= -\lim_{y \searrow 0} iyf(\alpha + iy), \quad \alpha \in \mathbb{R}. \end{aligned}$$

This shows that γ is uniquely determined and as $f(i) = a + i(b + \gamma(\mathbb{R}))$, also a and b are unique.

Conversely, as $\operatorname{Im}\left(\frac{1+zx}{x-z}\right) = \operatorname{Im}\left(\frac{1}{x-z}\right)(1+x^2) = \frac{\operatorname{Im}(z)(1+x^2)}{|x-z|^2} > 0$ for $z \in \mathbb{H}$, every function of the form (3.5) maps $\mathbb{H}$ holomorphically into $\mathbb{H} \cup \mathbb{R}$. □

The value b can be computed from f as follows.

THEOREM 3.13. *Let $f : \mathbb{H} \to \mathbb{H} \cup \mathbb{R}$ be holomorphic with Nevanlinna triple (a, b, γ) and let $c > 0$. Then*

$$\frac{f(z_n)}{z_n} \to b \quad \text{as } n \to \infty$$

for any sequence $(z_n)_n \subset \mathbb{H}$ with $|z_n| \to \infty$ and $|\mathrm{Re}(z_n)| < c\,\mathrm{Im}(z_n)$. (A non-tangential approach to ∞.)

PROOF. Write $z_n = a_n + ib_n$, $a_n \in \mathbb{R}, b_n > 0$. Then

$$\delta_n := \frac{f(z_n) - a - bz_n}{z_n} = \int_{\mathbb{R}} \frac{\frac{1}{z_n} + x}{x - z_n} \gamma(dx).$$

We need to show that δ_n converges to 0 as $n \to \infty$. We split the integrand into $\frac{\frac{1}{z_n}}{x-z_n} + \frac{x}{x-z_n}$. Then

$$\left| \frac{\frac{1}{z_n}}{x - z_n} \right| = \frac{1}{|z_n||x - z_n|} \leq \frac{1}{b_n^2} \quad \text{and}$$

$$\frac{x}{x - z_n} = \frac{x(x - a_n) + ixb_n}{(x - a_n)^2 + b_n^2} = \frac{x(x - a_n)}{(x - a_n)^2 + b_n^2} + \frac{ixb_n}{(x - a_n)^2 + b_n^2}.$$

With real calculus we find that the maximum of the real part is given by $\frac{1}{2} + \frac{1}{2}\sqrt{1 + a_n^2/b_n^2} \leq \frac{1}{2} + \frac{1}{2}\sqrt{1 + c^2}$, and its minimum by $\frac{1}{2} - \frac{1}{2}\sqrt{1 + a_n^2/b_n^2}$. Hence, the absolute value is bounded by $\frac{1}{2} + \frac{1}{2}\sqrt{1 + c^2}$.

The maximum of the imaginary part is given by

$$\frac{1}{2}(\sqrt{1 + a_n^2/b_n^2} + a_n/b_n) \leq \frac{1}{2}(\sqrt{1 + c^2} + c).$$

The dominated convergence theorem implies $\delta_n \to 0$ as $n \to \infty$. □

We summarize the inversion formulas for the Nevanlinna representation.

REMARK 3.14. Let $f : \mathbb{H} \to \mathbb{H} \cup \mathbb{R}$ be holomorphic with Nevanlinna triple (a, b, γ). Then

$$f(i) = a + i(b + \gamma(\mathbb{R})).$$

Furthermore, put $\tau(dx) = (1 + x^2)\gamma(dx)$. Then we have the Stieltjes–Perron inversion formula

(3.6)

$$\frac{1}{2}\tau(\{\alpha\}) + \frac{1}{2}\tau(\{\beta\}) + \tau((\alpha, \beta)) = \frac{1}{\pi} \lim_{y \searrow 0} \int_{\alpha}^{\beta} \mathrm{Im}(f(x + iy))\, dx, \quad \alpha, \beta \in \mathbb{R}, \alpha < \beta,$$
$$\tau(\{\alpha\}) = - \lim_{y \searrow 0} iyf(\alpha + iy).$$

Finally, b can be calculated by

$$b = \lim_{y \to \infty} \frac{f(iy)}{iy}. \tag{3.7}$$

We now obtain a very simple characterization of Cauchy transforms.

THEOREM 3.15. *Let $f : \mathbb{H} \to \mathbb{H}^-$ be holomorphic.*

(a) *$f = G_\mu$ for some $\mu \in \mathcal{P}(\mathbb{R})$ if and only if*

$$\lim_{y\to\infty} iyf(iy) = 1.$$

(b) *$f = G_\mu$ for some $\mu \in \mathcal{P}_c(\mathbb{R})$ if and only if f extends analytically to ∞ with the expansion*

$$f(z) = \frac{1}{z} + \frac{c_2}{z^2} + \frac{c_3}{z^3} + \dots$$

PROOF.

(a) Let $f = G_\mu$. Then

$$iyf(iy) = \int_{\mathbb{R}} \frac{iy}{iy - x}\mu(dx) = \int_{\mathbb{R}} \frac{y^2}{y^2 + x^2}\mu(dx) - i\int_{\mathbb{R}} \frac{xy}{y^2 + x^2}\mu(dx).$$

The dominated convergence theorem implies $\lim_{y\to\infty} iyf(iy) = 1$.

Now assume that $\lim_{y\to\infty} iyf(iy) = 1$. We have $-f(z) = a + bz + \int_{\mathbb{R}} \frac{1+zx}{x-z}\gamma(dx)$ for a finite, non-negative measure γ on $\mathbb{R}$ and some $a \in \mathbb{R}, b \geq 0$. Next we write

$$\frac{1 + zx}{x - z} = \left(\frac{1}{x - z} - \frac{x}{1 + x^2}\right)(1 + x^2).$$

Then

$$f(z) = -a - bz - \int_{\mathbb{R}} \left(\frac{1}{x - z} - \frac{x}{1 + x^2}\right)(1 + x^2)\gamma(dx). \tag{3.8}$$

Now we calculate $\mathrm{Re}(iyf(iy))$:

$$\mathrm{Re}(iyf(iy)) = by^2 - \int_{\mathbb{R}} \mathrm{Re}\left(\frac{ixy - y^2}{x^2 + y^2} - \frac{ixy}{1 + x^2}\right)(1 + x^2)\gamma(dx)$$

$$= by^2 + \int_{\mathbb{R}} \frac{y^2}{x^2 + y^2}(1 + x^2)\gamma(dx).$$

As $\mathrm{Re}(iyf(iy))/y^2 \to 0$, we have

$$b + \int_{\mathbb{R}} \frac{1 + x^2}{x^2 + y^2}\gamma(dx) \to b = 0 \quad \text{as} \quad y \to \infty.$$

So $\mathrm{Re}(iyf(iy)) = \int_{\mathbb{R}} \frac{y^2}{x^2+y^2}(1 + x^2)\gamma(dx) \to 1$, which shows that the measure $\mu(dx) := (1 + x^2)\gamma(dx)$ is finite with $\mu(\mathbb{R}) = 1$, i.e., μ is a probability measure. So we can write

$$f(z) = -a + \int_{\mathbb{R}} \frac{1}{z - x}\mu(dx) + \int_{\mathbb{R}} \frac{x}{1 + x^2}\mu(dx). \tag{3.9}$$

Now we calculate $\mathrm{Im}(iyf(iy))$:

$$\mathrm{Im}(iyf(iy)) = -ya + \int_{\mathbb{R}} \mathrm{Im}\left(\frac{iy}{iy - x}\right)\mu(dx) + \int_{\mathbb{R}} \frac{yx}{1 + x^2}\mu(dx)$$

$$= -ya - \int_{\mathbb{R}} \frac{yx}{x^2 + y^2}\mu(dx) + \int_{\mathbb{R}} \frac{yx}{1 + x^2}\mu(dx).$$

From $\mathrm{Im}(iyf(iy))/y \to 0$, we obtain

$$-a - \int_{\mathbb{R}} \frac{x}{x^2+y^2}\mu(dx) + \int_{\mathbb{R}} \frac{x}{1+x^2}\mu(dx) \to -a + \int_{\mathbb{R}} \frac{x}{1+x^2}\mu(dx) = 0$$

as $y \to \infty$. With (3.9) we get

$$f(z) = \int_{\mathbb{R}} \frac{1}{z-x}\mu(dx).$$

(b) If $f = G_\mu$ for $\mu \in \mathcal{P}_c(\mathbb{R})$, then we obtain the expansion $f(z) = 1/z + \dots$ by (3.2). Conversely, if f extends analytically to ∞ with $f(z) = 1/z + \dots$, then $\lim_{y\to\infty} iyf(iy) = 1$ and (a) implies that $f = G_\mu$ for some $\mu \in \mathcal{P}(\mathbb{R})$. As $\overline{f(z)} = f(\overline{z})$ and f extends analytically to $\hat{\mathbb{C}}\setminus[-M,M]$ for some $M > 0$, we see that $f(x) \in \mathbb{R}$ for all $x \in \mathbb{R}\setminus[-M,M]$. The Stieltjes–Perron inversion formula implies that $\mu \in \mathcal{P}_c(\mathbb{R})$.

□

COROLLARY 3.16. *Let $f : \mathbb{H} \to \mathbb{H}\cup\mathbb{R}$ be holomorphic with Nevanlinna triple (a, b, γ). Then*

$$\mathrm{Im}(f(z)) \geq \mathrm{Im}(z) \textit{ for all } z \in \mathbb{H} \textit{ if and only if } b \geq 1.$$

Furthermore, if $b \geq 1$ and $\mathrm{Im}(f(z_0)) = \mathrm{Im}(z_0)$ for some $z_0 \in \mathbb{H}$, then $f(z) = z + c$ for some $c \in \mathbb{R}$.

PROOF. If $b \geq 1$, then the function $f(z) - z$ is of the form (3.5) with Nevanlinna triple $(a, b-1, \gamma)$. Thus $f(z) - z$ maps $\mathbb{H}$ into $\mathbb{H}\cup\mathbb{R}$. Conversely, if $\mathrm{Im}(f(z)) \geq \mathrm{Im}(z)$ for all $z \in \mathbb{H}$, then $f(z) = z + g(z)$ for a holomorphic function $g : \mathbb{H} \to \mathbb{H}\cup\mathbb{R}$ with Nevanlinna triple (a', b', γ'), and we have $b = 1 + b'$ by (3.7), thus $b \geq 1$.

Now assume that $b \geq 1$ and $\mathrm{Im}(f(z_0)) = \mathrm{Im}(z_0)$ for some $z_0 \in \mathbb{H}$. Then $g(z) = f(z) - z$ is a holomorphic mapping into $\mathbb{H}\cup\mathbb{R}$ with $g(z_0) = c \in \mathbb{R}$. The open mapping theorem implies that g is constant. □

3.4. Convergence

Let $f : \mathbb{H} \to \mathbb{H}\cup\mathbb{R}$ be holomorphic with Nevanlinna triple (a, b, γ). We write

$$f(z) = a + bz + \int_{\mathbb{R}} \frac{1+xz}{x-z}\gamma(dx) = a + \int_{\hat{\mathbb{R}}} \frac{1+xz}{x-z}\hat{\gamma}(dx),$$

where $\hat{\gamma} = \gamma + b\delta_\infty$. Let us call $(a, \hat{\gamma})$ the Nevanlinna pair of f.

THEOREM 3.17. *Let $f, f_1, f_2, \dots$ be holomorphic mappings from $\mathbb{H}$ into $\mathbb{H}\cup\mathbb{R}$ with Nevanlinna pairs $(a, \hat{\gamma}), (a_1, \hat{\gamma}_1), \dots$ Then the following statements are equivalent:*

(1) *$f_n \to f$ locally uniformly in $\mathbb{H}$.*
(2) *$a_n \to a$ and $\hat{\gamma}_n$ converges weakly to $\hat{\gamma}$.*

PROOF. Assume that $a_n \to a$ and $\hat{\gamma}_n \to \hat{\gamma}$. The definition of weak convergence implies that $f_n(z) \to f(z)$ for all $z \in \mathbb{H}$. As the set of all holomorphic mappings from $\mathbb{H}$ into $\mathbb{H}\cup\hat{\mathbb{R}}$ forms a normal family, pointwise convergence already implies locally uniform convergence by Vitali's theorem, see [**62**, p.9].

Now assume that $f_n \to f$ locally uniformly. Then, by Remark 3.14, $f_n(i) = a_n + i\hat{\gamma}_n(\hat{\mathbb{R}}) \to a + i\hat{\gamma}(\hat{\mathbb{R}}) = f(i)$, and we see that $a_n \to a$ and $\hat{\gamma}_n(\hat{\mathbb{R}}) \to \hat{\gamma}(\hat{\mathbb{R}})$. Now

Helly's selection theorem yields a subsequence $(\hat{\gamma}_{n_k})_k$ which converges weakly to some finite, non-negative measure ρ on $\hat{\mathbb{R}}$. As $f_{n_k} \to f$, we conclude that $\rho = \hat{\gamma}$ by the Stieltjes–Perron inversion formula (3.6). Thus every convergent subsequence of $(\hat{\gamma}_n)_n$ has the same limit $\hat{\gamma}$, and we conclude that $\hat{\gamma}_n \to \hat{\gamma}$ as $n \to \infty$. □

The same proof applies to Cauchy transforms.

THEOREM 3.18. *Let μ and $\mu_1, \mu_2, \ldots$ be probability measures on $\mathbb{R}$. Then the following statements are equivalent:*

(1) *G_{μ_n} converges to G_μ locally uniformly in $\mathbb{H}$.*
(2) *μ_n converges weakly to μ.*

PROOF. It is clear that (2) implies (1). So assume that G_{μ_n} converges to G_μ locally uniformly in $\mathbb{H}$. Helly's selection theorem yields a subsequence $(\mu_{n_k})_k$ that converges to a probability measure ρ on $\hat{\mathbb{R}}$. Then $\int_{\mathbb{R}} \frac{1}{z-x}\rho(dx) = G_\mu$ and the Stieltjes–Perron inversion formula implies that μ is equal to the restriction of ρ to $\mathbb{R}$. As μ is also a probability measure, $\rho(\{\infty\}) = 0$, and thus, $\rho = \mu$. So every convergent subsequence of $(\mu_n)_n$ has the same limit μ, and we conclude that $\mu_n \to \mu$ as $n \to \infty$. □

3.5. Discrete semigroups

Consider a compositional semigroup $(F_n)_{n\in\mathbb{N}}$ of holomorphic self-mappings of $\mathbb{H}$, i.e., $F_n = F^{\circ n} = F \circ \cdots \circ F$ with $F = F_1$. Such semigroups can be classified by looking at the behavior of $F^{\circ n}$ as $n \to \infty$. This classification is usually stated for the unit disk $\mathbb{D}$, but we can simply pass from $\mathbb{H}$ to $\mathbb{D}$ via the Cayley transform $C : \mathbb{H} \to \mathbb{D}$. Note that $C \circ F_n \circ C^{-1} = (C \circ F \circ C^{-1})^{\circ n}$ is a semigroup on $\mathbb{D}$.

To state the result, we need two further notions. For a function $f : \mathbb{D} \to \mathbb{C}$ and $p \in \partial\mathbb{D}$, the existence of the non-tangential limit $\angle \lim_{z\to p} f(z) = c$ means that $\lim_{n\to\infty} f(z_n) = c$ for every sequence $(z_n) \subset \mathbb{D}$ which converges to p within a sector, i.e., the angle of the vector $p - z_n$ must be in $(\arg p - \varepsilon, \arg p + \varepsilon)$ for some $\varepsilon \in (0, \pi/2)$ and all n.

A horodisk in $\mathbb{D}$ at $p = 1$ is the image of a set of the form $\{z \in \mathbb{H} \mid \operatorname{Im}(z) > c\}$, $c > 0$, under the Cayley transform, see Figure 3.1. By rotating, we define horodisks in $\mathbb{D}$ at every other $p \in \partial\mathbb{D}$.

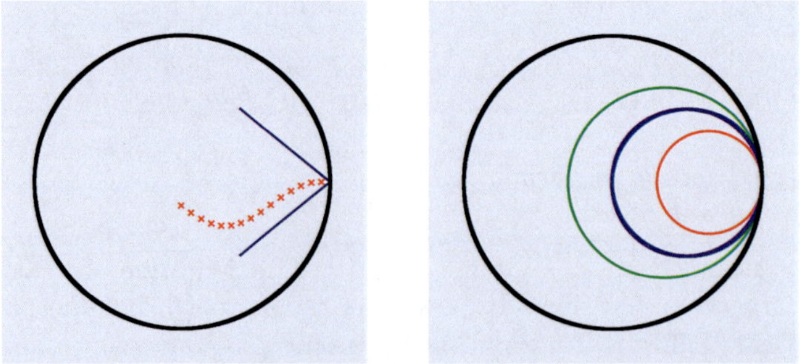

FIGURE 3.1. Left: non-tangential approach to 1. Right: horodisks at 1.

THEOREM 3.19. *Let $F : \mathbb{D} \to \mathbb{D}$ be a holomorphic self-map which is not an elliptic automorphism, i.e., F is not conjugated to a rotation $z \mapsto e^{i\alpha}z$, $\alpha \in \mathbb{R}$.*

(a) *If F has a fixed point $p \in \mathbb{D}$, then $F^{\circ n} \to p$ locally uniformly as $n \to \infty$ and $|F'(p)| < 1$.*
(b) *If F has no fixed points in $\mathbb{D}$, then there is a point $p \in \partial\mathbb{D}$ such that $F^{\circ n} \to p$ locally uniformly as $n \to \infty$. Furthermore, $\angle \lim_{z\to p} F(z) = p$ and $\angle \lim_{z\to p} F'(z) \in (0, 1]$.*

The point p in (a) or (b) is called the Denjoy–Wolff point *of F.*

(c) *Conversely, if there exists $p \in \partial\mathbb{D}$ with*
$$\angle \lim_{z\to p} F(z) = p \quad \textit{and} \quad \angle \lim_{z\to p} F'(z) \in (0, 1],$$
then p is the Denjoy–Wolff point of F.
(d) *If F has no fixed points in $\mathbb{D}$, then $p \in \partial\mathbb{D}$ is the Denjoy–Wolff point of F if and only if F maps every horodisk at p into itself.*

PROOF. The statements (a)-(c) can be found in [**175**, The Grand Iteration Theorem]. Statement (d) follows from Wolff's theorem, see [**175**, Section 5.3]. □

COROLLARY 3.20. *Let $f : \mathbb{H} \to \mathbb{H} \cup \mathbb{R}$ be holomorphic with Nevanlinna triple (a, b, γ) and assume that f is not the identity. Then ∞ is the Denjoy–Wolff point of f, i.e., $f^{\circ n} \to \infty$ as $n \to \infty$, if and only if $b \geq 1$.*

PROOF. If ∞ is the Denjoy–Wolff point of f, then Theorem 3.19 (d) implies that $\mathrm{Im}(f(z)) \geq \mathrm{Im}(z)$ for all $z \in \mathbb{H}$, which implies $b \geq 1$ by Corollary 3.16. Conversely, assume that $b \geq 1$. Then Corollary 3.16 implies $\mathrm{Im}(f(z)) \geq \mathrm{Im}(z)$ for all $z \in \mathbb{H}$. If f has a fixed point z_0, then $f(z) = z$ for all $z \in \mathbb{H}$ by Corollary 3.16. But as f is not the identity, f has no fixed points and Theorem 3.19 (d) implies that ∞ is the Denjoy–Wolff point of f. □

REMARK 3.21. A holomorphic self-mapping of $\mathbb{D}$ with a fixed point in $\mathbb{D}$ is called *elliptic*. By virtue of Theorem 3.19 (b), non-elliptic self-mappings are further classified as *hyperbolic* if $\angle \lim_{z\to p} F'(z) < 1$, and *parabolic* if $\angle \lim_{z\to p} F'(z) = 1$. Thus, the mapping f in Corollary 3.20, with $b \geq 1$, is non-elliptic because its Denjoy–Wolff point is at ∞, and it is parabolic if and only if $b = 1$.

3.6. Continuous semigroups

DEFINITION 3.22. Let $D \subset \mathbb{C}$ be a domain. A *continuous semigroup* on D is a family $(F_t)_{t\geq 0}$ of holomorphic self-mappings $F_t : D \to D$ such that

(1) $F_0(z) = z$ for all $z \in D$,
(2) $t \mapsto F_t$ is continuous with respect to locally uniform convergence,
(3) $F_{t+s} = F_t \circ F_s$ for all $s, t \geq 0$.

We will only need the case where $D \subsetneq \mathbb{C}$ is simply connected. Due to the Riemann mapping theorem we may assume that $D = \mathbb{D}$. (In fact, all other domains, even all other Riemann surfaces, lead to rather boring continuous semigroups, see [**1**, Section 1.4.3].)

A continuous semigroup of holomorphic self-mappings can be represented by a differential equation, [**35**, Theorem 1.1], see also [**1**, Theorem 1.4.11].

THEOREM 3.23. *Let $(F_t)_{t\geq 0}$ be a continuous semigroup on $\mathbb{D}$. Then $t \mapsto F_t(z)$ is differentiable for every $z \in \mathbb{D}$ and there exists a holomorphic function $G : \mathbb{D} \to \mathbb{C}$ such that*

$$\frac{\partial}{\partial t} F_t(z) = G(F_t(z)) \tag{3.10}$$

for all $z \in \mathbb{D}$ and $t \geq 0$.

PROOF. Let $K \subset \mathbb{D}$ be a compact subset. For $\alpha \in (0,1)$, the set $\cup_{0\leq t\leq \alpha} F_t(K)$ is compact and we denote by $\tilde{K} \subset \mathbb{D}$ its convex hull. For $c > 0$, we find $\delta \in (0, \alpha]$ such that $\sup_{z\in\tilde{K}} |F_t'(z) - 1| \leq c$ for all $0 \leq t \leq \delta$. Hence, for all $t \in [0, \delta]$ and $z \in K$ we have

$$|F_{2t}(z) - 2F_t(z) + z| = \left| \int_z^{F_t(z)} \frac{d}{dw}(F_t(w) - w) dw \right| \leq c|F_t(z) - z|, \tag{3.11}$$

where the integration is along the straight line segment from z to $F_t(z)$, which is contained in $\tilde{K}$. From $|F_{2t}(z) - 2F_t(z) + z| \geq 2|F_t(z) - z| - |F_{2t}(z) - z|$, we obtain

$$2|F_t(z) - z| \leq c|F_t(z) - z| + |F_{2t}(z) - z|, \quad \text{and thus,}$$

$$|F_t(z) - z| \leq \frac{1}{2 - c} |F_{2t}(z) - z| \leq 2^{-2/3} |F_{2t}(z) - z| \tag{3.12}$$

for all $z \in K$ and $t \in [0, \delta]$, where we assume that c is sufficiently close to 0. Let $k \in \mathbb{N}$ be such that $2^{-k} \leq \delta$ and put $M = 2^{2k/3} \sup\{|F_t(z) - z| | z \in K, t \in [2^{-k}, 1]\}$. Then, for all $t \in [0, 1]$ and $z \in K$:

$$|F_t(z) - z| \leq Mt^{2/3}. \tag{3.13}$$

This is clear if $t \in [2^{-k}, 1]$. If $t \in [2^{-k-1}, 2^{-k})$, then (3.12) implies

$$|F_t(z) - z| \leq 2^{-2/3} |F_{2t}(z) - z| \leq 2^{-2/3} M(2t)^{2/3} = Mt^{2/3},$$

and the inequality follows by induction for all $t \in (0, 2^{-k})$.

Now repeat the same argument for a compact subset $L \subset \mathbb{D}$ with $\tilde{K} \subset L$, which yields a constant $M_2 > 0$ such that

$$|F_t(z) - z| \leq M_2 t^{2/3}$$

for all $z \in L$ and $t \in [0, 1]$. The Cauchy inequalities give us a constant $M_3 > 0$ such that

$$|F_t'(z) - 1| \leq M_3 t^{2/3}$$

for all $z \in L$ and $t \in [0, 1]$. Now we look again at (3.11) and we use the previous estimate and (3.13) to get

$$|F_{2t}(z) - 2F_t(z) + z| = \left| \int_z^{F_t(z)} \frac{d}{dw}(F_t(w) - w) dw \right| \leq M_3 t^{2/3} |F_t(z) - z| \leq MM_3 t^{4/3}$$

for all $z \in K$ and $t \in [0, \alpha]$. Hence

$$\left| \frac{F_{2t}(z) - z}{2t} - \frac{F_t(z) - z}{t} \right| \leq (MM_3/2)t^{1/3}$$

for all $z \in K$ and $t \in (0, \alpha]$. Consequently, the limit

$$\lim_{n\to\infty} (F_{2^{-n}}(z) - z)/2^{-n}$$

exists uniformly on compact subsets of $\mathbb{D}$ and defines a holomorphic function $G : \mathbb{D} \to \mathbb{C}$. For $z_0 \in \mathbb{D}$ and $t_0 > 0$, the set $\{F_t(z_0) \,|\, 0 \leq t \leq t_0\}$ is compact and thus $(F_{t+2^{-n}}(z_0) - F_t(z_0))/2^{-n} = (F_{2^{-n}}(F_t(z_0)) - F_t(z_0))/2^{-n}$ converges uniformly on $[0, t_0]$ to $G(F_t(z_0))$ as $n \to \infty$. Due to (3.13), $t \mapsto F_t(z_0)$ is locally absolutely continuous. We conclude

$$F_t(z) = z + \int_0^t G(F_s(z))ds$$

for all $t \geq 0$ and $z \in \mathbb{D}$, which implies (3.10). □

EXAMPLE 3.24. For $a \in \mathbb{C}$ with $\mathrm{Re}(a) \leq 0$, the functions $F_t(z) = e^{at}z$ form a continuous semigroup on $\mathbb{D}$ with $G(z) = az$. If $\mathrm{Re}(a) < 0$, then $F_t(z) \to 0$ as $t \to \infty$ and if $\mathrm{Re}(a) = 0$, we simply have rotations of $\mathbb{D}$.

Note that Theorem 3.23 shows that the continuity of $t \mapsto F_t$, assumption (2), makes a semigroup automatically differentiable. Furthermore, $t \mapsto F_t(z)$ is the unique solution to the initial value problem

$$\frac{d}{dt}v(t) = G(v(t)), \quad v(0) = z. \tag{3.14}$$

This has a simple but interesting consequence. If we look at the initial value problem (3.14) for two different initial values $z_0, w_0 \in \mathbb{D}$, $z_0 \neq w_0$, then $F_t(z_0) \neq F_t(w_0)$ for all $t \geq 0$. Otherwise, if $F_T(z_0) = F_T(w_0)$ for some $T > 0$, we could solve the differential equation at time T and go backward in time:

$$\frac{d}{dt}v(t) = -G(v(t)), \quad v(0) = F_T(z_0).$$

This initial value problem would have two solutions for $t \in [0, T]$, namely $v(t) = F_{T-t}(z_0)$ and $v(t) = F_{T-t}(w_0)$, which would contradict the Picard–Lindelöf uniqueness theorem. So $z \mapsto F_t(z)$ is injective.

DEFINITION 3.25. An injective holomorphic function is also called *univalent*.

COROLLARY 3.26. *All elements F_t of a continuous semigroup are univalent functions.*

This statement can also be derived directly from Definition 3.22, see [**1**, Prop. 1.4.6]. Conversely, can any univalent $f : \mathbb{D} \to \mathbb{D}$ be embedded into a continuous semigroup? The answer is no (which follows, for example, from Theorem 8.19). However, any univalent $f : \mathbb{D} \to \mathbb{D}$ can be embedded into a generalization of a semigroup called *Loewner chain*, which will be studied in Chapter 6.

The holomorphic function $G : D \to \mathbb{C}$ is also called the *infinitesimal generator* of the semigroup. Since every semigroup is uniquely described by its generator, problems concerning semigroups can be translated into problems concerning infinitesimal generators. Therefore, it is of interest to describe and classify all possible infinitesimal generators on $\mathbb{D}$.

THEOREM 3.27.

(a) *Every infinitesimal generator G on $\mathbb{D}$ has the following form (Berkson–Porta formula):*

$$G(z) = (\tau - z)(1 - \overline{\tau}z)p(z), \tag{3.15}$$

where $\tau \in \overline{\mathbb{D}}$ and $p : \mathbb{D} \to \mathbb{C}$ is holomorphic with $\mathrm{Re}(p) \geq 0$. Conversely, every such function is the infinitesimal generator of a continuous semigroup on $\mathbb{D}$. If the corresponding semigroup $(F_t)_{t\geq 0}$ does not consist of elliptic automorphisms, then τ is the Denjoy–Wolff point of the semigroup, i.e., $F_t \to \tau$ locally uniformly as $t \to \infty$.

(b) *Assume that F_1 is not an elliptic automorphism. Then F_1 has a fixed point $p \in \mathbb{D}$ if and only if $G(p) = 0$. In this case,*

$$F_t'(p) = e^{tG'(p)} \quad \text{for all} \quad t \geq 0.$$

Furthermore, $p \in \partial\mathbb{D}$ is the Denjoy–Wolff point of F_1 if and only if

$$\angle \lim_{z\to p} G(z) = 0 \quad \text{and} \quad \angle \lim_{z\to p} G'(z) \in (-\infty, 0].$$

In this case, $\angle \lim_{z\to p} F_t'(z) = \angle \lim_{z\to p} e^{tG'(z)}$ for all $t \geq 0$.

PROOF. For (a), we refer to [**35**] or [**1**, Theorem 1.4.19], and (b) can be found in [**54**, Theorem 1]. □

EXAMPLE 3.28. $F_t(z) = e^{it}z$ is a continuous semigroup, where F_1 is an elliptic automorphism. We obtain the generator $G(z) = iz$, which corresponds to $\tau = 0$ and $p(z) \equiv -i$. The continuous semigroup $F_t(z) = e^{-t}z$ corresponds to $G(z) = -z$, i.e., $p(z) \equiv 1$ and $\tau = 0$, which is the Denjoy–Wolff point of the semigroup.

If $(F_t)_{t\geq 0}$ is a continuous semigroup on $\mathbb{D}$ with generator G and $C : E \to \mathbb{D}$, $E \subset \mathbb{C}$, is a conformal mapping, then $\hat{F}_t = C^{-1} \circ F_t \circ C$ is a continuous semigroup on E. We consider the upper half-plane $E = \mathbb{H}$, which is mapped conformally onto $\mathbb{D}$ by the Cayley transform $C : \mathbb{H} \to \mathbb{D}$, $C(z) = (z - i)/(z + i)$.

EXAMPLE 3.29. The mappings $(e^t w)_{t\geq 0}$ form a continuous semigroup on $\mathbb{H}$ with Denjoy–Wolff point ∞. Via conjugation by C, they correspond to the semigroup

$$\left(\frac{z + \tanh(t/2)}{1 + z\tanh(t/2)}\right)_{t\geq 0}$$

on $\mathbb{D}$, whose generator G is given by $G(z) = \frac{1}{2}(1 - z^2)$. G has the Berkson–Porta data $\tau = 1$ and $p(z) = \frac{1}{2}\frac{1+z}{1-z}$.

We obtain a formula for generators on the upper half-plane $\mathbb{H}$ by differentiation of $\hat{F}_t = C^{-1} \circ F_t \circ C$ with respect to t (at $t = 0$). The infinitesimal generator H of $\hat{F}_t$ is given by

$$\begin{aligned} H(z) = (C^{-1})'(C(z)) \cdot G(C(z)) &= \frac{-i(z+i)^2}{2} G(C(z)) \\ &= -(z+i)^2(\tau - C(z))(1 - \overline{\tau}C(z)) \cdot \frac{i}{2} p(C(z)). \end{aligned}$$

If $\tau = 1$, this expression reduces to $H(z) = 2ip(C(z))$. If $\tau \neq 1$, then the first three factors form a polynomial of degree 2 with zeros $\sigma := C^{-1}(\tau) \in \mathbb{H} \cup \mathbb{R}$ and $\overline{\sigma}$. In this case, we obtain $H(z) = (1-\tau)(1-\overline{\tau})(z-\sigma)(z-\overline{\sigma}) \cdot \frac{i}{2}p(C(z))$. Together with Theorem 3.27, we obtain the following representation.

LEMMA 3.30. *Every infinitesimal generator G on $\mathbb{H}$ has the form*

$$G(z) = q(z) \qquad \text{or} \qquad G(z) = (z - \sigma)(z - \overline{\sigma})q(z),$$

for a holomorphic function $q : \mathbb{H} \to \mathbb{H} \cup \mathbb{R}$ and $\sigma \in \mathbb{H} \cup \mathbb{R}$. Conversely, every such function is the infinitesimal generator of a continuous semigroup on $\mathbb{H}$. The first case corresponds to all semigroups whose Denjoy–Wolff point is ∞ unless $G = 0$. In the second case, σ is the Denjoy–Wolff point unless the semigroup consists of elliptic automorphisms.

3.7. Univalent functions on the unit disk

A holomorphic function f is injective in a neighborhood of a point p if and only if $f'(p) \neq 0$. Can the global injectivity of a holomorphic function also be described analytically? We define the class S of all univalent functions on $\mathbb{D}$ with a normalization for $f(0)$ and $f'(0)$.

DEFINITION 3.31.

$$S := \{f : \mathbb{D} \to \mathbb{C} \,|\, f \text{ is univalent}, f(0) = 0, f'(0) = 1\}.$$

[*Univalent* functions were also called "schlicht" ("simple") in German.]

If $g : \mathbb{D} \to \mathbb{C}$ is univalent, then $(g(z) - g(0))/g'(0) \in S$.

EXAMPLE 3.32. The Koebe function

$$k(z) = \frac{z}{(1-z)^2} = \sum_{n=1}^{\infty} n z^n$$

maps $\mathbb{D}$ conformally onto $\mathbb{C} \setminus (-\infty, -\frac{1}{4}]$, see Figure 3.2, as

$$k(z) = \frac{1}{4}\left(\left(\frac{1+z}{1-z}\right)^2 - 1\right),$$

and $(1+z)/(1-z)$ maps $\mathbb{D}$ conformally onto the right half-plane.

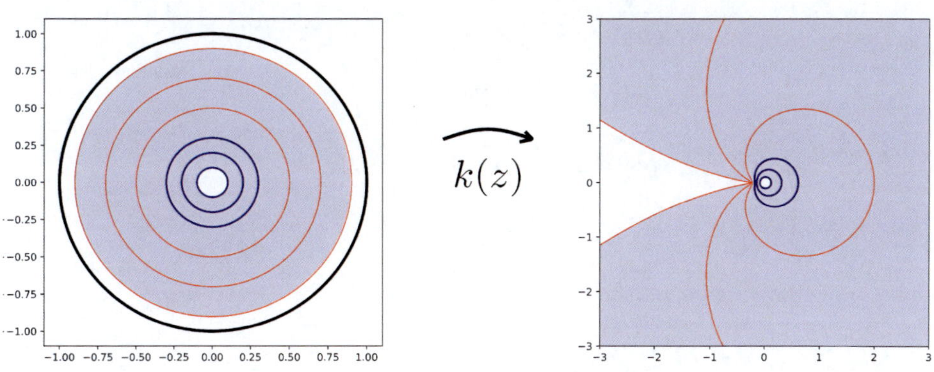

FIGURE 3.2. The Koebe function.

EXAMPLE 3.33. Let $f(z) = z + \sum_{n=2}^{\infty} a_n z^n \in S$. There are several transformations of f that generate new functions in S. For example, for $\alpha \in \mathbb{R}$, the rotation of f by α is given by

$$e^{-i\alpha} f(e^{i\alpha} z) = z + \sum_{n=2}^{\infty} a_n e^{i(n-1)\alpha} z^n \in S.$$

The square root transform $\sqrt{f(z^2)} \in S$ is defined by writing

$$\sqrt{f(z^2)} = g(z) := z\sqrt{1 + a_2 z^2 + a_3 z^4 + \dots}$$

(The function $z \mapsto 1 + a_2 z^2 + a_3 z^4 + \dots$ has no zeros in $\mathbb{D}$. Otherwise, f would have at least two zeros.) Then

$$\sqrt{f(z^2)} = z + \frac{a_2}{2} z^3 + c_5 z^5 + \dots$$

is an odd function. It is univalent, for $g(z_1) = g(z_2)$ implies $f(z_1^2) = f(z_2^2)$, and thus, $z_1 = \pm z_2$. If $z_1 = -z_2$, then $g(z_2) = -g(z_2)$, thus $g(z_2) = 0$, i.e., $z_2 = z_1 = 0$.

To better understand the class S, we first look at the class Σ of all univalent functions g defined in $\mathbb{C} \setminus \overline{\mathbb{D}}$ with the expansion

$$g(z) = z + \sum_{n=0}^{\infty} b_n z^{-n}$$

at ∞. If $f \in S$, then

$$\text{(3.16)} \qquad \frac{1}{f(1/z)} = \frac{z}{1 + a_2 z^{-1} + \dots} = z - a_2 + (a_2^2 - a_3)z^{-1} + \dots \in \Sigma.$$

THEOREM 3.34 (Grönwall's area theorem). *Let $g \in \Sigma$ and $E = \mathbb{C} \setminus g(\mathbb{C} \setminus \overline{\mathbb{D}})$. Then*

$$Area(E) = \pi \left(1 - \sum_{n=1}^{\infty} n|b_n|^2\right).$$

PROOF. For $r > 1$ we let $E(r) = \mathbb{C} \setminus g(\{z \in \mathbb{C} \,|\, |z| \geq r\})$ and $C_r = \{z \in \mathbb{C} \,|\, |z| = r\}$. Then $E(r)$ is bounded by the Jordan curve $g(C_r)$. Green's theorem and the observation $\overline{w}dw = (x - iy)(dx + idy) = xdx + ydy + ixdy - iydx$ imply

$$\begin{aligned}
Area(E(r)) &\underset{\text{Green's theorem}}{=} \frac{1}{2}\int_{g(C_r)} xdy - ydx = \frac{1}{2i}\int_{g(C_r)} ixdy - iydx \\
&= \frac{1}{2i}\int_{g(C_r)} xdx + ydy + ixdy - iydx = \frac{1}{2i}\int_{g(C_r)} \overline{w}dw \\
&= \frac{1}{2i}\int_{C_r} \overline{g(z)}g'(z)dz = \frac{1}{2i}\int_0^{2\pi} ire^{it}\overline{g(re^{it})}g'(re^{it})dt \\
&= \frac{1}{2}\int_0^{2\pi} (re^{-it} + \sum_{n=0}^{\infty} \overline{b_n} r^{-n} e^{int})(re^{it} - \sum_{n=1}^{\infty} nb_n r^{-n} e^{-int})dt \\
&= \pi(r^2 - \sum_{n=1}^{\infty} n|b_n|^2 r^{-2n}).
\end{aligned}$$

As $Area(E(r)) \geq 0$, we have $\sum_{n=1}^{N} n|b_n|^2 r^{-2n} \leq r^2$, and thus, $\sum_{n=1}^{N} n|b_n|^2 \leq 1$ for every $N \in \mathbb{N}$. Consequently, the sum $\sum_{n=1}^{\infty} n|b_n|^2$ converges, and we can take the limit $r \to 1$ to obtain

$$Area(E) = \pi \left(1 - \sum_{n=1}^{\infty} n|b_n|^2\right). \qquad \square$$

We can now prove that, in case of the class Σ, the univalence of functions does imply a coefficient bound.

COROLLARY 3.35. *Let* $g(z) = z + \sum_{n=0}^{\infty} b_n z^{-n} \in \Sigma$. *Then*

$$\sum_{n=1}^{\infty} n|b_n|^2 \leq 1.$$

In particular, $|b_1| \leq 1$.

PROOF. Follows directly from the area theorem, as $Area(E) \geq 0$. □

The following estimate will be useful in Chapter 8.

COROLLARY 3.36. *Let* $g(z) = z + \sum_{n=0}^{\infty} b_n z^{-n} \in \Sigma$ *and* $E = \mathbb{C} \setminus g(\mathbb{C} \setminus \overline{\mathbb{D}})$. *Then*

$$E \subset \{z \in \mathbb{C} \,|\, |z - b_0| \leq 2\}.$$

PROOF. For $w \in E$, the function $z^{-2}(g(z^2) - w)$ is holomorphic and $\neq 0$ in $\mathbb{C} \setminus \overline{\mathbb{D}}$. So $h(z) = z\sqrt{z^{-2}(g(z^2) - w)}$ is a holomorphic function in $\mathbb{C} \setminus \overline{\mathbb{D}}$ with expansion

$$h(z) = z + \frac{1}{2}(b_0 - w)z^{-1} + \dots$$

at ∞. If $h(z_1) = h(z_2)$, then $g(z_1^2) = g(z_2^2)$ and, since g is univalent, $z_1^2 = z_2^2$. If $z_1 = -z_2$, then, as h is an odd function, $h(-z_1) = -h(z_1) \neq h(z_1)$. Hence, h is univalent in $\mathbb{C} \setminus \overline{\mathbb{D}}$ and belongs to the class Σ. Corollary 3.35 implies that $|w - b_0| \leq 2$. □

COROLLARY 3.37. *Let* $f(z) = z + \sum_{n=2}^{\infty} a_n z^n \in S$. *Then*

$$|a_2^2 - a_3| \leq 1.$$

PROOF. Follows directly from Corollary 3.35 and (3.16). □

In 1916, L. Bieberbach proved the following inequality ([**39**]).

COROLLARY 3.38 (Bieberbach's theorem). *Let* $f(z) = z + \sum_{n=2}^{\infty} a_n z^n \in S$. *Then*

$$|a_2| \leq 2$$

and equality holds if and only if $f(z) = \frac{z}{(1-e^{i\alpha}z)^2} = e^{-i\alpha}k(e^{i\alpha}z)$, $\alpha \in \mathbb{R}$.

PROOF. According to Example 3.33, the mapping $\sqrt{f(z^2)}$ is in S, and with (3.16), the mapping $g(z) = (f(1/z^2))^{-1/2}$ is in Σ. As

$$g(z) = z - \frac{a_2}{2}z^{-1} + \dots,$$

it follows from Corollary 3.35 that $|a_2| \leq 2$. Equality holds if and only if $g(z) = z - e^{i\alpha}z^{-1}$, $\alpha \in \mathbb{R}$, which leads to $f(z) = \frac{z}{(1-e^{i\alpha}z)^2}$. □

Bieberbach then stated that $z + \sum_{n=2}^{\infty} a_n z^n \in S$ might imply

$$|a_n| \leq n \quad \text{for all } n \geq 2. \qquad \textbf{(Bieberbach conjecture)}$$

Note that the Koebe function $k(z) = \sum_{n=1}^{\infty} nz^n$ shows that there is no lower bound for $|a_n|$.

Bieberbach's theorem implies that S is compact, and thus, $|a_n|$ is indeed bounded for fixed $n \geq 2$ (see Appendix A). In Section 8.6, we will come back to this conjecture and see how Loewner managed to prove the case $n = 3$.

Another simple consequence of Bieberbach's theorem is the following result.

THEOREM 3.39 (Koebe's 1/4–theorem). *Let $f \in S$. Then*

$$\frac{1}{4}\mathbb{D} \subset f(\mathbb{D}).$$

PROOF. Let $w \notin f(\mathbb{D})$ and $g : \mathbb{D} \to \mathbb{C}$,

$$g(z) = \frac{wf(z)}{w - f(z)} = z + \left(a_2 + \frac{1}{w}\right) z^2 + \dots$$

Then g is univalent. Bieberbach's theorem for g gives $|a_2 + \frac{1}{w}| \leq 2$. Together with $|a_2| \leq 2$, we obtain

$$\frac{1}{|w|} \leq |-a_2| + \left|a_2 + \frac{1}{w}\right| \leq 4, \quad \text{so} \quad |w| \geq \frac{1}{4}. \qquad \square$$

3.8. Starlike and convex functions

Geometric properties of univalent functions can often be translated into analytic properties. Two classical examples are starlikeness and convexity of the image domains.

DEFINITION 3.40. A mapping $f \in S$ is called *starlike* if $f(\mathbb{D})$ is a starlike domain with respect to 0, i.e.,

$$w \in f(\mathbb{D}) \text{ implies } tw \in f(\mathbb{D}) \text{ for all } t \in [0,1].$$

The set of all starlike mappings is denoted by S^*. We call $f \in S$ *convex* if $f(\mathbb{D})$ is convex, i.e.,

$$a, b \in f(\mathbb{D}) \text{ implies } ta + (1-t)b \in \mathbb{D} \text{ for all } t \in [0,1].$$

Obviously, if $f \in S$ is convex, it is also starlike. For our analytic descriptions, we need the following class of holomorphic functions.

DEFINITION 3.41. The class $\mathcal{P}$ is defined as

$$\mathcal{P} = \{p : \mathbb{D} \to \mathbb{C} \,|\, p \text{ holomorphic}, \ \mathrm{Re}(p) > 0, \ p(0) = 1\}.$$

The starlike and convex functions can now be characterized via $\mathcal{P}$. We follow [**62**, Section 2.5].

THEOREM 3.42. *Let $f : \mathbb{D} \to \mathbb{C}$ be holomorphic with $f(0) = 0, f'(0) = 1$. Then $f \in S^*$ (in particular, f is univalent) if and only if*

$$z\frac{f'(z)}{f(z)} \in \mathcal{P}.$$

PROOF. Let $f \in S^*$. The domain $f(r\mathbb{D})$, $r \in (0,1)$, is starlike as well: First, for every $t \in (0,1)$, the mapping $tf(z)$ can also be written as $f(\omega(z))$ with $\omega : \mathbb{D} \to \mathbb{D}$, $\omega(0) = 0$. The Schwarz lemma implies $|\omega(z)| \leq |z|$. Hence,

$$t \cdot f(rz) = f(\omega(rz)) = f(r\omega_1(z)),$$

where $\omega_1(z) = \omega(rz)/r$, which implies $|\omega_1(z)| \leq |z|$.

Consequently, $f(r\partial\mathbb{D})$ is an analytic Jordan curve that surrounds a starlike domain. Hence, the mapping $\varphi \mapsto \arg f(re^{i\varphi})$ is monotonically increasing (see Figure 3.3) and

$$0 \leq \frac{\partial}{\partial\varphi} \arg f(re^{i\varphi}) = \frac{\partial}{\partial\varphi} \mathrm{Im} \log f(re^{i\varphi}) = \mathrm{Im}\left(ire^{i\varphi}\frac{f'(re^{i\varphi})}{f(re^{i\varphi})}\right) = \mathrm{Re}\left(z\frac{f'(z)}{f(z)}\right).$$

As $f(0) = 0$ and $f'(0) = 1$, the function $z\frac{f'(z)}{f(z)} = 1 + \dots$ is analytic in $\mathbb{D}$ and $\operatorname{Re}\left(z\frac{f'(z)}{f(z)}\right)$ is not constant 0. The maximum principle implies $\operatorname{Re}\left(z\frac{f'(z)}{f(z)}\right) > 0$ for all $z \in \mathbb{D}$, and we get $z\frac{f'(z)}{f(z)} \in \mathcal{P}$.

Conversely, assume $z\frac{f'(z)}{f(z)} \in \mathcal{P}$. Then $0 < \frac{\partial}{\partial\varphi} \arg f(re^{i\varphi})$. Let $r \in (0,1)$. Assume that the winding number of the curve $[0, 2\pi] \ni \varphi \mapsto f(re^{i\varphi})$ with respect to 0 is greater than 1. Then f has more than one zero in $\mathbb{D}$ because of the argument principle. This is impossible because $z\frac{f'(z)}{f(z)}$ is analytic and $f'(0) = 1$. It follows that the winding number is equal to 1 and the curve does not intersect itself. Hence, $f(r\mathbb{D})$ is a starlike domain and every $w \in f(r\mathbb{D})$ has exactly one preimage under f. We conclude that f is univalent and starlike. □

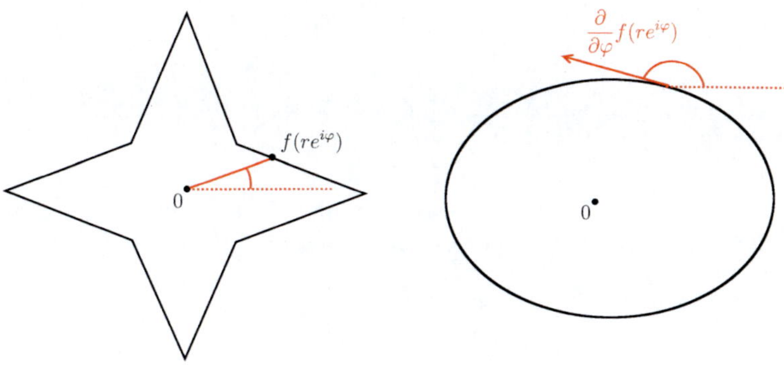

FIGURE 3.3. $\varphi \mapsto \arg f(re^{i\varphi})$ is increasing for a starlike function and $\varphi \mapsto \arg\left(\frac{\partial}{\partial\varphi}f(re^{i\varphi})\right)$ is increasing for a convex function.

THEOREM 3.43. *Let $f : \mathbb{D} \to \mathbb{C}$ be holomorphic with $f(0) = 0, f'(0) = 1$. Then f is convex (in particular, f is univalent) if and only if*

$$1 + z\frac{f''(z)}{f'(z)} \in \mathcal{P}.$$

PROOF. Let $f \in S$ be convex. The domain $f(r\mathbb{D})$, $r \in (0,1)$, is convex as well: Let $z_1, z_2 \in r\mathbb{D}$ with $|z_1| \le |z_2| \ne 0$, and $w_1 = f(z_1)$, $w_2 = f(z_2)$. We consider $w_0 = tw_1 + (1-t)w_2$, $t \in (0,1)$. Since f is convex, there exists a $z_0 \in \mathbb{D}$ with $f(z_0) = w_0$. We have to show that $|z_0| < r$. Let

$$g(z) = tf(z_1/z_2 \cdot z) + (1-t)f(z).$$

Then $g(z_2) = w_0$ and $g(\mathbb{D}) \subset f(\mathbb{D})$. Thus, we can consider the mapping

$$h(z) = f^{-1}(g(z)).$$

We have $h(0) = 0$ and $h(\mathbb{D}) \subset \mathbb{D}$. The Schwarz lemma implies $|h(z)| \le |z|$ and

$$|z_0| = |h(z_2)| \le |z_2| < r.$$

We conclude that $f(r\partial\mathbb{D})$ is an analytic Jordan curve that surrounds a convex domain. Hence, the mapping $\varphi \mapsto \arg\left(\frac{\partial}{\partial\varphi} f(re^{i\varphi})\right)$ is monotonically increasing, and we obtain

$$\begin{aligned}
0 &\leq \frac{\partial}{\partial\varphi} \arg\left(\frac{\partial}{\partial\varphi} f(re^{i\varphi})\right) = \frac{\partial}{\partial\varphi} \operatorname{Im}\left(\log(ire^{i\varphi} f'(re^{i\varphi}))\right) \\
&= \operatorname{Im}\left(\frac{-re^{i\varphi} f'(re^{i\varphi}) - r^2 e^{2i\varphi} f''(re^{i\varphi})}{ire^{i\varphi} f'(re^{i\varphi})}\right) = \operatorname{Im}\left(\frac{if'(re^{i\varphi}) + ire^{i\varphi} f''(re^{i\varphi})}{f'(re^{i\varphi})}\right) \\
&= \operatorname{Re}\left(\frac{f'(re^{i\varphi}) + re^{i\varphi} f''(re^{i\varphi})}{f'(re^{i\varphi})}\right) = \operatorname{Re}\left(\frac{f'(z) + zf''(z)}{f'(z)}\right) = \operatorname{Re}\left(1 + \frac{zf''(z)}{f'(z)}\right).
\end{aligned}$$

The maximum principle implies $1 + \frac{zf''(z)}{f'(z)} \in \mathcal{P}$.

Conversely, assume $1 + \frac{zf''(z)}{f'(z)} \in \mathcal{P}$. Then

$$\begin{aligned}
&\int_0^{2\pi} \frac{\partial}{\partial\varphi} \arg\left(\frac{\partial}{\partial\varphi} f(re^{i\varphi})\right) d\varphi = \int_0^{2\pi} \operatorname{Re}\left(1 + \frac{re^{i\varphi} f''(re^{i\varphi})}{f'(re^{i\varphi})}\right) d\varphi \\
&= \operatorname{Re}\left(\int_0^{2\pi} 1 + \frac{re^{i\varphi} f''(re^{i\varphi})}{f'(re^{i\varphi})} d\varphi\right) = 2\pi + \operatorname{Re}\left(\int_{r\partial\mathbb{D}} \frac{zf''(z)}{f'(z)} \frac{dz}{iz}\right) = 2\pi.
\end{aligned}$$

Thus, $f(r\partial\mathbb{D})$ is a closed curve that does not intersect itself, and the above calculation shows that it surrounds a convex domain. The argument principle implies that f is univalent, and thus, f is a convex function. □

3.9. Exercises

E 3.9.1. Let

$$f(z) = \frac{Az + B}{Cz + D}$$

be an automorphism of $\mathbb{H}$ with $A, B, C, D \in \mathbb{R}$, $AD - BC > 0$. Determine the Nevanlinna triple (a, b, γ) of f.

E 3.9.2. Consider three points $p_1, p_2, p_3 \in \partial\mathbb{D}$ and three points $q_1, q_2, q_3 \in \partial\mathbb{D}$, both in counter-clockwise order. Show that there exists a unique $f \in \operatorname{Aut}(\mathbb{D})$ with $f(p_j) = q_j$ for all $j = 1, 2, 3$.

E 3.9.3. The semicircle distribution $W(\mu, \sigma^2)$, also called Wigner's semicircle law, is given by the density

$$\frac{1}{2\pi\sigma^2}\sqrt{4\sigma^2 - (x-\mu)^2}, \quad x \in [\mu - 2\sigma, \mu + 2\sigma].$$

It has mean μ and variance σ^2.

(a) Let $G(z) = \frac{z - \sqrt{z^2 - 4}}{2}$, where the branch of the square root is chosen such that $\sqrt{\cdot}$ maps $\mathbb{C} \setminus [-4, \infty)$ into the upper half-plane. Prove that $G = G_\mu$ for some $\mu \in \mathcal{P}(\mathbb{R})$.
(b) Use the Stieltjes–Perron inversion formula to show that $\mu = W(0, 1)$.

E 3.9.4 (Plemelj formula). For a probability measure μ on $\mathbb{R}$, consider the following limits (which may or may not exist):

$$\hat{\mathcal{H}}_\mu(x) := \lim_{\varepsilon \searrow 0} \hat{\mathcal{H}}_{\varepsilon,\mu}(x), \quad \hat{\mathcal{H}}_{\varepsilon,\mu}(x) := \frac{1}{\pi} \operatorname{Re}\, G_\mu(x + i\varepsilon),$$

and

$$\mathcal{H}_\mu(x) := \lim_{\varepsilon \searrow 0} \mathcal{H}_{\varepsilon,\mu}(x), \quad \mathcal{H}_{\varepsilon,\mu}(x) := \frac{1}{\pi} \int_{|x-t|>\varepsilon} \frac{1}{x-t} \mu(dt) \quad \text{(Hilbert transform).}$$

Let μ be an absolutely continuous probability measure with compact support and continuous density $f(x)dx$. Let $x \in \mathbb{R}$. Show: $\mathcal{H}_\mu(x)$ exists if and only if $\hat{\mathcal{H}}_\mu(x)$ exists. If these limits exist, then $\mathcal{H}_\mu(x) = \hat{\mathcal{H}}_\mu(x)$.

E 3.9.5. Let $f : \mathbb{H} \to \mathbb{H}$ be holomorphic with Nevanlinna triple (a, b, γ) and $b > 0$. Show that there exists $C > 0$ such that $f(iy)$ belongs to the set $\{x + iy \in \mathbb{H} \,|\, |x| < Cy\}$ for all y large enough.

E 3.9.6. Let $f : \mathbb{D} \to \mathbb{C}$ be holomorphic with $\operatorname{Re} f \geq 0$ and $f(0) = 1$. Consider the power series expansion $f(z) = 1 + \sum_{n=1}^{\infty} c_n z^n$ and show that

$$|c_n| \leq 2 \quad \text{for all } n \geq 1.$$

E 3.9.7. Let $f \in S^*$. Then

$$(*) \qquad zf'(z) = p(z)f(z)$$

with $p \in \mathcal{P}$. We consider the power series $p(z) = 1 + \sum_{k=1}^{\infty} c_k z^k$, $f(z) = z + \sum_{k=2}^{\infty} a_k z^k$.

(a) Find a relation between the coefficients of these series by comparing coefficients in equation $(*)$.
(b) Use the estimate $|c_k| \leq 2$ for all $k \geq 1$ from Exercise 3.9.6 to prove the Bieberbach conjecture for $f \in S^*$. (Hint: proof by induction.)

E 3.9.8. Let $f(z) = z + \sum_{n=2}^{\infty} a_n z^n \in S$ be convex. Show that

$$|a_n| \leq 1 \quad \text{for all } n \geq 2.$$

(Hint: show that $zf'(z) \in S^*$.)

3.10. Solutions

S 3.10.1. Case 1: We assume that $C = 0$. Then $f(z) = \frac{A}{D}z + \frac{B}{D}$ and we immediately obtain

$$a = \frac{B}{D}, \quad b = \frac{A}{D}, \quad \gamma = 0.$$

Case 2: $C \neq 0$. Then f has a pole at $x_0 = -\frac{D}{C}$ and $f(\infty) \neq \infty$. Equation (3.7) implies that $b = 0$ and (3.6) implies that γ has an atom at x_0. We put $\gamma = M\delta_{x_0}$, $M > 0$, and now we have enough parameters to recover f. We have

$$\begin{aligned} a + \int_{\mathbb{R}} \frac{1 + xz}{x - z}\gamma(dx) &= a + M\frac{1 + x_0 z}{x_0 - z} = a + M\frac{C - Dz}{-D - Cz} \\ &= \frac{(MD + aC)z - MC + aD}{Cz + D}, \end{aligned}$$

which leads to $M = \frac{AD-BC}{C^2+D^2}$ and $a = \frac{AC+BD}{C^2+D^2}$. All in all

$$a = \frac{AC + BD}{C^2 + D^2}, \quad b = 0, \quad \gamma = \frac{AD - BC}{C^2 + D^2}\delta_{-D/C}.$$

S 3.10.2. Due to rotations, we may assume that $p_1 = q_1 = 1$. Now we transfer the problem to the upper half-plane via the Cayley transform $C : \mathbb{H} \to \mathbb{D}$. We thus need to find $g \in \mathrm{Aut}(\mathbb{H})$ such that

$$g(C^{-1}(p_1)) = C^{-1}(q_1), \quad g(C^{-1}(p_2)) = C^{-1}(q_2), \quad g(\infty) = \infty.$$

The last condition implies $g(z) = az + b$ for $a > 0, b \in \mathbb{R}$. Obviously, g is then uniquely determined by the other two equations.

S 3.10.3.

(a) The function $\sqrt{z^2 - 4}$ maps $\mathbb{H}$ holomorphically into itself. Let (a, b, γ) be its Nevanlinna triple. Then $b = 1$ by (3.7) and Corollary 3.16 implies that $G(z) := \frac{z - \sqrt{z^2-4}}{2}$ maps $\mathbb{H}$ holomorphically into $\mathbb{H}^-$. We have

$$y\frac{|iy - \sqrt{(iy)^2 - 4}|}{2} = y\frac{|y - \sqrt{y^2 + 4}|}{2} = \frac{4y}{2(y + \sqrt{y^2 + 4})} \to 1$$

as $y \to \infty$. So $G(z) = G_\mu$ for some $\mu \in \mathcal{P}(\mathbb{R})$ by Theorem 3.15.

(b) G can be extended continuously to $\mathbb{H} \cup \mathbb{R}$ with $G(x) = \frac{x - i\sqrt{4-x^2}}{2}$ for $x \in [-2, 2]$, $G(x) = \frac{x - \sqrt{x^2-4}}{2}$ for $x > 2$, and $G(x) = \frac{x + \sqrt{x^2-4}}{2}$ for $x < -2$ (with the usual square root for non-negative real numbers). Hence, $\mu(\{x\}) = 0$ for all $x \in \mathbb{R}$ by (3.4). For $a, b \in \mathbb{R}$, $a < b$, (3.3) gives

$$\begin{aligned} \mu((a, b)) &= -\frac{1}{\pi} \lim_{y \searrow 0} \int_a^b \mathrm{Im}(G_\mu(x + iy))\, dx = -\frac{1}{\pi} \int_a^b \mathrm{Im}(G_\mu(x))\, dx \\ &= -\frac{1}{\pi} \int_a^b \frac{-\sqrt{4 - x^2}}{2} \mathbf{1}_{[-2,2]}(x)\, dx = \int_a^b \frac{\sqrt{4 - x^2}}{2\pi} \mathbf{1}_{[-2,2]}(x)\, dx. \end{aligned}$$

Hence, $\mu = W(0, 1)$.

S 3.10.4. First, we consider the relevant integrals and change t to $t = x + \varepsilon u$, which gives

$$\begin{aligned}
&\int_{\mathbb{R}} \frac{1}{x-t+i\varepsilon} f(t)\, dt - \int_{|x-t|>\varepsilon} \frac{1}{x-t} f(t)\, dt \\
&= \int_{\mathbb{R}} \frac{f(x+\varepsilon u)\varepsilon du}{-\varepsilon u + i\varepsilon} - \int_{|\varepsilon u|>\varepsilon} \frac{f(x+\varepsilon u)\varepsilon du}{-\varepsilon u} \\
&= \int_{\mathbb{R}} \left(\frac{1}{i-u} + \mathbf{1}_{\{|u|>1\}}(u)\frac{1}{u} \right) f(x+\varepsilon u) du.
\end{aligned}$$

Denote the function in parentheses by $g(u)$. For $|u| \le 1$, we have $|g(u)| = \frac{1}{|i-u|} \le 1$, and if $|u| > 1$, then $|g(u)| = \frac{1}{|i-u||u|} \le \frac{1}{|u|^2}$. So g is integrable and a direct calculation yields $\int_{\mathbb{R}} g(u) du = -i\pi$. So $\int_{\mathbb{R}} \operatorname{Re}(g(u)) du = 0$.

Let $[a, b]$ be a compact interval containing the support of μ. Then $|f(x+\varepsilon u) - f(x)|$ is uniformly bounded by $2 \max_{t\in[a,b]} |f(t)|$ and the dominated convergence theorem implies that

$$\int_{\mathbb{R}} (f(x+\varepsilon u) - f(x)) g(u)\, du \to 0$$

as $\varepsilon \to 0$. Hence,

$$\operatorname{Re} \int_{\mathbb{R}} f(x+\varepsilon u) g(u)\, du = \int_{\mathbb{R}} f(x+\varepsilon u) \operatorname{Re}(g(u))\, du \to \int_{\mathbb{R}} f(x) \operatorname{Re}(g(u))\, du = 0$$

as $\varepsilon \to 0$. So, $\lim_{\varepsilon \searrow 0} \hat{\mathcal{H}}_{\varepsilon,\mu}(x)$ exists if and only if $\lim_{\varepsilon \searrow 0} \mathcal{H}_{\varepsilon,\mu}(x)$ exists, and if these limits exist, then they coincide.

S 3.10.5. We may assume that $b = 1$. Otherwise we regard $f(z)/b$.

The statement is clear if $\gamma(\mathbb{R}) = 0$. So assume that $\gamma(\mathbb{R}) > 0$. We have

$$\begin{aligned}
f(iy) &= a + iy + \int_{\mathbb{R}} \frac{1+iyx}{x-iy} \gamma(dx) = a + iy + \int_{\mathbb{R}} \frac{x + iy + iyx^2 - xy^2}{x^2+y^2} \gamma(dx) \\
&= \int_{\mathbb{R}} \frac{a(x^2+y^2)/\gamma(\mathbb{R}) + x - xy^2}{x^2+y^2} \gamma(dx) \\
&\quad + i \int_{\mathbb{R}} \frac{y^3/\gamma(\mathbb{R}) + y + yx^2(1+1/\gamma(\mathbb{R}))}{x^2+y^2} \gamma(dx).
\end{aligned}$$

For $y > 1$, the minimum of the integrand of the imaginary part is given by $\frac{y}{\gamma(\mathbb{R})} + \frac{1}{y}$. On the other hand, still assuming $y > 1$,

$$\begin{aligned}
\frac{|a(x^2+y^2)/\gamma(\mathbb{R}) + x - xy^2|}{x^2+y^2} &\le \frac{|a|}{\gamma(\mathbb{R})} + \frac{|x|}{x^2+y^2} + \frac{|x|y^2}{x^2+y^2} \\
&\le \frac{|a|}{\gamma(\mathbb{R})} + \frac{|x|}{x^2+1} + \frac{|x|y^2}{x^2+y^2} \le \frac{|a|}{\gamma(\mathbb{R})} + \frac{1}{2} + \frac{y}{2}.
\end{aligned}$$

($\frac{xy^2}{x^2+y^2}$ takes its maximum $\frac{y}{2}$ over $\mathbb{R}$ at $x = y$.) Hence, there exists $C > 0$ such that $|\operatorname{Re} f(iy)| < C \operatorname{Im}(f(iy))$ for all y large enough.

S 3.10.6. We have

$$\frac{u+w}{u-w} = 1 + \frac{2w}{u-w} = 1 + 2w/u \sum_{n=0}^{\infty} \left(\frac{w}{u}\right)^n$$

for $w \in \mathbb{D}$, $u \in \partial\mathbb{D}$, and the Herglotz representation formula gives

$$f(z) = 1 + 2\int_{\partial\mathbb{D}} \sum_{n=0}^{\infty} \left(\frac{w}{u}\right)^{n+1} \rho(du),$$

where ρ is a probability measure on $\partial\mathbb{D}$. Hence,

$$c_n = 2\int_{\partial\mathbb{D}} \frac{1}{u^n}\rho(du), \quad \text{and} \quad |c_n| \le 2\int_{\partial\mathbb{D}} \frac{1}{|u^n|}\rho(du) = 2.$$

S 3.10.7.

(a) From $zf'(z) = p(z)f(z)$, we obtain

$$na_n = a_n + c_{n-1} + \sum_{k=2}^{n-1} c_{n-k}a_k, \quad \text{or} \quad (n-1)a_n = c_{n-1} + \sum_{k=2}^{n-1} c_{n-k}a_k$$

for all $n \ge 2$.

(b) Induction basis ($n = 2$): Bieberbach's theorem gives $|a_2| \le 2$. Inductive step ($n-1 \to n$): Let $|a_k| \le k$ for all $k = 2, \ldots, n-1$. Then

$$|(n-1)a_n| \le |c_{n-1}| + \sum_{k=2}^{n-1} |c_{n-k}||a_k| \le 2 + \sum_{k=2}^{n-1} 2k = 2\sum_{k=1}^{n-1} k = (n-1)n,$$

so $|a_n| \le n$.

S 3.10.8. By Theorem 3.43, $1 + z\frac{f''(z)}{f'(z)} \in \mathcal{P}$. Thus, for $g(z) = zf'(z)$ we have $g(0) = 0, g'(0) = 1$ and

$$z\frac{g'(z)}{g(z)} = z\frac{f'(z) + zf''(z)}{zf'(z)} = 1 + \frac{zf''(z)}{f'(z)}.$$

By Theorem 3.42, $g \in S^*$. We have $zf'(z) = z + \sum_{n=2}^{\infty} na_n z^n$ and Exercise 3.9.7 implies $|na_n| \le n$.

CHAPTER 4

Quantum Probability Theory

4.1. The algebraization of probability theory

The path to quantum probability theory starts with the fundamental observation that working with a probability space $(\Omega, \mathcal{F}, \mathbb{P})$ is equivalent to working with the set of all its bounded complex-valued random variables $\mathcal{A} = \{X : \Omega \to \mathbb{C} \,|\, X \text{ measurable and bounded}\}$ together with their expectations, i.e., the function $\varphi : \mathcal{A} \to \mathbb{C}$, $\varphi(X) = \mathbb{E}[X]$:

$$(\Omega, \mathcal{F}, \mathbb{P}) \quad \Longleftrightarrow \quad (\mathcal{A}, \varphi).$$

Using complex numbers instead of $\mathbb{R}$ is not strictly necessary, but it makes things easier. Recall that the characteristic function of a real-valued random variable X is the expectation of the complex-valued random variable e^{itX}. The same holds for boundedness. Of course, we are typically also interested in unbounded random variables X, but their expectation might not exist. Instead, X can be replaced by the set of all bounded random variables of the form $f(X)$, where $f : \mathbb{C} \to \mathbb{C}$ is bounded and continuous.

Notions in $(\Omega, \mathcal{F}, \mathbb{P})$ can be translated into notions in $(\mathcal{A}, \varphi)$:

- An event $A \in \mathcal{F}$ can be represented by $\mathbf{1}_A \in \mathcal{A}$ and a sub-σ-algebra $\mathcal{G} \subset \mathcal{F}$ induces a subalgebra $\mathcal{B} \subset \mathcal{A}$, consisting of all $\mathcal{G}$-measurable elements in $\mathcal{A}$.
- The intersection and union of events corresponds to multiplication and addition:

$$\mathbf{1}_A \cdot \mathbf{1}_B = \mathbf{1}_{A \cap B}, \quad \mathbf{1}_A + \mathbf{1}_B - \mathbf{1}_A \cdot \mathbf{1}_B = \mathbf{1}_{A \cup B}.$$

- Real-valued $X_1, \ldots, X_n \in \mathcal{A}$ are *independent* if and only if (see Theorem 2.22) they satisfy the following *dependency*:

$$\mathbb{E}[X_1^{k_1} \cdots X_n^{k_n}] = \mathbb{E}[X_1^{k_1}] \cdots \mathbb{E}[X_n^{k_n}] \quad \text{for all } k_1, \ldots, k_n \in \mathbb{N}_0. \tag{4.1}$$

In $\mathcal{A}$, we can add and multiply elements and we also have scalar multiplication, i.e., we can form $X + Y, X \cdot Y, \lambda \cdot X$, where $\lambda \in \mathbb{C}$ and $X, Y \in \mathcal{A}$. These operations obey certain rules, in particular, the commutativity of the product: $X \cdot Y = Y \cdot X$.

The quantum probabilist is now confronted with the following task: Which algebraic properties does $(\mathcal{A}, \varphi)$ have, and how can they be formalized and generalized? This leads to more abstract probability spaces. In particular, the product may no longer be commutative, which is why quantum probability theory is also called "non-commutative probability theory".

Let us consider the space $\mathbb{C}^{2\times 2}$ of all 2×2-matrices as a possible version of non-commutative complex-valued random variables. Let $A, B \in \mathbb{C}^{2\times 2}$. When should we call A and B "independent"?

We look at a concrete example. Consider two simple diagonal matrices

$$A = \begin{pmatrix} \alpha_1 & 0 \\ 0 & \alpha_2 \end{pmatrix}, \quad B = \begin{pmatrix} \beta_1 & 0 \\ 0 & \beta_2 \end{pmatrix}, \quad \alpha_1, \alpha_2, \beta_1, \beta_2 \in \mathbb{R}. \tag{4.2}$$

In quantum mechanics, values of an observable quantity are described by the eigenvalues of an operator. By defining the "expectation" $\varphi(X) = \frac{1}{2}\operatorname{Tr}(X)$ on $\mathbb{C}^{2\times 2}$, which is the arithmetic average of the eigenvalues of X, we obtain $\varphi(A^n) = \frac{1}{2}\alpha_1^n + \frac{1}{2}\alpha_2^n$ and $\varphi(B^n) = \frac{1}{2}\beta_1^n + \frac{1}{2}\beta_2^n$. Hence, we might say that the distributions of A and B are

$$\mu_A = \frac{1}{2}\delta_{\alpha_1} + \frac{1}{2}\delta_{\alpha_2}, \quad \mu_B = \frac{1}{2}\delta_{\beta_1} + \frac{1}{2}\delta_{\beta_2}.$$

Are A and B independent in the sense of (4.1) with respect to φ? We have

$$\frac{1}{2}\operatorname{Tr}(AB) = \frac{1}{2}(\alpha_1\beta_1 + \alpha_2\beta_2), \quad \text{but} \quad \frac{1}{2}\operatorname{Tr}(A)\cdot\frac{1}{2}\operatorname{Tr}(B) = \frac{1}{4}(\alpha_1+\alpha_2)(\beta_1+\beta_2).$$

Both numbers are equal only in special cases, e.g., if $\alpha_1 = \alpha_2$. The convolution of the distributions is given by

$$\mu_A * \mu_B = \frac{1}{4}\delta_{\alpha_1+\beta_1} + \frac{1}{4}\delta_{\alpha_1+\beta_2} + \frac{1}{4}\delta_{\alpha_2+\beta_1} + \frac{1}{4}\delta_{\alpha_2+\beta_2}.$$

In general, these are four different eigenvalues, and $A+B$ obviously has at most only two eigenvalues. However, we can force the convolution to appear by embedding A and B into the higher dimensional space $\mathbb{C}^{4\times 4}$. Consider the following two tensor products:

$$A\otimes I = \begin{pmatrix} \alpha_1 & 0 & 0 & 0 \\ 0 & \alpha_1 & 0 & 0 \\ 0 & 0 & \alpha_2 & 0 \\ 0 & 0 & 0 & \alpha_2 \end{pmatrix}, \quad I\otimes B = \begin{pmatrix} \beta_1 & 0 & 0 & 0 \\ 0 & \beta_2 & 0 & 0 \\ 0 & 0 & \beta_1 & 0 \\ 0 & 0 & 0 & \beta_2 \end{pmatrix}.$$

With respect to the expectation $\frac{1}{4}\operatorname{Tr}(X)$ we have $\mu_{A\otimes I} = \mu_A, \mu_{I\otimes B} = \mu_B$ and now the convolution $*$ appears:

$$\mu_{A\otimes I + I\otimes B} = \mu_A * \mu_B.$$

4.2. Non-negative operators

We are now heading toward the definition of a more abstract probability space to obtain also non-commutative random variables. We start with a complex Hilbert space H. Recall that H is a vector space over the field $\mathbb{C}$ together with an inner product $\langle\cdot,\cdot\rangle$ (we use inner products that are linear in the second argument: $\langle v, \lambda w\rangle = \lambda\langle v, w\rangle = \langle\overline{\lambda} v, w\rangle$) which is complete with respect to the norm $\|x\| = \sqrt{\langle x, x\rangle}$, i.e., every Cauchy sequence in H converges to an element in H.

Definition 4.1. We let $B(H)$ be the set of all complex-linear and bounded mappings $A : H \to H$ and endow this space with the operator norm defined by

$$\|A\| = \sup_{v\in H, \|v\|=1} \|Av\|.$$

The identity operator will be denoted by I. For $X \in B(H)$ and $a \in \mathbb{C}$ we will sometimes simply write $X + a$ instead of $X + aI$. We recall that $B(H)$ endowed with the operator norm is a Banach space (a complete normed vector space). See, e.g., [**55**] for an introduction to functional analysis. We will be mainly interested in the following two classes of operators.

DEFINITION 4.2.

- $A \in B(H)$ is called *self-adjoint* if

$$\langle Av, w \rangle = \langle v, Aw \rangle \quad \text{for all } v, w \in H. \tag{4.3}$$

- We write $A \geq 0$ if $\langle v, Av \rangle \geq 0$ for all $v \in H$. An operator A with $A \geq 0$ is called *non-negative*. (In the literature, it is also called positive semidefinite.)

For any $A \in B(H)$, one can define the adjoint $A^* \in B(H)$, which has the property

$$\langle Av, w \rangle = \langle v, A^* w \rangle$$

for all $v, w \in H$. Thus A is self-adjoint if and only if $A = A^*$. The operation * can be thought of as playing the role of conjugation for complex numbers, and thus, self-adjoint elements correspond to real numbers.

EXAMPLE 4.3. If $H = \mathbb{C}^n$ with the usual inner product $\langle w, v \rangle = v_1 \overline{w_1} + \cdots + v_n \overline{w_n}$, then $B(H) = \mathbb{C}^{n \times n}$ is the space of all complex $n \times n$-matrices. For $A \in \mathbb{C}^{n \times n}$, the adjoint A^* is given by the conjugate transpose $A^* = \overline{A}^T$. The operator norm can also be expressed by

$$\|A\| = \sqrt{\lambda_{max}(A^* A)}, \quad A \in \mathbb{C}^{n \times n},$$

where λ_{max} is the largest eigenvalue of the non-negative matrix $A^* A$.

We would like to regard the self-adjoint elements of $B(H)$ as quantum random variables. The spectral theorem comes in handy for this purpose, as it assigns a projection-valued measure to each self-adjoint operator, see Remark 4.15. However, we will not use the spectral theorem and will instead follow a more elementary approach. First, we need to collect a few properties of non-negative operators.

LEMMA 4.4. *Let $A \in B(H)$, $A \geq 0$. Then A is self-adjoint. Furthermore, $A^n \geq 0$ for all $n \in \mathbb{N}$.*

PROOF. Let $v, w \in H$. We need to show that $\langle Av, w \rangle = \langle v, Aw \rangle$. We have (polarization identity):

$$\begin{aligned} \langle Av, w \rangle = {} & \frac{1}{4}(\langle A(v+w), v+w \rangle - \langle A(v-w), v-w \rangle \\ & - i \langle A(v+iw), v+iw \rangle + i \langle A(v-iw), v-iw \rangle), \quad \text{and} \end{aligned}$$

$$\begin{aligned} \overline{\langle v, Aw \rangle} = \langle Aw, v \rangle = {} & \frac{1}{4}(\langle A(v+w), v+w \rangle - \langle A(w-v), w-v \rangle \\ & - i \langle A(v-iw), v-iw \rangle + i \langle A(v+iw), v+iw \rangle). \end{aligned}$$

As $\langle x, Ax \rangle \geq 0$, we have $\langle x, Ax \rangle = \langle Ax, x \rangle$ for all $x \in H$, and we get $\langle Av, w \rangle = \langle v, Aw \rangle$.

For the second statement, consider an even exponent $n = 2m$ first. Then $\langle A^{2m}v, v\rangle = \langle A^m v, A^m v\rangle = \|A^m v\|^2 \geq 0$. If $n = 2m+1$ is odd, we have $\langle A^{2m+1}v, v\rangle = \langle AA^m v, A^m v\rangle \geq 0$. □

We write $A \leq B$ if $B - A \geq 0$.

LEMMA 4.5. *Assume that $X : H \to H$ is linear and self-adjoint (i.e., it satisfies (4.3)) with $-c \cdot I \leq X \leq c \cdot I$, $c > 0$. Then X is a bounded operator with $\|X\| \leq c$.*

PROOF. We have $0 \leq \langle (c \cdot I + X)v, v\rangle$ and $0 \leq \langle (c \cdot I - X)v, v\rangle$, in other words $-c\|v\|^2 \leq \langle Xv, v\rangle \leq c\|v\|^2$. For each $\lambda > 0$ we have

$$\begin{aligned} 4\|Xv\|^2 &= \langle X(\lambda v + 1/\lambda Xv), \lambda v + 1/\lambda Xv\rangle - \langle X(\lambda v - 1/\lambda Xv), \lambda v - 1/\lambda Xv\rangle \\ &\leq c\|\lambda v + 1/\lambda Xv\|^2 + c\|\lambda v - 1/\lambda Xv\|^2 = 2c\lambda^2\|v\|^2 + \frac{2c}{\lambda^2}\|Xv\|^2. \end{aligned}$$

For $\|Xv\| \neq 0$, the last expression is minimal for $\lambda^2 = \|Xv\|/\|v\|$ and, hence, $\|Xv\|^2 \leq c\|Xv\|\|v\|$, and thus, $\|Xv\| \leq c\|v\|$. Therefore, the operator X is bounded with $\|X\| \leq c$. □

LEMMA 4.6. *Let $(X_n)_{n\in\mathbb{N}}$ be a sequence of self-adjoint operators in $B(H)$ with $0 \leq X_n \leq X_{n+1} \leq I$. Then there exists a self-adjoint $X \in B(H)$ with $X_n v \to Xv$ as $n \to \infty$ for all $v \in H$.*

PROOF. For $m < n$ let $\Delta_{m,n} = X_n - X_m$. Then $0 \leq \Delta_{m,n} \leq I$ and $\|\Delta_{m,n}\| \leq 1$ because of Lemma 4.5. Let $v \in H$. Then

$$\|\Delta_{m,n}v\|^4 = \langle \Delta_{m,n}v, \Delta_{m,n}v\rangle^2 \leq \langle \Delta_{m,n}v, v\rangle \langle \Delta_{m,n}\Delta_{m,n}v, \Delta_{m,n}v\rangle .$$

(Use the Cauchy–Schwarz inequality for the linear form $L(v,w) = \langle \Delta_{m,n}v, w\rangle$.) Hence

$$\begin{aligned} \|\Delta_{m,n}v\|^4 &\leq (\langle X_n v, v\rangle - \langle X_m v, v\rangle) \langle \Delta_{m,n}\Delta_{m,n}v, \Delta_{m,n}v\rangle \\ &\leq (\langle X_n v, v\rangle - \langle X_m v, v\rangle)\|\Delta_{m,n}\Delta_{m,n}v\|\|\Delta_{m,n}v\| \\ &\leq (\langle X_n v, v\rangle - \langle X_m v, v\rangle)\|v\|^2. \end{aligned}$$

As $\langle X_n v, v\rangle$ is monotonically increasing and bounded, it is convergent. Thus, $X_n v$ is a Cauchy sequence in H and thus convergent with limit $Xv := \lim_{n\to\infty} X_n v$. X is a linear mapping $X : H \to H$ with $\langle Xv, w\rangle = \lim_{n\to\infty} \langle X_n v, w\rangle$. Hence, X is self-adjoint and $0 \leq X \leq I$. This implies that $X \in B(H)$ by Lemma 4.5. □

THEOREM 4.7. *(Square root of non-negative operators) Let $A \in B(H)$, $A \geq 0$.*

(1) *There exists a unique $\sqrt{A} \in B(H)$ with $\sqrt{A} \geq 0$ and $(\sqrt{A})^2 = A$.*
(2) *Any $C \in B(H)$ that commutes with A, i.e., $AC = CA$, also commutes with $\sqrt{A}$.*

PROOF. We may assume that $\|A\| \leq 1$, and then $I - A \geq 0$. Define the sequence $(X_n)_{n\in\mathbb{N}} \subset B(H)$ by

$$X_0 = 0, \quad X_{n+1} = \frac{1}{2}(I - A + X_n^2).$$

By induction we see that $X_n \leq I$ for all $n \in \mathbb{N}$, and that each X_n is self-adjoint. Furthermore, we prove that $X_n \geq 0$ and $X_n - X_{n-1} \geq 0$ for all $n \in \mathbb{N}$ as follows. Put $B = I - A$. Then $X_1 = X_1 - X_0 = \frac{1}{2}B \geq 0$. We prove by induction that $X_n = p_n(B)$ and $X_n - X_{n-1} = q_n(B)$ for polynomials $p_n, q_n : \mathbb{R} \to \mathbb{R}$ having non-negative coefficients. This is true for $n = 1$ with $p_1(x) = q_1(x) = \frac{1}{2}x$. We have $X_{n+1} = \frac{1}{2}(B + p_n(B)^2) = p_{n+1}(B)$ and

$$X_{n+1} - X_n = \frac{1}{2}(X_n^2 - X_{n-1}^2) = \frac{1}{2}q_n(B)(p_n(B) + p_{n-1}(B)) = q_{n+1}(B)$$

with polynomials $p_{n+1}, q_{n+1} : \mathbb{R} \to \mathbb{R}$ having non-negative coefficients. It follows that $X_n \geq 0$ and $X_n - X_{n-1} \geq 0$ for all $n \in \mathbb{N}$.

Lemma 4.6 implies that there exists $X \in B(H)$ with $Xv = \lim_{n\to\infty} X_n v$, $X \geq 0$ and $I - X \geq 0$. X satisfies $2X = I - A + X^2$, i.e., $(I - X)^2 = A$, and thus, we can define $\sqrt{A} = I - X$.

Furthermore, by induction we see that if $AC = CA$ for some $C \in B(H)$, then $CX_n = X_n C$ for all n, and thus, also $C\sqrt{A} = \sqrt{A}C$.

Uniqueness: Let $D \in B(H)$ be another square root of A, i.e., assume $D^2 = A$ and $D \geq 0$. As D commutes with A ($DA = DD^2 = AD$), it also commutes with $\sqrt{A}$. Let $v \in H$ and $w = (\sqrt{A} - D)v$. Then

$$\left\langle (\sqrt{A} + D)w, w \right\rangle = \left\langle ((\sqrt{A})^2 - D^2)v, w \right\rangle = 0.$$

As $\sqrt{A} \geq 0$ and $D \geq 0$, we must have $\left\langle \sqrt{A}w, w \right\rangle = 0$ and $\langle Dw, w\rangle = 0$. Now let $Y \geq 0$ be a square root of $\sqrt{A}$. Then $\langle Y^2 w, w\rangle = \|Yw\|^2 = 0$. So $Yw = 0$, and thus, $\sqrt{A}w = 0$. In the same way, we get $Dw = 0$. Consequently, $\|(\sqrt{A} - D)v\|^2 = \left\langle (\sqrt{A} - D)^2 v, v \right\rangle = \left\langle (\sqrt{A} - D)w, v \right\rangle = 0$. It follows that $D = \sqrt{A}$. □

We can use the square root for proving the following simple property.

LEMMA 4.8. *Let $A, B \in B(H)$, $A, B \geq 0$ and assume $AB = BA$. Then $AB \geq 0$.*

PROOF. With Theorem 4.7 (1) and (2) we have $AB = \sqrt{A}\sqrt{A}B = \sqrt{A}B\sqrt{A}$. Let $v \in H$. By Lemma 4.4, we obtain

$$\langle ABv, v\rangle = \left\langle \sqrt{A}B\sqrt{A}v, v \right\rangle = \left\langle B\sqrt{A}v, \sqrt{A}v \right\rangle \geq 0 \text{ as } B \geq 0. \qquad \square$$

4.3. Quantum probability spaces

DEFINITION 4.9. Let H be a complex Hilbert space. We define a *quantum probability space* as a pair $(B(H), \varphi)$, where φ is an *expectation* on $B(H)$, defined as a linear functional $\varphi : B(H) \to \mathbb{C}$ with

$$\varphi(A) \geq 0 \quad \text{whenever } A \geq 0 \quad \text{and} \quad \varphi(I) = 1.$$

The self-adjoint elements of $B(H)$ will be called (real, quantum) *random variables.*

EXAMPLE 4.10. In the case of $H = \mathbb{C}^n$, $B(H) = \mathbb{C}^{n\times n}$, we obtain a quantum probability space by defining the expectation as

$$\varphi(A) = \frac{1}{n}\operatorname{Tr}(A).$$

Indeed, $\varphi(I) = 1$ and as $\operatorname{Tr}(A)$ is the sum of all eigenvalues $\lambda_1, \dots, \lambda_n$ of A, we have $\operatorname{Tr}(A) \geq 0$ whenever $A \geq 0$. Furthermore, for any $m \in \mathbb{N}$ we have $\varphi(A^m) = (\lambda_1^m + \cdots + \lambda_n^m)/n$.

LEMMA 4.11. *Let $A \in B(H)$ be self-adjoint. Then $\varphi(A) \in \mathbb{R}$.*

PROOF. As A is self-adjoint, $A^2 \geq 0$ and we can take the square root $|A| := \sqrt{A^2}$ and define

$$P = \frac{1}{2}(|A| + A), \quad N = \frac{1}{2}(|A| - A).$$

Clearly, $A = P - N$. If we knew that both P and N were non-negative, then $\varphi(A) = \varphi(P) - \varphi(N)$ would be the difference of two non-negative numbers and thus real.

As A commutes with A^2, A commutes with $|A|$ by Theorem 4.7. In particular, $0 = |A|^2 - A^2 = (|A| + A)(|A| - A)$.

Let $Q \in B(H)$ be the projection onto the kernel of P. Then $Q \geq 0$, $I - Q \geq 0$, and $PQ = 0$. From $0 = \langle PQQPv, v\rangle = \langle QPv, QPv\rangle = \|QPv\|^2$ for all $v \in H$, we see that $QP = 0$ as well.

As $0 = (|A| + A)(|A| - A)$, the image of $|A| - A$ is contained in the kernel of $|A| + A$, and thus, $QN = N$. Again, we also have $NQ = N$. ($0 = \langle (Q - I)NN(Q - I)v, v\rangle = \langle N(Q - I)v, N(Q - I)v\rangle = \|N(Q - I)v\|^2$ for all $v \in H$.)

Hence

$$Q|A| = Q(P + N) = N, \quad |A|Q = (P + N)Q = N.$$

Thus $A = |A| - 2N = (I - 2Q)|A|$ and we can write

$$P = \frac{1}{2}(|A| + (I - 2Q)|A|) = (I - Q)|A|, \quad N = \frac{1}{2}(|A| - (I - 2Q)|A|) = Q|A|.$$

As Q and $|A|$ commute, Lemma 4.8 implies that $Q|A|$ and $(I - Q)|A|$ are non-negative, thus N and P are non-negative. □

LEMMA 4.12. *Let $A \in B(H)$. Then $\varphi(A^*) = \overline{\varphi(A)}$.*

PROOF. As $A + A^*$ is self-adjoint, we have $\varphi(A + A^*) \in \mathbb{R}$ by Lemma 4.11. So $\operatorname{Im}(\varphi(A)) = -\operatorname{Im}(\varphi(A^*))$. Furthermore, also $iA - iA^*$ is self-adjoint, which gives $\varphi(iA - iA^*) \in \mathbb{R}$, i.e., $i(\varphi(A) - \varphi(A^*)) \in \mathbb{R}$, and thus, $\operatorname{Re}(\varphi(A)) = \operatorname{Re}(\varphi(A^*))$. Thus $\varphi(A^*) = \overline{\varphi(A)}$. □

Let $X \in B(H)$ be self-adjoint. As we interpret $\varphi(X)$ as the expectation of the quantum random variable X, we could go further and regard $\varphi(X^2) - \varphi(X)^2$ as its variance, and more generally, $\varphi(X^m)$, $m \in \mathbb{N}$, as the mth moment of X. So we are already close to assigning a probability distribution $\mu_X \in \mathcal{P}(\mathbb{R})$ to X. We would require μ_X to satisfy

$$\varphi(X^m) = \int_{\mathbb{R}} x^m \mu_X(dx), \quad m \in \mathbb{N}.$$

But can we construct a probability measure whose moments are the given numbers $\varphi(X^m)$?[1] In Example 4.10, this measure exists and is given by $\mu_X = \frac{1}{n}\delta_{\lambda_1} + \cdots + \frac{1}{n}\delta_{\lambda_n}$. For the general case, we can show that if μ_X exists, then it has compact support with $\operatorname{supp}\mu_X \subset [-\|X\|, \|X\|]$: Assume $\|X\| = 1$. By Exercise 4.6.2, we

[1]The question of whether a given sequence $m_1, m_2, \ldots$ can be regarded as moments of a probability measure is also known as the *moment problem*. The literature contains several results on this more general problem, see [**171**].

have $|\varphi(X^{2n})| \leq \|X^{2n}\| \leq \|X\|^{2n} = 1$. Suppose there exists $p \in \operatorname{supp} \mu_X$ with $|p| > 1$. Then we find n large enough such that $\int_{\mathbb{R}} x^{2n} \mu_X(dx) > 1$, a contradiction.

To prove the existence of μ_X, we now follow [**178**], which uses the following lemma.

LEMMA 4.13. *Let $n, N \in \mathbb{N}$ with $N \geq n$. There exist functions $\alpha_{k,n}(N)$, $k = 0, \ldots, n$, such that*

$$\sum_{k=0}^{N} \binom{N}{k} x^k (1-x)^{N-k} (k/N)^n = x^n + \sum_{k=0}^{n} \alpha_{k,n}(N) x^k$$

for all $x \in \mathbb{R}$ and $\alpha_{k,n} = \mathcal{O}(1/N)$.

PROOF. The polynomial $x(x-1)\cdots(x-n+1) = x^n + \ldots$ has degree n and with $c_{n,n} := 1$, $x^n - c_{n,n}x(x-1)\cdots(x-n+1)$ has degree $n-1$. So we find $c_{n-1,n}$ such that $x^n - c_{n,n}x(x-1)\cdots(x-n+1) - c_{n-1,n}x(x-1)\cdots(x-n+2)$ has degree $n-2$. Inductively, we find constants $c_{k,n} \in \mathbb{R}$, $k = 0, \ldots, n$, $c_{n,n} = 1$, such that

$$x^n = \sum_{k=0}^{n} c_{k,n} x(x-1)\ldots(x-k+1).$$

Then

$$\left(x\frac{\partial}{\partial x}\right)^n = \sum_{k=0}^{n} c_{k,n} x^k \left(\frac{\partial}{\partial x}\right)^k \tag{4.4}$$

as operators on the set of all real polynomials, because $\left(x\frac{\partial}{\partial x}\right)^n (x^m) = m^n x^m$, $m \in \mathbb{N}_0$, and $x^k (\frac{\partial}{\partial x})^k (x^m) = m(m-1)\ldots(m-k+1)x^m$. Note that this is also true if $m < k$, in which case both expressions are 0.

Now let $N \geq n$. For fixed $y \in \mathbb{R}$, we have

$$\left(x\frac{\partial}{\partial x}\right)^n (x+y)^N = \left(x\frac{\partial}{\partial x}\right)^n \left[\sum_{k=0}^{N} \binom{N}{k} x^k y^{N-k}\right] = \sum_{k=0}^{N} \binom{N}{k} k^n x^k y^{N-k}$$
$$= N^n \sum_{k=0}^{N} \binom{N}{k} x^k y^{N-k} (k/N)^n.$$

On the other hand, using relation (4.4) yields

$$\begin{aligned}\left(x\frac{\partial}{\partial x}\right)^n (x+y)^N &= \sum_{k=0}^{n} c_{k,n} x^k \left(\frac{\partial}{\partial x}\right)^k (x+y)^N \\ &= \sum_{k=0}^{n} c_{k,n} N(N-1)\cdots(N-k+1) x^k (x+y)^{N-k}.\end{aligned}$$

Hence

$$\sum_{k=0}^{N} \binom{N}{k} x^k y^{N-k} (k/N)^n = \sum_{k=0}^{n} \frac{c_{k,n}}{N^{n-k}} \cdot 1 \cdot (1-1/N)\cdots(1-(k-1)/N) x^k (x+y)^{N-k}.$$

Putting $y = 1 - x$ yields

$$\sum_{k=0}^{N} \binom{N}{k} x^k (1-x)^{N-k} (k/N)^n = \sum_{k=0}^{n} \frac{c_{k,n}}{N^{n-k}} (1 - 1/N) \cdots (1 - (k-1)/N) x^k.$$

The coefficient of x^n in the right sum is

$$c_{n,n}(1 - 1/N) \cdots (1 - (n-1)/N) =: 1 + \alpha_{n,n}(N).$$

Then $\alpha_{n,n}(N) = \mathcal{O}(1/N)$ as $c_{n,n} = 1$. For $k = 0, \dots, n-1$, let $\alpha_{k,n}(N)$ be the coefficient of x^k and then we also have $\alpha_{k,n}(N) = \mathcal{O}(1/N)$. □

THEOREM 4.14. *If $X \in B(H)$ is self-adjoint, then there exists a unique Borel probability measure $\mu_X \in \mathcal{P}_c(\mathbb{R})$ such that*

$$\varphi(p(X)) = \int_{\mathbb{R}} p(x) \mu_X(dx)$$

for all polynomials $p : \mathbb{R} \to \mathbb{R}$. (If $p(x) = a_0 + a_1 x + \cdots + a_n x^n$, then $p(X) := a_0 I + a_1 X + \cdots + a_n X^n$.) We call μ_X the distribution *of the quantum random variable X.*

PROOF. Concerning uniqueness, assume that $\mu, \nu \in \mathcal{P}_c(\mathbb{R})$ yield the same moments $\varphi(X^m)$, $m \in \mathbb{N}$. Then the Cauchy transforms G_μ and G_ν both have the same power series expansion at ∞. So $G_\mu = G_\nu$, and thus, $\mu = \nu$.

If $X = 0$, then we can choose $\mu_X = \delta_0$. So assume $X \neq 0$. Then $\|X\| > 0$ and we have $-\|X\| I \leq X \leq \|X\| I$. We switch to $Y = X/(2\|X\|) + I/2$. Then $0 \leq Y \leq I$.

We define a linear functional τ on the space of all real polynomials, regarded as a subspace of $C([0,1], \mathbb{R})$, by

$$\tau(p) = \varphi(p(Y)).$$

Step 1: If $p \geq 0$ on $[0,1]$, then $\tau(p) \geq 0$:

Let $f \in C([0,1], \mathbb{R})$. For $N \in \mathbb{N}$, let $B_{f,N}$ be the Bernstein polynomial

$$B_{f,N}(x) = \sum_{k=0}^{N} \binom{N}{k} f(k/N) \, x^k (1-x)^{N-k}.$$

Assume that $f \geq 0$. Then we also have $B_{f,N} \geq 0$ on $[0,1]$ and

$$\tau(B_{f,N}) = \sum_{k=0}^{N} \binom{N}{k} f(k/N) \, \varphi(Y^k (I - Y)^{N-k}).$$

We have $0 \leq Y \leq I$, and thus, also $Y^k (I - Y)^{N-k} \geq 0$ for all $k = 0, \dots, N$ by Lemma 4.8, and $\varphi(Y^k (I - Y)^{N-k}) \geq 0$ by definition of φ. Consequently, $\tau(B_{f,N}) \geq 0$.

Now let p be a polynomial of degree $m \in \mathbb{N}$ which is non-negative on $[0,1]$. Let $N \geq m$ and $n \leq m$. Then $B_{x^n,N}$ is the expression on the left side in Lemma 4.13 and we have $B_{x^n,N} = x^n + \dots$

Write $p(x) = \sum_{n=0}^{m} a_n x^n$. Lemma 4.13 yields functions $A_{k,m}(N)$, $k = 0, \ldots, m$, with $A_{k,m}(N) = \mathcal{O}(1/N)$ such that

$$B_{p,N}(x) = p(x) + \sum_{k=0}^{m} A_{k,m}(N) x^k.$$

As $p \geq 0$ on $[0,1]$, we have $\tau(B_{p,N}) \geq 0$, and thus,

$$\begin{aligned}\tau\left(p + \sum_{k=0}^{m} A_{k,m}(N)x^k\right) &= \tau(p) + \tau\left(\sum_{k=0}^{m} A_{k,m}(N)x^k\right)\\ &= \tau(p) + \sum_{k=0}^{m} A_{k,m}(N)\tau(x^k) \geq 0.\end{aligned}$$

As we can choose N arbitrarily large, we must have $\tau(p) \geq 0$.

Alternatively, we could use the fact that every polynomial p that is positive on $[0,1]$ can be written as

$$p(x) = \sum_{j,k}^{N} \alpha_{j,k} x^j (1-x)^k$$

for some $N \in \mathbb{N}$ with $\alpha_{j,k} \geq 0$ for all $j,k = 0, \ldots, N$. This is a theorem due to Bernstein, see [**171**, Proposition 3.4]. Thus

$$\tau(p) = \sum_{j,k}^{N} \alpha_{j,k} \varphi\left(Y^j (I-Y)^k\right) \geq 0.$$

If $p \geq 0$ on $[0,1]$, then $p + 1/n > 0$ on $[0,1]$ for all $n \in \mathbb{N}$ and $\tau(p + 1/n) = \tau(p) + 1/n \geq 0$. Thus $\tau(p) \geq 0$.

Step 2: The Riesz extension theorem (by Marcel Riesz) implies that τ has an extension to a linear functional on $C([0,1], \mathbb{R})$ with $\tau(f) \geq 0$ whenever $f \geq 0$. (See [**162**] or [**92**, p. 220], which gives a more general version called Krein's extension theorem.)

Step 3: The Riesz representation theorem (by Frigyes Riesz, see [**165**]) then tells us that the functional τ can be represented by an integral, i.e., there exists a Borel probability measure $\mu \in \mathcal{P}(\mathbb{R})$ with support contained in $[0,1]$ such that

$$\tau(f) = \int_{[0,1]} f(x) \mu(dx).$$

(The function $\Phi(x) = \tau(\mathbf{1}_{(-\infty,x]})$ defines a cumulative distribution function for μ.) □

REMARK 4.15. The spectral theorem says that X can be represented as

$$X = \int_{\mathbb{R}} x E(dx)$$

for a projection-valued measure E. E can also be seen as the "non-commutative distribution" of X. The spectral theorem is a powerful tool, which immediately gives us the probability distribution μ_X of X with respect to φ by the formula

$$\mu_X(B) = \varphi(E(B)).$$

For a polynomial p we have $p(X) = \int_{\mathbb{R}} p(x)E(dx)$, and thus,

$$\varphi(p(X)) = \int_{\mathbb{R}} p(x)\varphi(E(dx)) = \int_{\mathbb{R}} p(x)\mu_X(dx).$$

REMARK 4.16. The Weierstraß approximation theorem states that every continuous $f : [0,1] \to \mathbb{R}$ can be approximated uniformly by polynomials. A short proof is given by the fact that $B_{f,n}$ converges uniformly to f as $n \to \infty$. This proof is due to S. Bernstein [**36**].

Let $b_{k,n}(x) = \binom{n}{k} x^k (1-x)^{n-k}$, $0 \le k \le n$. Then

$$\sum_{k=0}^{n} b_{k,n}(x) = (x + 1 - x)^n = 1$$

and

$$\begin{aligned} |B_{f,n}(x) - f(x)| &= \left| \sum_{k=0}^{n} f\left(\frac{k}{n}\right) b_{k,n}(x) - f(x) b_{k,n}(x) \right| \\ &\le \sum_{k=0}^{n} \left| f\left(\frac{k}{n}\right) - f(x) \right| b_{k,n}(x). \end{aligned}$$

Let $\varepsilon > 0$. As f is uniformly continuous on the compact set $[0,1]$, we find $\delta > 0$ such that $|x - y| < \delta$, $x, y \in [0,1]$, implies $|f(x) - f(y)| < \varepsilon$. Let $M = \sup_{x \in [0,1]} |f(x)|$. Then

$$\begin{aligned} |B_{f,n}(x) - f(x)| &\le \sum_{|x-k/n|<\delta} |f(k/n) - f(x)|\, b_{k,n}(x) \\ &+ \sum_{|x-k/n|\ge\delta} |f(k/n) - f(x)|\, b_{k,n}(x) \\ &\le \sum_{|x-k/n|<\delta} \varepsilon b_{k,n}(x) + 2M \sum_{|x-k/n|\ge\delta} b_{k,n}(x) \le \varepsilon + 2M \sum_{|x-k/n|\ge\delta} b_{k,n}(x) \\ &\le \varepsilon + 2M \sum_{k=0}^{n} \delta^{-2}(x - k/n)^2 b_{k,n}(x) \\ &\underset{\text{see Exercise 4.6.3}}{=} \varepsilon + 2M\delta^{-2} \frac{x(1-x)}{n} \le \varepsilon + 2M\delta^{-2}\frac{1}{4n}. \end{aligned}$$

Consequently, $B_{f,n}$ converges to f uniformly on $[0,1]$ as $n \to \infty$.

EXAMPLE 4.17. If H is a Hilbert space and $v \in H$ with $\|v\| = 1$, then

$$\varphi(X) = \langle v, Xv \rangle$$

defines an expectation. In quantum mechanics, such a vector v is also called a *state* and due to this example, expectations are also called *states* in quantum probability theory.

EXAMPLE 4.18. We have seen that $B(H) = \mathbb{C}^{n\times n}$ together with $\varphi(A) = \frac{1}{n}\operatorname{Tr}(A)$ becomes a quantum probability space. Other examples of expectations can be obtained as follows. Let $\rho \in \mathbb{C}^{n\times n}$ be non-negative with $\operatorname{Tr}(\rho) = 1$, a so-called *density matrix*. We now obtain an expectation via

$$\varphi(A) = \operatorname{Tr}(A\rho).$$

We have $\varphi(I) = 1$ and if $A \geq 0$, then $\mathrm{Tr}(A\rho) = \mathrm{Tr}(\sqrt{A}\sqrt{A}\rho) = \mathrm{Tr}(\sqrt{A}\rho\sqrt{A})$. As $\sqrt{A}\rho\sqrt{A} \geq 0$, we have $\mathrm{Tr}(\sqrt{A}\rho\sqrt{A}) \geq 0$.

If $\rho = \frac{1}{n}I$, then $\mathrm{Tr}(A\rho) = \frac{1}{n}\mathrm{Tr}(A)$. For $v \in \mathbb{C}^n$ with $\|v\| = 1$ and $\rho = v\overline{v}^T$, we have $\mathrm{Tr}(A\rho) = \mathrm{Tr}(Av\overline{v}^T) = \mathrm{Tr}(\overline{v}^T Av) = \langle v, Av\rangle$. In fact, all expectations on $\mathbb{C}^{n\times n}$ can be written as $\mathrm{Tr}(A\rho)$ for a density matrix ρ, see Exercise 4.6.7.

EXAMPLE 4.19. Let $(\Omega, \mathcal{F}, \mathbb{P})$ be a classical probability space. Then $H = L^2(\Omega, \mathcal{F}, \mathbb{P})$ is a Hilbert space, see Remark 2.9. Let $\mathbf{1}_\Omega \in H$ be the function constant 1 and consider the quantum probability space $(B(H), \varphi)$ with

$$\varphi(A) = \langle \mathbf{1}_\Omega, A(\mathbf{1}_\Omega)\rangle = \int_\Omega (A(\mathbf{1}_\Omega))(\omega)d\mathbb{P}(\omega).$$

Then every bounded classical random variable $X : \Omega \to \mathbb{R}$ can be identified with a random variable $A_X \in B(H)$, namely as the multiplication operator $A_X(v) = X \cdot v$, $v \in H$. The distribution of A_X is equal to the distribution μ of X as

$$\varphi(p(A_X)) = \int_\Omega (p(X))(\omega)d\mathbb{P}(\omega) = \int_\mathbb{R} p(x)\mu(dx) = \mathbb{E}[p(X)],$$

for any polynomial $p : \mathbb{R} \to \mathbb{R}$.

EXAMPLE 4.20. We can combine the two previous examples to obtain random matrices as quantum random variables. Let $(\Omega, \mathcal{F}, \mathbb{P})$ be a classical probability space and let $N \in \mathbb{N}$. We take the Hilbert space $H = L^2(\Omega, \mathcal{F}, \mathbb{P}) \otimes \mathbb{C}^N$. Let $E_{j,k} \in \mathbb{C}^{N\times N}$ be the matrix with the (j,k)-entry equal to 1 and all other entries equal to 0. We construct an expectation on $B(L^2(\Omega, \mathcal{F}, \mathbb{P}) \otimes \mathbb{C}^N)$ by defining φ for $A \otimes E_{j,k}$ as $\varphi(A \otimes E_{j,k}) = \frac{1}{N}\int_\Omega (A(\mathbf{1}_\Omega))(\omega)d\mathbb{P}(\omega)$ if $j = k$ and 0 otherwise.

A bounded random matrix can be seen as a bounded random variable of the form $(X_{j,k})_{1\leq j,k\leq N} : \Omega \to \mathbb{C}^{N\times N}$, which can be identified with the operator $X = \sum_{j,k} A_{X_{j,k}} \otimes E_{j,k}$, and thus,

$$\varphi(X) = \sum_{j,k} \varphi(A_{X_{j,k}} \otimes E_{j,k}) = \frac{1}{N}\sum_j \mathbb{E}[X_{j,j}] = \frac{1}{N}\mathbb{E}[\mathrm{Tr}(X)].$$

4.4. Independence

Classical independence can be identified algebraically with the tensor product of quantum probability spaces. This leads to the question of whether other products of quantum probability spaces might yield reasonable notions of independence. Note that the *independence* of classical random variables can be reformulated as an algebraic relation, i.e., a *dependency*, between them. The non-commutative theory, with its rich historical development, allows for a variety of types of "independence", i.e., constraints of different types among quantum random variables. We refer to the articles [**2**, **6**].

In this chapter, we consider the following notions of independence: *classical* (or *tensor*) independence, *Boolean* independence, *free* independence, and *monotone* and *anti-monotone* independence. Boolean independence appeared in the work of W. von Waldenfels in the 1970s ([**3**]). In the 1980s, D. Voiculescu discovered the important notion of free independence with impressive applications to the theory of random matrices. Around the year 2000, N. Muraki introduced monotone and anti-monotone independence for a tuple of quantum random variables.

Due to the works of Speicher, Schürmann, Ben Ghorbal, and Muraki, these five types of independence can be described through one axiomatic framework, see [**180**], [**30**], [**142**], or [**18**, Section "The Five Universal Independences"]. In the subsequent chapters of this book, the primary focus is on monotone independence, with a connection to free independence explored in Chapter 9.

DEFINITION 4.21 (Independence). Let $(B(H), \varphi)$ be a quantum probability space and let $X_1, \ldots, X_n \in B(H)$ be random variables. We define five different notions of independence for $X_1, \ldots, X_n$. For $j = 1, \ldots, n$, let

$$\mathcal{A}_j = \{p(X_j) \,|\, p : \mathbb{R} \to \mathbb{R} \text{ is a polynomial with } p(0) = 0\}.$$

(1) Tensor independence: $X_1, \ldots, X_n$ are *tensor independent* if

$$\varphi(Y_1 \cdots Y_m) = \prod_{j=1,\ldots,n} \varphi\Big(\prod_{\substack{k=1,\ldots,m \\ Y_k \in \mathcal{A}_j}} Y_k \Big)$$

for all $m \in \mathbb{N}$ and $Y_k \in \mathcal{A}_{j_k}$.

(2) Free independence: $X_1, \ldots, X_n$ are *freely independent* if

$$\varphi(Y_1 \cdots Y_m) = 0 \quad \text{whenever} \quad \varphi(Y_1) = \cdots = \varphi(Y_m) = 0,$$

for all $m \in \mathbb{N}$ and $Y_k \in \mathcal{A}_{j_k}$, where $j_k \neq j_{k+1}$ for all $k = 1, \ldots, m-1$.

(3) Boolean independence: $X_1, \ldots, X_n$ are *Boolean independent* if

$$\varphi(Y_1 \cdots Y_m) = \varphi(Y_1) \cdots \varphi(Y_m)$$

for all $m \in \mathbb{N}$ and $Y_k \in \mathcal{A}_{j_k}$, where $j_k \neq j_{k+1}$ for all $k = 1, \ldots, m-1$.

(4) Monotone independence: The tuple $(X_1, \ldots, X_n)$ is *monotonically independent* if

$$\varphi(Y_1 \cdots Y_k \cdots Y_m) = \varphi(Y_k)\varphi(Y_1 \cdots Y_{k-1} Y_{k+1} \cdots Y_m)$$

for all $m \in \mathbb{N}$ and $Y_l \in \mathcal{A}_{j_l}$, whenever $j_{k-1} < j_k$ and $j_k > j_{k+1}$ (and if $k = 1$ or $k = n$, the first or second inequality respectively can be ignored).

(5) Anti-monotone independence: $(X_1, \ldots, X_n)$ is *anti-monotonically* independent if $(X_n, \ldots, X_1)$ is monotonically independent.

Note that tensor independence corresponds to classical independence. Within the argument of φ, we can regroup as if $X_1, \ldots, X_n$ commute, and split $\varphi(Y_1 \cdots Y_m)$ into a product just as in (4.1).

Monotone and anti-monotone independence also depend on the order of the random variables. So (X, Y) might be monotonically independent while (Y, X) is not.

REMARK 4.22. The reason why the polynomials in $\mathcal{A}_j$ have to satisfy $p(0) = 0$ is to exclude the case of independence of X with the identity, which would be too strict, see, e.g., Exercise 4.6.4.

EXAMPLE 4.23. Let X and Y be two random variables. In all cases of independence of X and Y, we have

$$\varphi(XY) = \varphi(YX) = \varphi(X)\varphi(Y).$$

Thus, X and Y are uncorrelated. In the case of free independence, note that $\varphi((X-\varphi(X))(Y-\varphi(Y))=0$ and $\varphi((Y-\varphi(Y))(X-\varphi(X))=0$. Hence, with the solution to Exercise 2.8.6, we also have $\mathrm{Var}(X+Y)=\mathrm{Var}(X)+\mathrm{Var}(Y)$, where $\mathrm{Var}(A):=\varphi((A-\varphi(A))^2)$.

EXAMPLE 4.24. Let X and Y be two random variables. If (X,Y) is monotonically independent, we have

$$\varphi(XYX)=\varphi(Y)\varphi(X^2) \quad \text{and} \quad \varphi(YXY)=\varphi(X)\varphi(Y)^2,$$

while the case of Boolean independence gives

$$\varphi(XYX)=\varphi(Y)\varphi(X)^2, \quad \varphi(YXY)=\varphi(X)\varphi(Y)^2,$$

and the case of tensor independence yields

$$\varphi(XYX)=\varphi(Y)\varphi(X^2), \quad \varphi(YXY)=\varphi(X)\varphi(Y^2).$$

REMARK 4.25. The notion of monotone independence was introduced by N. Muraki [**139–141**] based on earlier works on monotone Fock spaces [**59, 124, 137, 138**]. In [**141**] Muraki defines monotone independence by the following two stronger conditions:

(i) For all $i,j,k\in\{1,\dots,n\}$ with $i<j>k$ and any $X\in\mathcal{A}_i$, $Y\in\mathcal{A}_j$, $Z\in\mathcal{A}_k$ we have
$$XYZ=\varphi(Y)XZ.$$

(ii) For any $r,s\in\mathbb{N}_0$, $i_1,\dots,i_r,j,k_1\dots,k_s\in\{1,\dots,n\}$ with $i_1>\cdots>i_r>j<k_s<\cdots<k_1$ and any $X_1\in\mathcal{A}_{i_1},\dots,X_r\in\mathcal{A}_{i_r}$, $Y\in\mathcal{A}_j$, $Z_1\in\mathcal{A}_{k_1},\dots,Z_s\in\mathcal{A}_{k_s}$, we have
$$\varphi(X_1\cdots X_rYZ_s\cdots Z_1)=\varphi(X_1)\cdots\varphi(X_r)\varphi(Y)\varphi(Z_s)\cdots\varphi(Z_1).$$

The conditions (i) and (ii) imply monotone independence as defined in Definition 4.21. [**69**, Remark 3.2 (c)] gives a condition under which these definitions are equivalent if $\varphi=\langle v,\cdot v\rangle$.

Similar to the tensor products $A\otimes I$ and $I\otimes B$ of the matrices (4.2), we can construct models for the five types of independence. Let H_1 and H_2 be Hilbert spaces with expectations $\varphi_1: B(H_1)\to\mathbb{C}$, $\varphi_1(A)=\langle v_1,Av_1\rangle$, and $\varphi_2: B(H_2)\to\mathbb{C}$, $\varphi_2(A)=\langle v_2,Av_2\rangle$ for some unit vectors $v_1\in H_1$ and $v_2\in H_2$. We can now construct a new probability space. Consider the tensor product $H=H_1\otimes H_2$. The inner product on H is defined as $\langle(a\otimes b),(c\otimes d)\rangle=\langle a,c\rangle\cdot\langle b,d\rangle$, and by linear and continuous extension. Let $v=v_1\otimes v_2$ and $\varphi: B(H)\to\mathbb{C}$, $\varphi(A)=\langle v,Av\rangle$.

THEOREM 4.26. *Let $X\in B(H_1)$ and $Y\in B(H_2)$ be self-adjoint. Denote by I_j the identity on H_j and by P_j the projection in H_j onto the space spanned by v_j, $j=1,2$. Then*

- *$X\otimes I_2$ and $I_1\otimes Y$ are tensor independent.*
- *$X\otimes P_2$ and $P_1\otimes Y$ are Boolean independent.*
- *$(X\otimes P_2, I_1\otimes Y)$ is monotonically independent.*
- *$(X\otimes I_2, P_1\otimes Y)$ is anti-monotonically independent.*

PROOF. We only consider the Boolean case (see Section "The Five Universal Independences" in [**18**]).

Let $k \in \mathbb{N}$. Then $(X \otimes P_2)^k = X^k \otimes P_2$ and $(P_1 \otimes Y)^k = P_1 \otimes Y^k$. Thus $\langle v, (X^k \otimes P_2)v \rangle = \langle v_1, X^k v_1 \rangle \langle v_2, P_2 v_2 \rangle = \langle v_1, X^k v_1 \rangle$ and also $\langle v, (P_1 \otimes Y^k)v \rangle = \langle v_2, Y^k v_2 \rangle$.

Instead of arbitrary polynomials $p(X), p(Y)$, it suffices to consider powers of X and Y. Thus, for $k_1, \ldots, k_m, l_1, \ldots, l_m \in \mathbb{N}$, we obtain

$$\begin{aligned}(X^{k_1} \otimes P_2) \cdot (P_1 \otimes Y^{l_1}) \cdots (X^{k_m} \otimes P_2) \cdot (P_1 \otimes Y^{l_m}) \\ = (X^{k_1} P_1 \cdots X^{k_m} P_1 \otimes P_2 Y^{l_1} \cdots P_2 Y^{l_m}).\end{aligned}$$

Hence,

$$\begin{aligned}&\langle (v_1 \otimes v_2), (X^{k_1} P_1 \cdots X^{k_m} P_1 \otimes P_2 Y^{l_1} \cdots P_2 Y^{l_m})(v_1 \otimes v_2) \rangle \\ &= \langle v_1, X^{k_1} P_1 \cdots X^{k_m} P_1 v_1 \rangle \cdot \langle v_2, P_2 Y^{l_1} \cdots P_2 Y^{l_m} v_2 \rangle \\ &= \langle P_1 X^{k_1} v_1, X^{k_2} P_1 \cdots X^{k_m} P_1 v_1 \rangle \cdot \langle P_2 Y^{l_1} v_2, Y^{l_2} P_2 \cdots Y^{l_m} v_2 \rangle \\ &= \langle v_1, X^{k_1} v_1 \rangle \cdots \langle v_1, X^{k_m} v_1 \rangle \cdot \langle v_2, Y^{l_1} v_2 \rangle \cdots \langle v_2, Y^{l_m} v_2 \rangle . \qquad \square\end{aligned}$$

REMARK 4.27. Free independence is modeled in a different way by using a free product for Hilbert spaces, see [**15**], [**192**].

EXAMPLE 4.28. Consider the two matrices A and B from (4.2). We choose $v_1 = v_2 = (1, 0)^T$. Then $\mu_A = \delta_{\alpha_1}$ and $\mu_B = \delta_{\beta_1}$. Let $v = v_1 \otimes v_2 = (1, 0, 0, 0)^T$. The following two matrices are monotonically independent in $(\mathbb{C}^{4\times 4}, \langle v, \cdot v \rangle)$:

$$A \otimes P_2 = \begin{pmatrix} \alpha_1 & 0 & 0 & 0 \\ 0 & 0 & 0 & 0 \\ 0 & 0 & \alpha_2 & 0 \\ 0 & 0 & 0 & 0 \end{pmatrix}, \quad I_1 \otimes B = \begin{pmatrix} \beta_1 & 0 & 0 & 0 \\ 0 & \beta_2 & 0 & 0 \\ 0 & 0 & \beta_1 & 0 \\ 0 & 0 & 0 & \beta_2 \end{pmatrix}.$$

Boolean independence is realized by

$$A \otimes P_2 = \begin{pmatrix} \alpha_1 & 0 & 0 & 0 \\ 0 & 0 & 0 & 0 \\ 0 & 0 & \alpha_2 & 0 \\ 0 & 0 & 0 & 0 \end{pmatrix}, \quad P_1 \otimes B = \begin{pmatrix} \beta_1 & 0 & 0 & 0 \\ 0 & \beta_2 & 0 & 0 \\ 0 & 0 & 0 & 0 \\ 0 & 0 & 0 & 0 \end{pmatrix}.$$

REMARK 4.29. Let us consider the monotone case in Theorem 4.26 from a quantum mechanical perspective. Assume we make a quantum measurement in H_2. Then the state v_2 changes to some $w \in H_2$, $\|w\| = 1$. Suppose $\langle w, v_2 \rangle = 0$. (For example, take $w = (0, 1)^T$ in the previous example.) Then $v = v_1 \otimes v_2$ changes to $v' = v_1 \otimes w$ ($v' = (0, 1, 0, 0)^T$), and

$$\langle v', (A \otimes P_2)^n v' \rangle = 0 \text{ for all } n \in \mathbb{N}.$$

Conversely, let us change v_1 to some $w \in H_1$, $\|w\| = 1$, with $\langle w, v_1 \rangle = 0$. (For example, take $w = (0, 1)^T$ in the previous example.) Then v changes to $v' = w \otimes v_2$ ($v' = (0, 0, 1, 0)^T$), and

$$\langle v', (I_1 \otimes B)^n v' \rangle = \langle v_2, B^n v_2 \rangle \text{ for all } n \in \mathbb{N}.$$

We see that the measurement in H_2 "annihilates" the observable $A \otimes P_2$, while the measurement in H_1 leaves $I_1 \otimes B$ unaffected.

4.5. Random matrices

An important application of the non-commutative versions of independence, in particular of free independence, concerns random matrices. There are several strong theorems on the asymptotic behavior of random matrices as their size goes to ∞.

Let $N \in \mathbb{N}$ and consider (classical) independent random variables $(X_j)_{1\le j\le N} \cup (Y_{j,k})_{1\le j<k\le N}$ such that each X_j has a $\mathcal{N}(0,2)$ and each $Y_{j,k}$ has a $\mathcal{N}(0,1)$ distribution. We define the $N \times N$-random matrix

$$A_N = [A_{N,j,k}]_{1\le j,k\le N},$$

where

$$A_{N,j,k} = Y_{j,k} \text{ for } j<k,\, A_{N,j,k} = Y_{k,j} \text{ for } k<j, \text{ and } A_{N,j,j} = X_j.$$

Then A_N is a symmetric Gaussian random matrix. The variances are chosen such that if $O \in \mathbb{R}^{N\times N}$ is orthogonal ($OO^T = I$), then $O^T A_N O$ has the same distribution as A_N, see [**9**, Section 2.5]. For this reason, the distribution of A_N is also called the *Gaussian Orthogonal Ensemble* (GOE).

We are interested in the behavior of the eigenvalues of A_N as $N \to \infty$. The first moment of A_N with respect to $\frac{1}{N}\operatorname{Tr}$ is given by

$$\frac{1}{N}\operatorname{Tr}(A_N) = \frac{1}{N}\sum_{j=1}^{N} X_j.$$

This random variable converges to 0 almost surely as $N \to \infty$, due to the strong law of large numbers. For the second moment, we have

$$\frac{1}{N}\operatorname{Tr}(A_N^2) = \frac{1}{N}\sum_{j,k=1}^{N}(A_{N,j,k})^2.$$

This sum involves $N(N+1)/2$ independent terms, but it is scaled only by $1/N$ and so it will not converge. This suggests that we should scale A_N by $1/\sqrt{N}$ in order to obtain a limiting distribution for the eigenvalues.

So, define

$$B_N := A_N/\sqrt{N}$$

and denote by $\lambda_1 \le \cdots \le \lambda_N$ the eigenvalues of B_N. We write $B_N = [b_{j,k}]_{j,k=1,\dots,N}$, where we suppress the dependence of $b_{j,k}$ on N to simplify notation. We have

$$\mathbb{E}[b_{j,k}] = 0, \quad \mathbb{E}[b_{j,j}^2] = \tfrac{2}{N}, \quad \mathbb{E}[b_{j,k}^2] = \tfrac{1}{N} \text{ if } j \neq k.$$

The simulation from Figure 4.1 suggests that the empirical spectrum

$$\mu_N := \sum_{j=1}^{N} \delta_{\lambda_j}/N$$

has a deterministic limit, namely a semicircle distribution $W(0,1)$. This was first shown by E. Wigner [**194, 195**], who used random matrices for models in physics, and $W(0,1)$ is therefore also called *Wigner's semicircle law.*

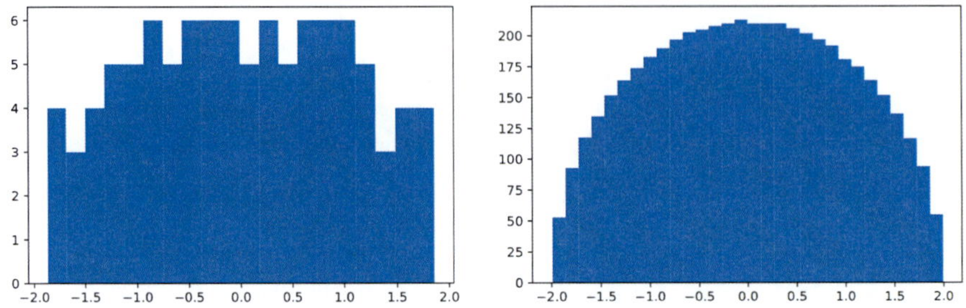

FIGURE 4.1. Histograms for the scaled eigenvalues of simulated GOE random matrices for $N = 100$ (left) and $N = 5000$ (right).

Note that μ_N is a random probability distribution. One can show that μ_N converges weakly to $W(0,1)$ almost surely, see [**186**, Theorem 2.4.2]. We prove a weaker statement by regarding the averaged empirical distribution $\alpha_N := \mathbb{E}[\mu_N]$, which is defined by

$$\alpha_N(B) = \mathbb{E}\left[\mu_N(B)\right]$$

for every Borel subset $B \subset \mathbb{R}$.

THEOREM 4.30. *Let A_N be an $N \times N$ GOE random matrix. Let μ_N be the empirical spectrum of $A_N/\sqrt{N}$. Then*

$$\alpha_N = \mathbb{E}[\mu_N] \to W(0,1)$$

with respect to weak convergence as $N \to \infty$.

In other words, if N is large, we expect that the number of eigenvalues of $A_N/\sqrt{N}$ in some interval $[a,b]$ is equal to

$$\frac{N}{2\pi}\int_a^b \sqrt{4-x^2}\,\mathbf{1}_{[-2,2]}(x)dx.$$

We follow [**66**, **101**], and we begin with the following lemma.

LEMMA 4.31 (Stein's lemma). *Let $X_1, \ldots, X_n$ be independent Gaussian random variables with mean 0 and variances $\mathrm{Var}(X_k) = \sigma_k^2 > 0$. Let $f : \mathbb{R}^n \to \mathbb{R}$ be a continuously differentiable function such that f and its partial derivatives are all of polynomial growth. Then, for all $k = 1, \ldots, n$, we have*

$$\mathbb{E}[f(X_1,\ldots,X_n)X_k] = \sigma_k^2 \cdot \mathbb{E}\left[\frac{\partial f}{\partial x_k}(X_1,\ldots,X_n)\right].$$

PROOF. Partial integration gives us

$$\begin{aligned}
&\int_{\mathbb{R}} x_k f(x_1,\dots,x_n) e^{-(x_1^2/\sigma_1^2+\dots+x_n^2/\sigma_n^2)/2} dx_k \\
&= \int_{\mathbb{R}} f(x_1,\dots,x_n) \frac{\partial}{\partial x_k} \left[-\sigma_k^2 e^{-(x_1^2/\sigma_1^2+\dots+x_n^2/\sigma_n^2)/2} \right] dx_k \\
&= -\lim_{R\to\infty} \left[\sigma_k^2 f(x_1,\dots,x_n) e^{-(x_1^2/\sigma_1^2+\dots+x_n^2/\sigma_n^2)/2} \right]_{x_k=-R}^{R} \\
&+ \int_{\mathbb{R}} \frac{\partial}{\partial x_k} f(x_1,\dots,x_n) \sigma_k^2 e^{-(x_1^2/\sigma_1^2+\dots+x_n^2/\sigma_n^2)/2} dx_k \\
&= \int_{\mathbb{R}} \frac{\partial}{\partial x_k} f(x_1,\dots,x_n) \sigma_k^2 e^{-(x_1^2/\sigma_1^2+\dots+x_n^2/\sigma_n^2)/2} dx_k.
\end{aligned}$$

Thus

$$\begin{aligned}
&\mathbb{E}[f(X_1,\dots,X_n)X_k] = \int_{\mathbb{R}^n} x_k f(x_1,\dots,x_n) e^{-(x_1^2/\sigma_1^2+\dots+x_n^2/\sigma_n^2)/2} dx_1 \dots dx_n \\
&= \int_{\mathbb{R}^n} \frac{\partial}{\partial x_k} f(x_1,\dots,x_n) \sigma_k^2 e^{-(x_1^2/\sigma_1^2+\dots+x_n^2/\sigma_n^2)/2} dx_1 \dots dx_n \\
&= \sigma_k^2 \cdot \mathbb{E}\left[\frac{\partial f}{\partial x_k}(X_1,\dots,X_n) \right]. \qquad \square
\end{aligned}$$

We will also need the following theorem. A proof can be found in [**44**, Section 3.7].

THEOREM 4.32 (Gaussian Poincaré Inequality). *Let $X_1,\dots,X_n$ be independent Gaussian random variables with mean 0 and variances* $\mathrm{Var}(X_k) = \sigma_k^2 > 0$. *Let $f : \mathbb{R}^n \to \mathbb{R}$ be continuously differentiable. Then*

$$\mathrm{Var}(f(X_1,\dots,X_n)) \le \max_{k=1,\dots,n} \sigma_k^2 \cdot \mathbb{E}[\|\nabla f(X_1,\dots,X_n)\|^2].$$

PROOF OF THEOREM 4.30. We need to show that

$$G_{\alpha_N}(z) \to G_{W(0,1)}(z) = \frac{z-\sqrt{z^2-4}}{2} \tag{4.5}$$

(see Exercise 3.9.3) locally uniformly on $\mathbb{H}$ as $N\to\infty$. We can write G_{α_N} as

$$G_{\alpha_N}(z) = \int_{\mathbb{R}} \frac{1}{z-x} \mathbb{E}[\mu_N](dx) = \mathbb{E}\left[\int_{\mathbb{R}} \frac{1}{z-x} \mu_N(dx) \right] = \mathbb{E}[G_{\mu_N}(z)],$$

$z \in \mathbb{H}$. As $G_{\mu_N}(z) = \frac{1}{N}\operatorname{Tr}\left((zI-B_N)^{-1}\right)$, we have

$$G_{\alpha_N}(z) = \frac{1}{N}\mathbb{E}\left[\operatorname{Tr}\left((zI-B_N)^{-1} \right) \right].$$

Let $R = (zI-B_N)^{-1}$. (To simplify notation, we suppress the dependence of R on N and z.) Then $(zI-B_N)R = zR - B_N R = I$, and thus, $R = \frac{1}{z}I + \frac{1}{z}B_N R$. Hence $\frac{1}{N}\operatorname{Tr}(R) = \frac{1}{z} + \frac{1}{zN}\operatorname{Tr}(B_N R)$ and

$$G_{\alpha_N}(z) = \frac{1}{z} + \frac{1}{zN}\mathbb{E}[\operatorname{Tr}(B_N R)]. \tag{4.6}$$

We have $\operatorname{Tr}(B_N R) = \sum_{j,k=1}^{N} b_{j,k}[R]_{k,j}$. Now we can apply Stein's lemma to obtain

$$\begin{aligned}\mathbb{E}[\operatorname{Tr}(B_N R)] &= \mathbb{E}\left[\sum_{j,k=1}^{N} b_{j,k}[R]_{k,j}\right] = \sum_{j,k=1}^{N} \mathbb{E}\left[b_{j,k}[R]_{k,j}\right] \\ &= \sum_{j,k=1}^{N} \operatorname{Var}(b_{j,k})\mathbb{E}\left[\frac{\partial}{\partial b_{j,k}}[R]_{k,j}\right].\end{aligned} \tag{4.7}$$

(Here we denote by $\frac{\partial}{\partial b_{j,k}}[R]$ the partial derivative of the function $[x_{j,k}]_{j,k=1,\dots,n} \mapsto (zI - [x_{j,k}]_{j,k=1,\dots,n})^{-1}$, where $x_{k,j} = x_{j,k}$, with respect to $x_{j,k}$. The components of this function and its partial derivatives can be shown to be of polynomial growth by Cramer's rule.)

Next we calculate the partial derivatives. From $R(zI - B_N) = I$, we get

$$\frac{\partial R}{\partial b_{j,k}}(zI - B_N) - R\frac{\partial B_N}{\partial b_{j,k}} = \frac{\partial R}{\partial b_{j,k}}(zI - B_N) - RE_{j,k} = 0,$$

where we denote by $E_{j,k}$ the symmetric matrix whose elements are 0 except for two 1s at positions (j,k) and (k,j) (one 1 if $j = k$). Multiplying by R from the right yields

$$\frac{\partial R}{\partial b_{j,k}} = RE_{j,k}R.$$

So, if $j = k$, then

$$\left[\frac{\partial R}{\partial b_{j,j}}\right]_{j,j} = [RE_{j,j}R]_{j,j} = ([R]_{j,j})^2.$$

If $j \neq k$, then

$$\left[\frac{\partial R}{\partial b_{j,k}}\right]_{k,j} = [RE_{j,k}R]_{k,j} = ([R]_{k,j})^2 + [R]_{j,j}[R]_{k,k}.$$

Consequently,

$$\begin{aligned}&\sum_{j,k=1}^{N} \operatorname{Var}(b_{j,k})\mathbb{E}\left[\frac{\partial}{\partial b_{j,k}}[R]_{k,j}\right] = \sum_{j=1}^{N} \frac{2}{N}\mathbb{E}\left[([R]_{j,j})^2\right] \\ &+ \sum_{j,k=1, j\neq k}^{N} \frac{1}{N}\mathbb{E}\left[([R]_{k,j})^2 + [R]_{j,j}[R]_{k,k}\right] = \\ &\frac{1}{N}\sum_{j,k=1}^{N} \mathbb{E}\left[([R]_{k,j})^2\right] + \frac{1}{N}\sum_{j,k=1}^{N} \mathbb{E}\left[[R]_{j,j}[R]_{k,k}\right] = \frac{1}{N}\mathbb{E}\left[\operatorname{Tr}(R^2)\right] + \frac{1}{N}\mathbb{E}\left[\operatorname{Tr}(R)^2\right].\end{aligned} \tag{4.8}$$

With (4.6) and (4.7), we get

$$G_{\alpha_N}(z) = \frac{1}{z} + \frac{1}{z}\cdot\frac{1}{N^2}\mathbb{E}\left[\operatorname{Tr}(R^2)\right] + \frac{1}{z}\cdot\frac{1}{N^2}\mathbb{E}\left[\operatorname{Tr}(R)^2\right]. \tag{4.9}$$

The eigenvalues of R^2 are $\frac{1}{(z-\lambda_k)^2}$. Thus

$$\frac{1}{N^2}\left|\mathbb{E}\left[\operatorname{Tr}(R^2)\right]\right| \le \frac{1}{N^2}\mathbb{E}\left[\left|\operatorname{Tr}(R^2)\right|\right] = \frac{1}{N^2}\mathbb{E}\left[\left|\sum_{k=1}^{N}\frac{1}{(z-\lambda_k)^2}\right|\right]$$
$$\le \frac{1}{N^2}\mathbb{E}\left[\sum_{k=1}^{N}\frac{1}{|z-\lambda_k|^2}\right] \le \frac{1}{N^2}\mathbb{E}\left[\sum_{k=1}^{N}\frac{1}{\operatorname{Im}(z)^2}\right] = \frac{1}{N}\frac{1}{\operatorname{Im}(z)^2} \to 0$$

as $N \to \infty$. The second expectation in (4.9) can be written as

$$\begin{aligned}
&\frac{1}{N^2}\mathbb{E}\left[\operatorname{Tr}(R)^2\right] = \mathbb{E}\left[(\operatorname{Tr}(R)/N)^2\right] \\
&= \mathbb{E}\left[\operatorname{Tr}(R)/N\right]^2 + \mathbb{E}\left[(\operatorname{Tr}(R)/N - \mathbb{E}\left[\operatorname{Tr}(R)/N\right])^2\right] \\
&= G_{\alpha_N}(z)^2 + \mathbb{E}\left[(\operatorname{Tr}(R)/N - \mathbb{E}\left[\operatorname{Tr}(R)/N\right])^2\right] \\
&= G_{\alpha_N}(z)^2 + \mathbb{E}\left[\operatorname{Re}[(\operatorname{Tr}(R)/N - \mathbb{E}\left[\operatorname{Tr}(R)/N\right])^2]\right] \\
&+ i\mathbb{E}\left[\operatorname{Im}[(\operatorname{Tr}(R)/N - \mathbb{E}\left[\operatorname{Tr}(R)/N\right])^2]\right].
\end{aligned}$$

We want to show that the last two expressions converge to 0. As $\operatorname{Re}(z^2) \le \operatorname{Re}(z)^2$ for any $z \in \mathbb{C}$, we have

$$\begin{aligned}
&\mathbb{E}\left[\operatorname{Re}[(\operatorname{Tr}(R)/N - \mathbb{E}\left[\operatorname{Tr}(R)/N\right])^2]\right] \\
&\le \mathbb{E}\left[(\operatorname{Re}[\operatorname{Tr}(R)/N - \mathbb{E}\left[\operatorname{Tr}(R)/N\right]])^2\right] = \operatorname{Var}(\operatorname{Re}(\operatorname{Tr}(R)/N)).
\end{aligned}$$

From our previous calculations, we obtain for $j \neq k$:

$$\begin{aligned}
&\frac{\partial}{\partial b_{j,k}}\frac{1}{N}\operatorname{Tr}(R) = \frac{1}{N}\sum_{l=1}^{N}\left[\frac{\partial R}{\partial b_{j,k}}\right]_{l,l} = \frac{1}{N}\sum_{l=1}^{N}(R_{l,j}R_{k,l} + R_{j,l}R_{l,k}) \\
&= \frac{1}{N}([R^2]_{k,j} + [R^2]_{j,k}) = \frac{2}{N}[R^2]_{j,k},
\end{aligned}$$

where we used that R is symmetric (as the inverse of a symmetric matrix), and if $j = k$:

$$\frac{\partial}{\partial b_{j,j}}\frac{1}{N}\operatorname{Tr}(R) = \frac{1}{N}\sum_{l=1}^{N}\left[\frac{\partial R}{\partial b_{j,j}}\right]_{l,l} = \frac{1}{N}\sum_{l=1}^{N}(R_{j,l}R_{l,j}) = \frac{1}{N}[R^2]_{j,j}.$$

By Theorem 4.32,

$$\begin{aligned}
&\operatorname{Var}\left(\operatorname{Re}\left(\operatorname{Tr}(R)/N\right)\right) \le \frac{2}{N}\mathbb{E}\left[\sum_{j,k=1}^{N}\left(\frac{\partial}{\partial b_{j,k}}\operatorname{Re}\left(\operatorname{Tr}(R)/N\right)\right)^2\right] \\
&= \frac{2}{N}\mathbb{E}\left[\sum_{j,k=1}^{N}\left(\operatorname{Re}\left(\frac{\partial}{\partial b_{j,k}}\operatorname{Tr}(R)/N\right)\right)^2\right] \le \frac{8}{N^3}\mathbb{E}\left[\sum_{j,k=1}^{N}\left(\operatorname{Re}\left([R^2]_{j,k}\right)\right)^2\right] \\
&\le \frac{8}{N^3}\mathbb{E}\left[\sum_{j,k=1}^{N}\left|[R^2]_{j,k}\right|^2\right] = \frac{8}{N^3}\mathbb{E}\left[\|R^2\|_F^2\right] \le \frac{8}{N^3}\mathbb{E}\left[\|R\|_F^4\right] \le \frac{8}{N^3}N^2\mathbb{E}\left[\|R\|^4\right],
\end{aligned}$$

where $\|\cdot\|_F$ denotes the Frobenius norm and $\|\cdot\|$ the operator norm of R (with respect to the Euclidean norm). We used that $\|\cdot\|_F$ is sub-multiplicative ($\|XY\|_F \leq \|X\|_F\|Y\|_F$) and that the relation $\|X\|_F \leq \sqrt{N}\|X\|$ holds for every $X \in \mathbb{C}^{N\times N}$.[2]

As R is a normal operator ($RR^* = R^*R$), $\|R\|$ is equal to its spectral radius, i.e., $\|R\| = \max_{k=1,\ldots,N} |z-\lambda_k|^{-1} \leq 1/\operatorname{Im}(z)$. Therefore,

$$Var\left(\operatorname{Re}\left(\operatorname{Tr}(R)/N\right)\right) \leq \frac{8}{N}\frac{1}{\operatorname{Im}(z)^4} \to 0.$$

Similarly, we show that $Var\left(\operatorname{Im}\left(\operatorname{Tr}(R)/N\right)\right) \to 0$ as well.

Now, Helly's selection theorem implies that there is a subsequence $(\alpha_{n_k})_k$ which converges weakly to a probability measure α on $\hat{\mathbb{R}}$. From (4.9), we obtain that

$$G_\alpha(z) = \frac{1}{z} + \frac{1}{z}G_\alpha(z)^2,$$

which leads to $G_\alpha(z) = \frac{z-\sqrt{z^2-4}}{2} = G_{W(0,1)}$, and thus, $\alpha = W(0,1)$. Therefore, every convergent subsequence of $(\alpha_N)_{N\in\mathbb{N}}$ has the same limit $W(0,1)$, and we conclude that $\alpha_N \to W(0,1)$ as $N\to\infty$. $\square$

REMARK 4.33. Theorem 4.30 remains valid under the more general assumption that the variances $\operatorname{Var}(A_{n,j,j}) = \sigma_j^2$ of the diagonal entries are all bounded by a constant $C>0$, independent of j and N. The proof can be adapted as follows. We now have $\operatorname{Var}(b_{j,j}) = \sigma_j^2/N \leq C/N$ and equation (4.8) becomes:

$$\sum_{j,k=1}^{N} \operatorname{Var}(b_{j,k})\mathbb{E}\left[\frac{\partial}{\partial b_{j,k}}[R]_{k,j}\right] = \sum_{j=1}^{N}\frac{\sigma_j^2}{N}\mathbb{E}\left[([R]_{j,j})^2\right]$$

$$+\sum_{j,k=1,j\neq k}^{N}\frac{1}{N}\mathbb{E}\left[([R]_{k,j})^2 + [R]_{j,j}[R]_{k,k}\right] =$$

$$\sum_{j=1}^{N}\frac{\sigma_j^2-2}{N}\mathbb{E}\left[([R]_{j,j})^2\right] + \frac{1}{N}\mathbb{E}\left[\operatorname{Tr}(R^2)\right] + \frac{1}{N}\mathbb{E}\left[\operatorname{Tr}(R)^2\right].$$

It suffices to show that the first term is bounded by a constant independent of N. We have

$$\left|\sum_{j=1}^{N}\frac{\sigma_j^2-2}{N}\mathbb{E}\left[([R]_{j,j})^2\right]\right| \leq \sum_{j=1}^{N}\frac{\sigma_j^2+2}{N}\left|\mathbb{E}\left[([R]_{j,j})^2\right]\right| \leq \frac{C+2}{N}\mathbb{E}\left[\sum_{j=1}^{N}|[R]_{j,j}|^2\right]$$

$$\leq \frac{C+2}{N}\mathbb{E}\left[\sum_{j,k=1}^{N}|[R]_{j,k}|^2\right] = \frac{C+2}{N}\mathbb{E}\left[\|R\|_F^2\right] \leq (C+2)\mathbb{E}\left[\|R\|^2\right] \leq \frac{C+2}{\operatorname{Im}(z)^2}.$$

REMARK 4.34. It can be shown that the largest eigenvalue of B_N converges to 2 almost surely and its scaling limit can be described by a Tracy–Widom distribution, see [**17, 187**].

---

[2] Consider the singular value decomposition $X = UDV$, where U, V are unitary and D is diagonal with diagonal entries $\sigma_1,\ldots,\sigma_N \geq 0$. Then $\|X\|_F^2 = \operatorname{Tr}(X^*X) = \operatorname{Tr}(D^*D) = \sigma_1^2+\cdots+\sigma_N^2 \leq N(\max_{k=1,\ldots,N}\sigma_k)^2 = N\|X\|^2$.

REMARK 4.35. The Hermitian case is quite similar: Let $N \in \mathbb{N}$ and consider (classical) independent random variables $(X_{j,k})_{1\le j\le k\le N} \cup (Y_{j,k})_{1\le j<k\le N}$ such that each $X_{j,j}$ has a $\mathcal{N}(0,1)$ and each $X_{j,k}$, $j<k$, and each $Y_{j,k}$ has a $\mathcal{N}(0,1/2)$ distribution. The $N\times N$-random matrix $A_N = [Z_{j,k}]_{1\le j,k\le N}$ with

$$Z_{j,k} = X_{j,k} + iY_{j,k} \text{ for } j<k, Z_{j,k} = X_{k,j} - iY_{k,j} \text{ for } k<j, \text{ and } Z_{j,j} = X_{j,j},$$

is a Hermitian Gaussian random matrix. [The variances are chosen such that if $C \in \mathbb{C}^{N\times N}$ is unitary ($CC^* = I$), then C^*A_NC has the same distribution as A_N, see [**9**, Section 2.5]. For this reason, the distribution of A_N is also called a *Gaussian Unitary Ensemble* (GUE).] The empirical measure of $A_N/\sqrt{N}$ converges to $W(0,1)$ as $N\to\infty$, see [**186**, Theorem 2.4.2].

Another important result concerns singular values. Let A_N be an $m\times N$ random matrix such all entries are *iid* with distribution $\mathcal{N}(0,\sigma^2)$. We let $m = m(N)$ be dependent on N. Let $\lambda_1,\dots,\lambda_m$ be the eigenvalues of $S_N = \frac{1}{N}A_NA_N^T$, and consider the random measure $\mu_N = \sum_{k=1}^m \delta_{\lambda_k}/m$. As S_N is non-negative, $\lambda_k \ge 0$ for all k. Marchenko and Pastur found the limiting distribution as $N\to\infty$.

THEOREM 4.36 ([**129**]). *Assume that* $m(N)/N \to \lambda > 0$. *Then* μ_N *converges weakly almost surely to the* Marchenko–Pastur law μ_{MP} *given by*

$$\max[(1-1/\lambda),0]\delta_0 + \frac{\sqrt{(b-x)(x-a)}}{2\pi\sigma^2\lambda x}dx,$$

where $b = \sigma^2(1+\sqrt{\lambda})^2$, $a = \sigma^2(1-\sqrt{\lambda})^2$ *and* $x\in[a,b]$ *if* $\lambda\ne 1$, $x\in(0,b]$ *if* $\lambda = 1$.

Figure 4.2 shows examples of the density of μ_{MP}. The Cauchy transform of μ_{MP} is given by

$$G_{\mu_{MP}}(z) = \frac{z+\sigma^2(\lambda-1) - \sqrt{(z-\sigma^2(\lambda+1))^2 - 4\lambda\sigma^4}}{2\lambda\sigma^2 z}.$$

The distributions μ_{MP} and $W(0,1)$ have a deeper significance within free probability theory. While Wigner's semicircle law can be viewed as the free probability analog of the normal distribution, due to the free central limit theorem (Theorem 5.40), μ_{MP} is the free probability counterpart of the Poisson distribution (Exercise 5.9.9).

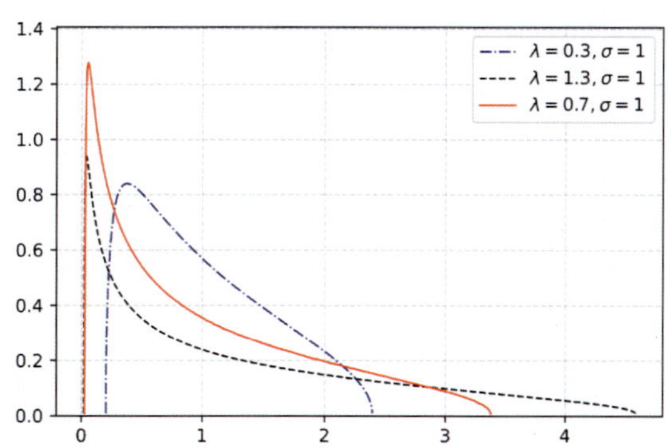

FIGURE 4.2. Density of the Marchenko–Pastur distribution for some values of λ.

The connection between random matrices and free probability theory goes beyond formal analogies. Voiculescu [**191**] discovered that classically independent Gaussian random matrices are asymptotically freely independent with respect to the expectation $\frac{1}{N}\mathbb{E}[\mathrm{Tr}(X)]$, see also [**186**, Proposition 2.5.21], [**132**, Chapter 1].

THEOREM 4.37. *For every $N \in \mathbb{N}$, let $A_{N,1}, \ldots, A_{N,n}$ be n independent $N \times N$ GOE (or GUE) random matrices, and put $B_{N,k} = A_{N,k}/\sqrt{N}$, $k = 1, \ldots, n$. Let $\varphi_N(X) = \frac{1}{N}\mathbb{E}[\mathrm{Tr}(X)]$.*

Then $B_{N,1}, \ldots, B_{N,n}$ are asymptotically freely independent:

$$\varphi_N((p_1(B_{N,i_1}) - \varphi_N(p_1(B_{N,i_1}))) \cdots (p_m(B_{N,i_m}) - \varphi_N(p_m(B_{N,i_m})))) \to 0 \tag{4.10}$$

as $N \to \infty$, for any polynomials $p_1, \ldots, p_m$ and indices $i_1, \ldots, i_m \in \{1, \ldots, n\}$ with $i_j \neq i_{j+1}$.

Thus, for large N, the classically independent matrices $A_{N,1}/\sqrt{N}, \ldots, A_{N,n}/\sqrt{N}$ can be treated as freely independent $W(0,1)$-distributed random variables via the expectation $\frac{1}{N}\mathbb{E}[\mathrm{Tr}(X)]$.

Let us consider a further result of this type. The set of all unitary $N \times N$-matrices $\mathcal{U}(N) = \{U \in \mathbb{C}^{N \times N} \mid UU^* = I\}$ forms a group and its normalized Haar measure is the unique Borel probability measure on $\mathcal{U}(N)$ that is invariant under the group operation. If $M \in \mathbb{C}^{N \times N}$ and U is a Haar unitary $N \times N$-random matrix, then UMU^* can be thought of as a random rotation of M.

THEOREM 4.38 (Theorem 23.14 in [**144**]). *Let $(A_N)_{N \in \mathbb{N}}$ and $(B_N)_{N \in \mathbb{N}}$ be sequences of (deterministic) $N \times N$-matrices such that $(A_N)_N$ and $(B_N)_N$ converge in distribution with respect to $X \to \frac{1}{N}\mathrm{Tr}(X)$ as $N \to \infty$. Furthermore, let $(U_N)_{N \in \mathbb{N}}$ be a sequence of Haar unitary $N \times N$-random matrices. Then A_N and $U_N B_N U_N^*$ are asymptotically freely independent in the sense of* (4.10) *as $N \to \infty$.*

There are numerous additional results on the limiting distributions of random matrices and the asymptotic free independence of random matrices, as discussed in the books [**9**], [**67**], [**130**], [**132**], [**144**], [**186**].

Random matrix theory finds applications in a wide range of fields, including physics, multivariate statistics, signal processing [**56**], machine learning [**57**], and more [**7**].

4.6. Exercises

E 4.6.1. Let $A, B \in B(H)$. Prove the Cauchy–Schwarz inequality

$$|\varphi(A^*B)|^2 \le \varphi(B^*B)\varphi(A^*A).$$

E 4.6.2. Let $(B(H), \varphi)$ be a quantum probability space and let $A \in B(H)$. Show that

$$|\varphi(A)| \le \|A\|.$$

E 4.6.3. With the notation from Remark 4.16, show that

$$\sum_{k=0}^{n}(x - k/n)^2 b_{k,n}(x) = \frac{x(1-x)}{n}.$$

E 4.6.4. Let $(B(H), \varphi)$ be a quantum probability space and denote by $I \in B(H)$ the identity.

(a) For which self-adjoint $X \in B(H)$ is (X, I) monotonically independent?
(b) For which self-adjoint $X \in B(H)$ is (I, X) monotonically independent?

E 4.6.5. Let $(B(H), \varphi)$ be a quantum probability space with $\varphi(AB) = \varphi(BA)$ for all $A, B \in B(H)$. Let $X, Y \in B(H)$ be random variables. Prove that if (X, Y) is monotonically independent, then the distribution of X or of Y is a point measure.

E 4.6.6. Let X and Y be random variables in a quantum probability space $(B(H), \varphi)$. Assume that X and Y are both classically and freely independent. Show that the distribution of X or of Y has to be a point measure.

E 4.6.7. Show that any expectation φ on $B(\mathbb{C}^{N\times N})$ can be written as $\varphi(X) = \mathrm{Tr}(X\rho)$ for a density matrix ρ.

E 4.6.8. Let A_N be an $N \times N$ GOE random matrix and $B_N = A_N/\sqrt{N}$. Use

$$\int_0^{\pi/2} \sin^{2k}(t)dt = \frac{\pi}{2}4^{-k}\binom{2k}{k} \tag{4.11}$$

to show that

$$\tfrac{1}{N}\mathbb{E}[\mathrm{Tr}(B_N^{2k})] \to \tfrac{1}{k+1}\tbinom{2k}{k} \quad \text{and} \quad \tfrac{1}{N}\mathbb{E}[\mathrm{Tr}(B_N^{2k-1})] \to 0$$

for all $k \in \mathbb{N}$ as $N \to \infty$.

4.7. Solutions

S 4.7.1. Let $A, B \in B(H)$ and let $c \in \mathbb{C}$ with $|c| = 1$. Then

$$\begin{aligned}&\varphi((A - tcB)^*(A - tcB)) = \varphi(A^*A) - t\varphi(cA^*B + \overline{c}B^*A) + t^2\varphi(B^*B)\\ &= \varphi(A^*A) - 2t\,\mathrm{Re}(\varphi(cA^*B)) + t^2\varphi(B^*B).\end{aligned}$$

Here we used that $\varphi(X + X^*) = 2\,\mathrm{Re}(\varphi(X))$ (as $\varphi(X^*) = \overline{\varphi(X)}$, Lemma 4.12). We know that $t \mapsto \varphi((A - tcB)^*(A - tcB)) \geq 0$ for all $t \in \mathbb{R}$. So the determinant of the quadratic function has to be ≤ 0, which gives

$$\mathrm{Re}(c\varphi(A^*B))^2 \leq \varphi(B^*B)\varphi(A^*A).$$

We now choose c such that $c\varphi(A^*B) \in [0, \infty)$, i.e., $c\varphi(A^*B) = |\varphi(A^*B)|$. Then

$$|\varphi(A^*B)|^2 \leq \varphi(B^*B)\varphi(A^*A).$$

S 4.7.2. Let $A \in B(H)$ be self-adjoint. Then we have $-\|A\|I \leq A \leq \|A\|I$ because

$$\langle v, (\|A\|I \pm A)v\rangle = \|A\|\|v\|^2 \pm \langle v, Av\rangle \geq \|Av\|\|v\| \pm \langle v, Av\rangle \geq |\langle v, Av\rangle| \pm \langle v, Av\rangle \geq 0,$$

for all $v \in H$, where we used the Cauchy–Schwarz inequality $|\langle v, Av\rangle| \leq \|Av\|\|v\|$. Thus,

$$0 \leq \varphi(\|A\|I \pm A) = \|A\| \pm \varphi(A)$$

and $|\varphi(A)| \leq \|A\|$.

Now, consider an arbitrary $A \in B(H)$. With Exercise 4.6.1, we obtain

$$|\varphi(A)|^2 = |\varphi(I \cdot A)|^2 \leq \varphi(A^*A) \leq \|A^*A\| \leq \|A\|^2.$$

(For $v \in H$ we have $|\langle v, A^*Av\rangle| = |\langle Av, Av\rangle| = \|Av\|^2 \leq \|A\|^2\|v\|^2$. So $0 \leq A^*A \leq \|A\|^2 I$ and $\|A^*A\| \leq \|A\|^2$ by Lemma 4.5.)

S 4.7.3. We have

(1)

$$\sum_{k=0}^{n} b_{k,n}(x) = 1.$$

(2)

$$\begin{aligned}&\sum_{k=0}^{n} kb_{k,n}(x) = \sum_{k=0}^{n} k\binom{n}{k}x^k(1-x)^{n-k}\\ &= \sum_{k=1}^{n} n\binom{n-1}{k-1}x^k(1-x)^{n-k} = n\sum_{j=0}^{n-1}\binom{n-1}{j}x^{j+1}(1-x)^{n-j-1} = xn.\end{aligned}$$

(3)

$$\begin{aligned}\sum_{k=0}^{n} k^2 b_{k,n}(x) &= \sum_{k=0}^{n} k^2 \binom{n}{k} x^k (1-x)^{n-k} = \sum_{k=1}^{n} nk \binom{n-1}{k-1} x^k (1-x)^{n-k}\\ &= nx \sum_{j=0}^{n-1} (j+1) \binom{n-1}{j} x^j (1-x)^{n-1-j}\\ &= nx + nx \sum_{j=0}^{n-1} j \binom{n-1}{j} x^j (1-x)^{n-1-j}\\ &= nx + nx(n-1)x,\end{aligned}$$

where we used (2) in the last step.

Now we expand $(x - k/n)^2 = x^2 - 2kx/n + k^2/n^2$ and obtain

$$\begin{aligned}&\sum_{k=0}^{n} (x - k/n)^2 b_{k,n}(x) = x^2 - 2x^2 + (nx + nx(n-1)x)/n^2 =\\ &= x^2 - 2x^2 + x/n + x^2 - x^2/n = x(1-x)/n.\end{aligned}$$

S 4.7.4.

(a) As $X^{n_0} I^{m_1} \cdots I^{m_k} X^{n_k} = X^{n_0} \cdots X^{n_k}$, we see that (X, I) is monotonically independent for every self-adjoint $X \in B(H)$.

(b) Assume that (I, X) is monotonically independent. Then

$$\varphi(X^n) = \varphi(X^{n-1} I X) = \varphi(X^{n-1}) \varphi(X),$$

and inductively, we see that $\varphi(X^n) = \varphi(X)^n$ for all $n \in \mathbb{N}$. Hence, X has the distribution δ_c for some $c \in \mathbb{R}$. Conversely, if the distribution of X is δ_c, then $\varphi(X^n) = \varphi(X)^n$ for all $n \in \mathbb{N}$, and we see that (I, X) is monotonically independent.

S 4.7.5. We may assume that $\varphi(X) = \varphi(Y) = 0$. (Otherwise, consider $X - \varphi(X)$ and $Y - \varphi(Y)$.) We have

$$\varphi(X^2Y^2) = \varphi(X^2)\varphi(Y^2) \quad \text{and} \quad \varphi(X^2Y^2) = \varphi(YXXY) = \varphi(X^2)\varphi(Y)^2 = 0.$$

If $\varphi(X^2) = 0$, then the distribution of X is δ_0. Otherwise, $\varphi(Y^2) = 0$ and the distribution of Y is δ_0.

S 4.7.6. Let $A = X - \varphi(X)$ and $B = Y - \varphi(Y)$. Then $\varphi(A) = \varphi(B) = 0$ and free independence of X and Y implies that

$$\varphi(ABAB) = 0.$$

On the other hand, the classical independence implies that

$$\varphi(ABAB) = \varphi(A^2)\varphi(B^2).$$

Hence, $\varphi(A^2) = 0$ or $\varphi(B^2) = 0$. So, X or Y has variance 0, which implies that the distribution of X or of Y has to be a point measure.

S 4.7.7. Let φ be an expectation. Then φ is uniquely determined by the numbers $c_{j,k} = \varphi(E_{j,k})$, where $E_{j,k}$ is the matrix which contains only 0s except for the (j,k)-entry, which is 1. Write $X = \sum_{j,k} x_{j,k} E_{j,k}$. Then

$$\varphi(X) = \sum_{j,k} x_{j,k} \varphi(E_{j,k}) = \sum_{j,k} x_{j,k} c_{j,k}.$$

Define $\rho = \sum_{j,k} c_{j,k} e_k \overline{e_j}^T$, where e_k is the k-th unit vector in $\mathbb{C}^n$. Then $c_{j,k} = \overline{e_k}^T \rho e_j$ and

$$\varphi(X) = \sum_{j,k} x_{j,k} c_{j,k} = \sum_{j,k} \overline{e_j}^T X e_k \overline{e_k}^T \rho e_j = \mathrm{Tr}(X\rho).$$

As $\varphi(I) = 1$, we see that $\mathrm{Tr}(\rho) = 1$. As $\mathrm{Tr}(\rho v \overline{v}^T) \geq 0$ for every $v \in \mathbb{C}^N$, we see that $\langle v, \rho v \rangle \geq 0$. Thus, ρ is non-negative, and consequently, ρ is a density matrix.

S 4.7.8. We have

$$\mathbb{E}[\mathrm{Tr}(B_N^n)/N] = \mathbb{E}\left[\int_{\mathbb{R}} x^n \mu_N(dx)\right] = \int_{\mathbb{R}} x^n \alpha_N(dx) \to \frac{1}{2\pi}\int_{-2}^{2} x^n \sqrt{4-x^2}\,dx.$$

If n is odd, we integrate an odd function and the result is 0. Now let $n = 2k$. Then

$$\frac{1}{2\pi}\int_{-2}^{2} x^{2k}\sqrt{4-x^2}\,dx = \frac{1}{\pi}\int_{0}^{2} x^{2k}\sqrt{4-x^2}\,dx =$$

$$\frac{1}{\pi}\int_0^{\pi/2} (2\sin(t))^{2k} \cdot \sqrt{4-4\sin(t)^2} \cdot 2\cos(t)dt = \frac{2^{2k+2}}{\pi}\int_0^{\pi/2} \sin(t)^{2k}\cos(t)^2 dt$$

$$= \frac{2^{2k+2}}{\pi}\left(\int_0^{\pi/2} \sin(t)^{2k} dt - \int_0^{\pi/2} \sin(t)^{2k+2} dt\right)$$

$$= \frac{2^{2k+2}}{\pi}\left(\frac{\pi}{2}4^{-k}\binom{2k}{k} - \frac{\pi}{2}4^{-k-1}\binom{2k+2}{k+1}\right) = 2\binom{2k}{k} - \frac{1}{2}\binom{2k+2}{k+1}$$

$$= \binom{2k}{k}\left(2 - \frac{(2k+2)(2k+1)k!}{2(k+1)(k+1)!}\right) = \binom{2k}{k}\left(2 - \frac{2k+1}{k+1}\right) = \frac{1}{k+1}\binom{2k}{k}.$$

CHAPTER 5

Convolutions and Additive Processes

In this chapter, we will examine the convolution induced by monotone independence. We prove the monotone central limit theorem and show that monotonically infinitely divisible distributions can be characterized by certain semigroups of holomorphic functions. We also discuss Boolean and free independence. In the case of free independence, we refer to the literature for proofs.

Furthermore, we define additive processes as quantum stochastic processes with independent increments and use a certain class of classical Markov processes to construct additive processes in the case of monotone independence.

5.1. F-transform

In monotone probability theory, the role of the characteristic function is taken by the F-transform, which is the multiplicative inverse of the Cauchy transform.

DEFINITION 5.1. For $\mu \in \mathcal{P}(\mathbb{R})$, the F-transform F_μ is defined as the holomorphic function

$$F_\mu : \mathbb{H} \to \mathbb{H}, \quad F_\mu(z) = \frac{1}{G_\mu(z)} = \left(\int_{\mathbb{R}} \frac{1}{z-x} \mu(dx) \right)^{-1}.$$

EXAMPLE 5.2. For $\mu = \frac{1}{2}\delta_a + \frac{1}{2}\delta_b$, $a, b \in \mathbb{R}$, we obtain the rational function

$$F_\mu(z) = \frac{(z-a)(z-b)}{z-(a+b)/2}.$$

In case $a = b$, we have $F_{\delta_a}(z) = z - a$, which is a simple translation of the upper half-plane.

The arcsine distribution will play a special role for monotone independence.

EXAMPLE 5.3. The arcsine distribution $Arc(\mu, \sigma^2)$ with mean μ and variance σ^2 is given by the density (see Figure 5.1)

$$\frac{1}{\pi\sqrt{2\sigma^2 - (x-\mu)^2}}, \quad x \in \left(\mu - \sqrt{2}\sigma, \mu + \sqrt{2}\sigma \right).$$

Similarly to the semicircle distribution, see Exercise 3.9.3, we can show that

$$G_{Arc(\mu,\sigma^2)}(z) = \frac{1}{\sqrt{(z-\mu)^2 - 2\sigma^2}},$$

and thus,

$$F_{Arc(\mu,\sigma^2)}(z) = \sqrt{(z-\mu)^2 - 2\sigma^2}.$$

This function is a conformal mapping from $\mathbb{H}$ onto $\mathbb{H}$ minus the vertical line segment from 0 to $\sqrt{2}\sigma i$, see Figure 5.2.

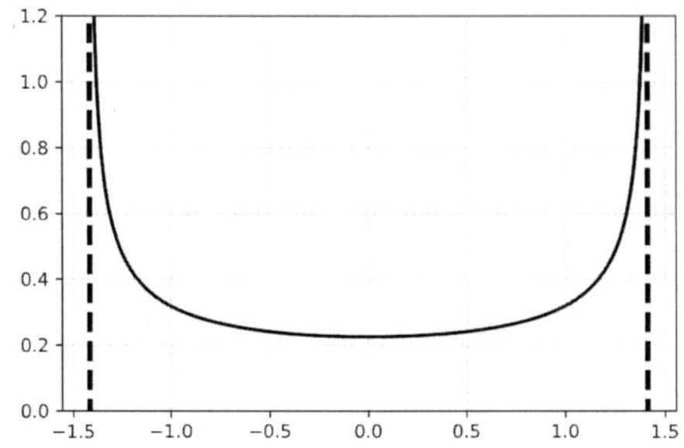

FIGURE 5.1. Density of the arcsine distribution $Arc(0,1)$.

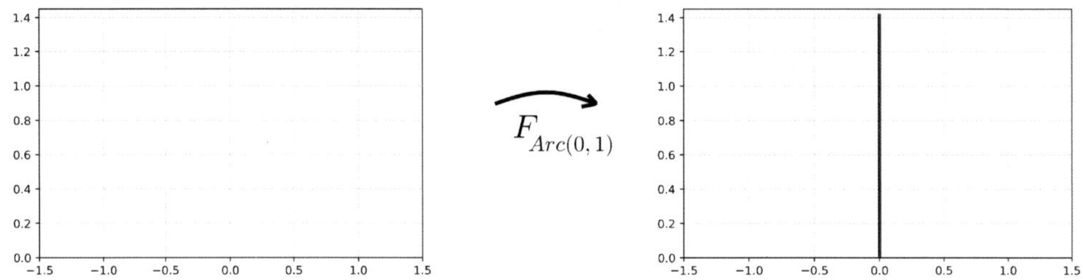

FIGURE 5.2. F-transform of the arcsine distribution $Arc(0,1)$.

REMARK 5.4. All odd moments of $Arc(0,1)$ are clearly equal to 0, and for even moments, we obtain, by using (4.11):

$$\int_{-\sqrt{2}}^{\sqrt{2}} \frac{x^{2n}}{\pi\sqrt{2-x^2}}dx = 2\int_0^{\sqrt{2}} \frac{x^{2n}}{\pi\sqrt{2-x^2}}dx \underbrace{=}_{x=\sqrt{2}\sin(y)} \frac{2}{\pi}2^n \int_0^{\pi/2} \sin(y)^{2n}dy$$

$$= \frac{2}{\pi}2^n\frac{\pi}{2}4^{-n}\binom{2n}{n} = 2^{-n}\binom{2n}{n} = \frac{(2n)!}{2^n n!}\cdot\frac{1}{n!} = \frac{1\cdot 3\cdot 5\cdot\dots\cdot(2n-1)}{n!}.$$

EXAMPLE 5.5. Let $\mu = W(0,1)$ be the standard semicircle distribution. Then Exercise 3.9.3 shows that

$$F_\mu(z) = \frac{2}{z-\sqrt{z^2-4}}.$$

As F_μ extends continuously to $\mathbb{H}\cup\mathbb{R}$, $F_\mu(\mathbb{H})$ is the unbounded complement of the curve $\{F_\mu(x)\,|\,x\in(-2,2)\}$ in $\mathbb{H}$. The solution to Exercise 3.9.3 shows that

$$\mathrm{Re}(G_\mu(x)) = \tfrac{x}{2} \text{ and } \mathrm{Im}(G_\mu(x)) = \tfrac{-\sqrt{4-x^2}}{2} \text{ for } x\in[-2,2].$$

Thus, for $x\in[-2,2]$, we have

$$F_\mu(x) = \frac{1}{\frac{x}{2}+i\frac{-\sqrt{4-x^2}}{2}} = \frac{2}{x-i\sqrt{4-x^2}} = \frac{2x+2i\sqrt{4-x^2}}{4} = \frac{x}{2}+i\sqrt{1-\left(\frac{x}{2}\right)^2},$$

and we see that the curve is the semicircle $(\partial\mathbb{D})\cap\mathbb{H}$, see Figure 5.3.

F-transforms of probability measures can be characterized as follows.

THEOREM 5.6. *Let $F:\mathbb{H}\to\mathbb{H}\cup\mathbb{R}$ be holomorphic. Then the following statements are equivalent:*

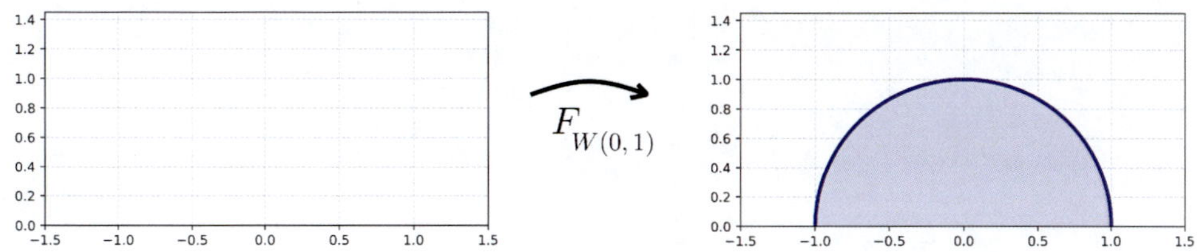

FIGURE 5.3. F-transform of the semicircle distribution $W(0,1)$.

(a) *There exists a probability measure* μ *on* $\mathbb{R}$ *such that* $F = F_\mu$.
(b) F *has the Nevanlinna representation*

$$F(z) = z + a + \int_{\mathbb{R}} \frac{1 + xz}{x - z} \gamma(dx), \tag{5.1}$$

where $a \in \mathbb{R}$ *and* γ *is a finite, non-negative measure on* $\mathbb{R}$.
(c) $\lim_{y\to\infty} \frac{F(iy)}{iy} = 1$.
(d) F *is the identity, or the Denjoy–Wolff point of* F *is parabolic and at* ∞.

If these equivalent conditions hold, then $\operatorname{Im}(F(z)) \geq \operatorname{Im}(z)$ *for all* $z \in \mathbb{H}$.

PROOF. The equivalence between (a) and (c) follows from Theorem 3.15, the equivalence between (b) and (c) follows from Theorem 3.12, and the equivalence between (c) and (d) follows from Corollary 3.20 and Remark 3.21.

Finally, if $F = F_\mu$, then Corollary 3.16 implies that $\operatorname{Im}(F(z)) \geq \operatorname{Im}(z)$ for all $z \in \mathbb{H}$. □

In the case of finite first and second moments, F_μ has another useful representation. These functions were studied, for example, in [**127**] in the context of free probability theory and, in the same year, in [**75**] in the context of Loewner theory.

THEOREM 5.7. *Let* $\mu \in \mathcal{P}(\mathbb{R})$ *and assume that the first and second moment of* μ *exist. Then* F_μ *can be represented as*

$$F_\mu(z) = z - a - \int_{\mathbb{R}} \frac{1}{z - x} \tau(dx), \tag{5.2}$$

where $a \in \mathbb{R}$ *is the first moment of* μ *and* τ *is a non-negative measure on* $\mathbb{R}$ *with* $\tau(\mathbb{R}) = \operatorname{Var}(\mu)$. *Conversely, if there exists* $a \in \mathbb{R}$ *and a finite, non-negative measure* τ *such that* (5.2) *holds, then the first and second moment of* μ *exist.*

Furthermore, μ *has compact support if and only if* τ *has compact support.*

PROOF. First, assume that the first and second moments m_1 and m_2 of μ exist. If $\operatorname{Var}(\mu) = 0$, then the statement is clearly true. So assume that $\operatorname{Var}(\mu) = m_2 - m_1^2 > 0$. Then

$$\begin{aligned} y^3 \left[G_\mu(iy) - \frac{1}{iy} - \frac{m_1}{(iy)^2} - \frac{m_2}{(iy)^3} \right] &= y^3 \left[\int_{\mathbb{R}} \frac{1}{iy - x} - \frac{1}{iy} - \frac{x}{(iy)^2} - \frac{x^2}{(iy)^3} \mu(dx) \right] \\ &= \int_{\mathbb{R}} \frac{x^3 y}{x^2 + y^2} - \frac{ix^4}{x^2 + y^2} \mu(dx) \to 0 \end{aligned}$$

as $y \to \infty$. So

$$iy\left(iy - m_1 - \frac{1}{G_\mu(iy)}\right) = iy\frac{iyG_\mu(iy) - m_1G_\mu(iy) - 1}{G_\mu(iy)}$$

$$= \frac{\frac{m_2-m_1^2}{iy} + o(1/y)}{\frac{1}{iy} + \frac{m_1}{(iy)^2} + \frac{m_2}{(iy)^3} + o(1/y^3)} = m_2 - m_1^2 + o(1).$$

Due to Theorem 5.6, the function $z - m_1 - 1/G_\mu(z)$ maps $\mathbb{H}$ into $\mathbb{H}^- \cup \mathbb{R}$. Theorem 3.15 implies that $(z - m_1 - \frac{1}{G_\mu(z)})/(m_2 - m_1^2)$ is the Cauchy transform of a probability measure $\hat{\tau}$. We put $\tau = (m_2 - m_1^2) \cdot \hat{\tau} = \mathrm{Var}(\mu) \cdot \hat{\tau}$ and obtain (5.2).

Conversely, assume that F_μ has the form (5.2) and that $\tau(\mathbb{R}) > 0$. Then

$$(iy)^3\left[G_\mu(iy) - \frac{1}{iy} - \frac{a}{(iy)^2} - \frac{\tau(\mathbb{R}) + a^2}{(iy)^3}\right]$$

$$= \frac{(iy)^3}{iy - a - \int_\mathbb{R}\frac{1}{iy-x}\tau(dx)} - (iy)^2 - aiy - (\tau(\mathbb{R}) + a^2)$$

$$= \frac{(iy)^2}{1 - \left(a/iy + \int_\mathbb{R}\frac{1}{iy(iy-x)}\tau(dx)\right)} - (iy)^2 - aiy - (\tau(\mathbb{R}) + a^2)$$

$$= (iy)^2\left(1 + a/iy + \int_\mathbb{R}\frac{1}{iy(iy-x)}\tau(dx) + a^2/(iy)^2 + o(1/y^2)\right)$$

$$- (iy)^2 - aiy - (\tau(\mathbb{R}) + a^2)$$

$$= \int_\mathbb{R}\frac{iy}{iy - x} - 1\tau(dx) + o(1) = \int_\mathbb{R}\frac{x}{iy - x}\tau(dx) + o(1) \to 0$$

as $y \to \infty$. Let $m = \tau(\mathbb{R}) + a^2$. We have

$$\mathrm{Re}\left[(iy)^3\left(G_\mu(iy) - \frac{1}{iy} - \frac{a}{(iy)^2} - \frac{m}{(iy)^3}\right)\right] = \int_\mathbb{R}\left(\frac{x^2}{1 + x^2/y^2} - m\right)\mu(dx) \to 0.$$

The function $x \mapsto \frac{x^2}{1+x^2/y^2}$ is monotonically increasing with respect to y, and the monotone convergence theorem implies that

$$m = \lim_{y\to\infty}\int_\mathbb{R}\frac{x^2}{1 + x^2/y^2}\mu(dx) = \int_\mathbb{R} x^2\mu(dx).$$

As $m < \infty$, the second, and thus, also the first moment of μ exists.

Finally, μ has compact support if and only if F_μ extends analytically to ∞, which is equivalent to τ having compact support. □

REMARK 5.8. By Theorem 3.15, we conclude that $\mu \in \mathcal{P}(\mathbb{R})$ has compact support with mean 0 and variance σ^2 if and only if F_μ extends analytically to ∞ with the expansion

$$F_\mu(z) = z - \frac{\sigma^2}{z} + \ldots$$

For $\mu \in \mathcal{P}_c(\mathbb{R})$, the F-transform F_μ extends analytically to ∞ with power series expansion $F_\mu(z) = z + a_0 + \frac{a_1}{z} + \ldots$, so F_μ can be inverted in a neighborhood of

∞. In general, we only have $F_\mu(\infty) = \infty$ and $F'_\mu(\infty) \neq 0$ in the sense of a non-tangential approach to ∞. However, in this general case, we can invert F_μ in a "non-tangential neighborhood" of ∞.

We follow [**132**, Section 3.6]. For $\alpha, \beta > 0$, define

$$\Gamma_{\alpha,\beta} = \{z \in \mathbb{H} \,|\, |\mathrm{Re}(z)| < \alpha \,\mathrm{Im}(z), |z| > \beta\}, \quad \text{see Figure 5.4.}$$

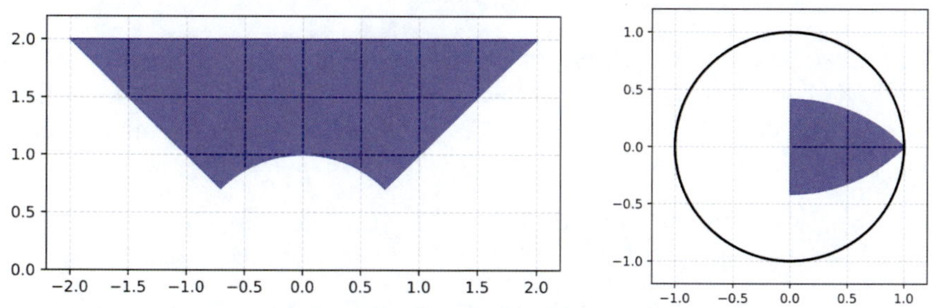

FIGURE 5.4. The domain $\Gamma_{\alpha,\beta}$ for $\alpha = \beta = 1$ (left) and its image in $\mathbb{D}$ with respect to the Cayley transform (right).

LEMMA 5.9. *Let $\mu \in \mathcal{P}(\mathbb{R})$ and let $0 < \alpha_1 < \alpha_2$, $\beta_2 > 0$. Then there exists $\beta_1 > \beta_2$ such that $F_\mu(\Gamma_{\alpha_1,\beta_1}) \subset \Gamma_{\alpha_2,\beta_2}$.*

PROOF. Consider the domain $\Gamma_{\alpha_1,\beta_2+1}$, which is a subset of $\Gamma_{\alpha_2,\beta_2}$. We find $\varepsilon > 0$ such that all $z \in \partial\Gamma_{\alpha_1,\beta_2+1}$ with $|z| = \beta_2 + 1$ satisfy $\{w \in \mathbb{C} \,|\, |w - z| < \varepsilon(\beta_2+1) = \varepsilon|z|\} \subset \Gamma_{\alpha_2,\beta_2}$. By scaling with $r \geq 1$, we obtain that $\{w \in \mathbb{C} \,|\, |w-z| < \varepsilon|z|\} \subset \Gamma_{\alpha_2,\beta_2}$ for all $z \in \overline{\Gamma_{\alpha_1,\beta_2+1}}$.

Next, there exists $\beta_1 > \beta_2 + 1$ such that $|F_\mu(z) - z| < \varepsilon|z|$ for all $z \in \Gamma_{\alpha_1,\beta_1}$, due to Theorem 3.13. Thus $F_\mu(\Gamma_{\alpha_1,\beta_1}) \subset \Gamma_{\alpha_2,\beta_2}$. □

LEMMA 5.10. *Let $\mu \in \mathcal{P}(\mathbb{R})$ and $\alpha_1 > 0$. Then there exist $\alpha_2, \beta_1, \beta_2 > 0$ such that $\Gamma_{\alpha_2,\beta_2} \subset F_\mu(\Gamma_{\alpha_1,\beta_1})$, and for each $w \in \Gamma_{\alpha_2,\beta_2}$, there exists a unique $z \in \Gamma_{\alpha_1,\beta_1}$ with $F_\mu(z) = w$.*

PROOF. Denote by L_1 and L_2 the two half-lines of $\partial\Gamma_{\alpha_1,1}$. Let $d(z)$ be the distance of z to the imaginary axis. Then $d(z) = c|z|$ for a constant $c > 0$ and all $z \in L_1 \cup L_2$. So $|z - iy| \geq c|z|$ for all $z \in L_1 \cup L_2$ and all $y > 0$.

We find $\alpha_2 > 0$, with $\alpha_2 < \alpha_1$, such that $|z - w| > \frac{c}{2}|z|$ for all $z \in L_1 \cup L_2$ and all $w \in \Gamma_{\alpha_2,1}$.

Let $\varepsilon \in (0,1)$ with $\varepsilon < \frac{c}{2}$. Due to Theorem 3.13, we find $\beta_1 > 1$ such that $|F_\mu(z) - z| < \varepsilon|z|$ for all $z \in \overline{\Gamma_{\alpha_1,\beta_1}}$.

Finally, we find $\beta_2 > \beta_1$ such that $|z - w| > \frac{c}{2}|z|$ for all $z \in \partial\Gamma_{\alpha_1,\beta_1}$ and all $w \in \Gamma_{\alpha_2,\beta_2}$.

Fix $w \in \Gamma_{\alpha_2,\beta_2}$. Then we find $r > \beta_2$ large enough such that $|z - w| > \frac{c}{2}|z|$ for all z on the closed curve $C = \partial(\Gamma_{\alpha_1,\beta_1} \cap r\mathbb{D})$, and the point w belongs to the interior of C. For all $z \in C$, we have

$$|F_\mu(z) - w - (z - w)| < \varepsilon|z| < \frac{c}{2}|z| < |z - w|.$$

Rouché's theorem implies that there exists exactly one z_0 in the interior of C with $F_\mu(z_0) = w$. We can take r arbitrarily large, so z_0 is the only point within $\Gamma_{\alpha_1,\beta_1}$ with $F_\mu(z_0) = w$. □

Hence, we can define a right compositional inverse of F_μ on some $\Gamma_{\alpha_2,\beta_2}$, i.e., $(F_\mu \circ F_\mu^{-1})(z) = z$ for all $z \in \Gamma_{\alpha_2,\beta_2}$. Let $\alpha = \alpha_2/2$. By Lemma 5.9, we find $\beta > \beta_2$ such that $F_\mu(\Gamma_{\alpha,\beta}) \subset \Gamma_{\alpha_2,\beta_2}$. So F_μ^{-1} is the left compositional inverse of F_μ restricted to $\Gamma_{\alpha,\beta}$. We have:

$$(F_\mu \circ F_\mu^{-1})(z) = z \quad \text{and} \quad (F_\mu^{-1} \circ F_\mu)(z) = z \quad \text{for all } z \in \Gamma_{\alpha,\beta}. \tag{5.3}$$

5.2. Monotone convolution

Let X be a random variable in a quantum probability space with distribution $\mu \in \mathcal{P}_c(\mathbb{R})$. For $z \in \mathbb{H}$ with $|z| > \|X\|$ large enough, the series

$$Y = \frac{1}{z}\sum_{n=0}^{\infty}(X/z)^n$$

converges and $(zI - X)Y = Y(zI - X) = I$. So $(zI - X)^{-1} = Y$.[1] Thus,

$$\begin{aligned} \varphi((zI - X)^{-1}) &= \frac{1}{z}\sum_{n=0}^{\infty}\varphi(X^n)/z^n = \frac{1}{z}\sum_{n=0}^{\infty}\int_{\mathbb{R}} x^n\mu(dx)/z^n \\ &= \int_{\mathbb{R}} \frac{1}{z-x}\mu(dx) = G_\mu(z) \end{aligned} \tag{5.4}$$

for all $z \in \mathbb{H}$ with $|z| > \|X\|$.

THEOREM 5.11 (Monotone convolution). *Let (X, Y) be monotonically independent with distributions $\mu, \nu \in \mathcal{P}_c(\mathbb{R})$. Let α be the distribution of $X + Y$. Then*

$$F_\alpha(z) = (F_\mu \circ F_\nu)(z) \quad \textit{for all } z \textit{ in } \mathbb{H}.$$

PROOF. For $z \in \mathbb{H}$ with $|z|$ large enough, we have

$$(z - (X + Y))^{-1} = \sum_{n=0}^{\infty}(X + Y)^n/z^{n+1}$$

$$= \sum_{n=0}^{\infty}\sum_{k=0}^{n}\sum_{\substack{j_0+j_1+\cdots+j_k=n-k \\ 0\le j_0,\ldots,j_k}} Y^{j_0}XY^{j_1}X\ldots XY^{j_k}/z^{n+1}.$$

[1]For $X \in B(H)$, the spectrum $\sigma(X)$ of X is defined as the set of all $z \in \mathbb{C}$ for which $X - zI$ does not have an inverse (in $B(H)$). One can show that $\sigma(X) \subset \mathbb{R}$ if X is self-adjoint. So $X - zI$ is invertible for every $z \in \mathbb{H}$ in this case.

Now we take the expectation and use the monotone independence of (X, Y):

$$
\begin{aligned}
G_\alpha(z) &= \varphi((z-(X+Y))^{-1}) \\
&= \sum_{n=0}^{\infty}\sum_{k=0}^{n} \sum_{\substack{j_0+j_1+\cdots+j_k=n-k \\ 0\le j_0,\dots,j_k}} \varphi(Y^{j_0}XY^{j_1}X\cdots XY^{j_k})/z^{n+1} \\
&= \sum_{n=0}^{\infty}\sum_{k=0}^{n} \sum_{\substack{j_0+j_1+\cdots+j_k=n-k \\ 0\le j_0,\dots,j_k}} \varphi(X^k)\varphi(Y^{j_0})\cdots\varphi(Y^{j_k})/z^{n+1}. \qquad (5.5)
\end{aligned}
$$

Furthermore, we have

$$
G_\nu(z)(1-G_\nu(z)X)^{-1} = \sum_{k=0}^{\infty} G_\nu(z)^{k+1}X^k = \sum_{k=0}^{\infty}\left(\sum_{n=0}^{\infty}\varphi(Y^n)/z^{n+1}\right)^{k+1} X^k.
$$

Taking the expectation yields $G_\mu(1/G_\nu(z))$, so
(5.6)

$$
G_\mu(1/G_\nu(z)) = \varphi(G_\nu(z)(1-G_\nu(z)X)^{-1}) = \sum_{k=0}^{\infty}\left(\sum_{n=0}^{\infty}\varphi(Y^n)/z^{n+1}\right)^{k+1} \varphi(X^k).
$$

By comparing the coefficients in (5.5) and (5.6), we see that both sums are identical. Note that all the expansions only hold when $|z|$ is large enough. So $G_\alpha(z) = G_\mu(1/G_\nu(z))$ in a neighborhood of ∞. The identity theorem for holomorphic functions implies that $G_\alpha(z) = G_\mu(1/G_\nu(z))$ for all $z \in \mathbb{H}$. □

Let $\mu, \nu \in \mathcal{P}(\mathbb{R})$. Then $F_\mu \circ F_\nu$ is again an F-transform, which can be seen as follows. Due to Exercise 3.9.5, the curve $y \mapsto F_\nu(iy)$ belongs to $\{x+iy \in \mathbb{H} \,|\, |x| < Cy\}$ for all y large enough and some $C>0$, and $\mathrm{Im}(F_\nu(iy)) \to \infty$. With Theorem 3.13, we obtain

$$
\lim_{y\to\infty} \frac{(F_\mu\circ F_\nu)(iy)}{iy} = \lim_{y\to\infty} \frac{F_\mu(F_\nu(iy))}{F_\nu(iy)}\frac{F_\nu(iy)}{iy} = 1\cdot 1 = 1, \qquad (5.7)
$$

and Theorem 5.6 implies that $F_\mu \circ F_\nu$ is an F-transform. We can thus make the following definition.

DEFINITION 5.12 (Monotone convolution). For $\mu, \nu \in \mathcal{P}(\mathbb{R})$ we define the *monotone convolution* $\mu \rhd \nu \in \mathcal{P}(\mathbb{R})$ via

$$
F_{\mu\rhd\nu} = F_\mu \circ F_\nu.
$$

A probability measure $\mu \in \mathcal{P}(\mathbb{R})$ is called *monotonically infinitely divisible* if, for every $n \in \mathbb{N}$, there exists $\mu_n \in \mathcal{P}(\mathbb{R})$ such that $\mu = \mu_n \rhd \cdots \rhd \mu_n$ (n-fold convolution).

The following statements are clear for compactly supported measures if we think of adding monotonically independent random variables.

THEOREM 5.13. *Let $\mu_1, \mu_2 \in \mathcal{P}(\mathbb{R})$ with finite variances and first moments a_1 and a_2. Then $\mu_1 \rhd \mu_2$ has the finite variance* $\mathrm{Var}(\mu_1)+\mathrm{Var}(\mu_2)$ *and its first moment is $a_1 + a_2$.*

PROOF. Let τ_1 and τ_2 be the measures from Theorem 5.7 corresponding to μ_1 and μ_2. Then

$$\begin{aligned}(F_{\mu_1} \circ F_{\mu_2})(z) &= F_{\mu_2}(z) - a_1 - \int_{\mathbb{R}} \frac{1}{F_{\mu_2}(z) - x} \tau_1(dx) \\ &= z - (a_1 + a_2) - \int_{\mathbb{R}} \frac{1}{z-x} \tau_2(dx) - \int_{\mathbb{R}} \frac{1}{F_{\mu_2}(z) - x} \tau_1(dx).\end{aligned}$$

We may assume that $\tau_1(\mathbb{R}) > 0$. Then the last integral is equal to

$$\tau_1(\mathbb{R}) G_{\tau_1/\tau_1(\mathbb{R})} \circ F_{\mu_2} = \tau_1(\mathbb{R}) G_{\tau_1/\tau_1(\mathbb{R}) \rhd \mu_2}.$$

With $\rho := \tau_1(\mathbb{R})(\tau_1/\tau_1(\mathbb{R}) \rhd \mu_2)$, we have

$$(F_{\mu_1} \circ F_{\mu_2})(z) = z - (a_1 + a_2) - \int_{\mathbb{R}} \frac{1}{z-x} [\tau_2 + \rho](dx).$$

By Theorem 5.7, $\mu_1 \rhd \mu_2$ has finite variance $\mathrm{Var}(\mu_1 \rhd \mu_2) = \tau_2(\mathbb{R}) + \rho(\mathbb{R}) = \tau_2(\mathbb{R}) + \tau_1(\mathbb{R}) = \mathrm{Var}(\mu_2) + \mathrm{Var}(\mu_1)$, and the first moment of $\mu_1 \rhd \mu_2$ is given by $a_1 + a_2$. □

EXAMPLE 5.14. Let us take $\mu = \nu = \frac{1}{2}\delta_{-1} + \frac{1}{2}\delta_1$. The classical convolution yields

$$\mu * \mu = \frac{1}{4}\delta_{-2} + \frac{1}{2}\delta_0 + \frac{1}{4}\delta_2.$$

Let us compare this to monotone convolution. We have $G_\mu(z) = \frac{1}{2(z+1)} + \frac{1}{2(z-1)}$ and

$$F_\mu(z) = z - \frac{1}{z}.$$

Thus, we obtain

$$F_{\mu \rhd \mu}(z) = F_\mu(F_\mu(z)) = \frac{(z^2 - z - 1)(z^2 + z - 1)}{z(z^2 - 1)},$$

$$G_{\mu \rhd \mu}(z) = \frac{z(z^2 - 1)}{(z^2 - z - 1)(z^2 + z - 1)}.$$

The poles of $G_{\mu \rhd \mu}$ are given by $\pm(\frac{\sqrt{5}}{2} - \frac{1}{2}) = \pm 0.61\ldots$ and $\pm(\frac{\sqrt{5}}{2} + \frac{1}{2}) = \pm 1.61\ldots$, and thus,

$$\mu \rhd \mu = m_1 \delta_{-\sqrt{5}/2 - 1/2} + m_2 \delta_{-\sqrt{5}/2 + 1/2} + m_2 \delta_{\sqrt{5}/2 - 1/2} + m_1 \delta_{\sqrt{5}/2 + 1/2}$$

with masses $m_1 = \frac{1}{4} + \frac{\sqrt{5}}{20} = 0.36\ldots$ and $m_2 = \frac{1}{4} - \frac{\sqrt{5}}{20} = 0.13\ldots$

EXAMPLE 5.15. For $s, t > 0$, let $\mu = Arc(0, s)$ and $\nu = Arc(0, t)$. Then, by Example 5.3,

$$F_{\mu \rhd \nu}(z) = F_\mu(F_\nu(z)) = \sqrt{(\sqrt{z^2 - 2t})^2 - 2s} = \sqrt{z^2 - 2(s+t)},$$

and we have $\mu \rhd \nu = Arc(0, s+t)$. Furthermore, $Arc(0, t)$ is monotonically infinitely divisible, since $Arc(0, t) = Arc(0, t/n) \rhd \cdots \rhd Arc(0, t/n)$ for every $n \in \mathbb{N}$.

EXAMPLE 5.16. Let us consider the convolution with a point measure. For $x \in \mathbb{R}$, we have

$$F_{\delta_x \rhd \mu}(z) = F_\mu(z) - x \quad \text{and} \quad F_{\mu \rhd \delta_x}(z) = F_\mu(z - x).$$

While $\mu \rhd \delta_x$ simply shifts the measure μ, the convolution $\delta_x \rhd \mu$ has a quite different effect in general.

Assume that the first and second moment of μ exist. Let m_1 be the first moment and let τ be the measure from representation (5.2). For $x = -m_1$, we get

$$F_{\delta_x \triangleright \mu}(z) = F_\mu(z) + m_1 = z - \int_{\mathbb{R}} \frac{1}{z-x}\tau(dx) \quad \text{and}$$

$$F_{\mu \triangleright \delta_x}(z) = F_\mu(z+m_1) = z - \int_{\mathbb{R}} \frac{1}{z-(x-m_1)}\tau(dx) = z - \int_{\mathbb{R}} \frac{1}{z-x}\tau'(dx),$$

with $\tau'(B) = \tau(B+m_1)$.

EXAMPLE 5.17. We compute the convolution $\delta_x \triangleright Arc(0,4)$ for $x=-1$. We have $F_{\delta_{-1}\triangleright Arc(0,4)}(z) = \sqrt{z^2-8}+1$. This function extends continuously to $\mathbb{H}\cup\mathbb{R}$ and it has a zero at $z=-3$. (The square root is chosen such that $\sqrt{x^2-8}<0$ when $x<-\sqrt{8}$.) So $\delta_{-1}\triangleright Arc(0,4)$ has an atom at $x=-3$, see Figure 5.5.

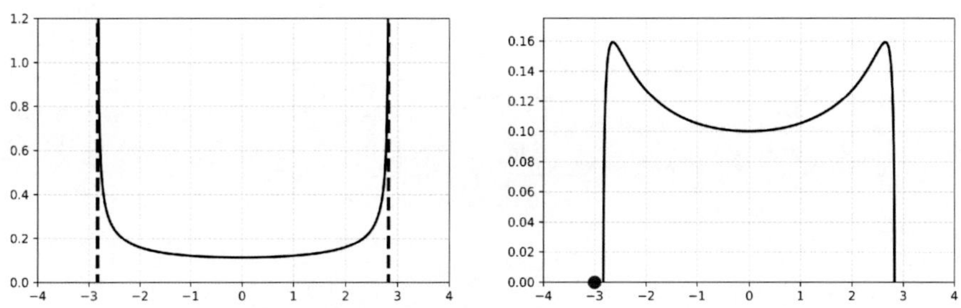

FIGURE 5.5. Left: The density of the arcsine distribution $Arc(0,4)$. Right: the distribution $\delta_{-1} \triangleright Arc(0,4)$ with density in $[-2\sqrt{2}, 2\sqrt{2}]$ and its atom at -3.

LEMMA 5.18. *Let $\mu \in \mathcal{P}(\mathbb{R})$. For $\lambda > 0$, denote by μ_λ the distribution $\mu(\cdot/\lambda)$. Then*

$$F_{\mu_\lambda}(z) = \lambda F_\mu(z/\lambda) \quad \textit{for all } z \in \mathbb{H}.$$

PROOF. We have $G_{\mu_\lambda}(z) = G_{\mu_1}(z/\lambda)/\lambda$, and thus, $F_{\mu_\lambda}(z) = \lambda F_{\mu_1}(z/\lambda)$. □

We can now prove the central limit theorem for monotone independence, which has been proven by Muraki in [**140**]. We follow the proof of [**193**, Theorem 3.1, (2)→(1)].

THEOREM 5.19 (Monotone central limit theorem). *Let $X_1, X_2, \ldots$ be a sequence of monotonically independent identically distributed random variables with mean c and variance $\sigma^2 > 0$. Let μ_n be the distribution of $S_n = (X_1+\cdots+X_n-nc)/(\sigma\sqrt{n})$. Then μ_n converges weakly to the arcsine distribution $Arc(0,1)$, given by the density*

$$\frac{1}{\pi\sqrt{2-x^2}}, \quad x \in (-\sqrt{2}, \sqrt{2}).$$

PROOF. Let $Y = (X_1 - c)/\sigma$ and denote its distribution by ν. By Lemma 5.18 and Theorem 5.11, we have

$$F_{\mu_n}(z) = \frac{1}{\sqrt{n}} F_{\nu^{\triangleright n}}(\sqrt{n}z) = \left(\frac{1}{\sqrt{n}} F_\nu\left(\sqrt{n}z\right)\right)^{\circ n}.$$

We need to show that $F_{\mu_n}(z) \to \sqrt{z^2-2}$ by Example 5.3 and Theorem 3.18. This can also be stated as

$$\psi_1 \circ F_{\mu_n} \circ \psi_2 \to z - 2$$

with $\psi_1(z) = z^2$ and $\psi_2(z) = \sqrt{z}$. Let $F_n(z) := \frac{1}{\sqrt{n}} F_\nu(\sqrt{n}z)$ and write $(\psi_1 \circ F_n \circ \psi_2)(z) = z + R_n(\psi_2(z))$, $z \in \mathbb{C} \setminus [0,\infty)$. We will show that

$$\lim_{n\to\infty} \sum_{j=0}^{n-1} R_n(F_n^{\circ j}(iy)) = -2 \tag{5.8}$$

for all $y > 1$. Vitali's theorem implies locally uniform convergence of the sum in $\mathbb{H}$, and hence,

$$\begin{aligned}\psi_1 \circ F_{\mu_n} \circ \psi_2 &= \psi_1 \circ F_n^{\circ n} \circ \psi_2 = (\psi_1 \circ F_n \circ \psi_2)^{\circ n} \\ &= (z + R_n(\psi_2(z)))^{\circ n} = z + \sum_{j=0}^{n-1} R_n(F_n^{\circ j}(\sqrt{z})) \to z - 2.\end{aligned}$$

By Theorem 5.7, we can write $F_\nu(z) = z + \int_{\mathbb{R}} \frac{1}{x-z}\tau(dx)$ for a finite, non-negative measure τ with compact support and $\tau(\mathbb{R}) = 1$ (the variance of ν). We have

$$F_n(z) = z + \frac{1}{\sqrt{n}} \int_{\mathbb{R}} \frac{1}{x - z\sqrt{n}} \tau(dx),$$

and for $z \in \mathbb{H}$, we obtain

$$\begin{aligned}|F_n(z) - z| &= \left| \frac{1}{\sqrt{n}} \int_{\mathbb{R}} \frac{1}{x - z\sqrt{n}} \tau(dx) \right| \\ &\le \frac{1}{\sqrt{n}} \int_{\mathbb{R}} \left| \frac{1}{x - \sqrt{n}z} \right| \tau(dx) \le \frac{\tau(\mathbb{R})}{\sqrt{n}} \frac{1}{\operatorname{Im}(z\sqrt{n})} = \frac{1}{n \operatorname{Im}(z)}.\end{aligned}$$

We have

$$R_n(z) = F_n(z)^2 - z^2 = \frac{2}{\sqrt{n}} \int_{\mathbb{R}} \frac{z}{x - \sqrt{n}z} \tau(dx) + \left(\frac{1}{\sqrt{n}} \int_{\mathbb{R}} \frac{1}{x - \sqrt{n}z} \tau(dx) \right)^2.$$

Let $y > 1$. Then $\operatorname{Im}(F_n^{\circ j}(iy)) > 1$ for all $j = 0, \dots, n-1$ and

$$\sum_{j=0}^{n-1} \left(\frac{1}{\sqrt{n}} \int_{\mathbb{R}} \frac{1}{x - \sqrt{n} F_n^{\circ j}(iy)} \tau(dx) \right)^2 \le n \cdot \frac{1}{n^2} \to 0 \tag{5.9}$$

as $n \to \infty$.

Let $y > 1$ and put $z_j = F_n^{\circ j}(iy)$, $j = 0, \dots, n-1$. Then also $\operatorname{Im}(z_j) > 1$ for all $j = 0, \dots, n-1$. We have

$$\begin{aligned}&\left| 2 + \sum_{j=0}^{n-1} \frac{2}{\sqrt{n}} \int_{\mathbb{R}} \frac{z_j}{x - \sqrt{n}z_j} \tau(dx) \right| = \left| \sum_{j=0}^{n-1} \frac{2}{\sqrt{n}} \int_{\mathbb{R}} \frac{1}{\sqrt{n}} + \frac{z_j}{x - \sqrt{n}z_j} \tau(dx) \right| \\ &= \left| \sum_{j=0}^{n-1} \frac{2}{\sqrt{n}} \int_{\mathbb{R}} \frac{x/\sqrt{n}}{x - \sqrt{n}z_j} \tau(dx) \right| \le \sum_{j=0}^{n-1} \frac{2}{\sqrt{n}} \int_{\mathbb{R}} \frac{|x|/\sqrt{n}}{|x - \sqrt{n}z_j|} \tau(dx) \\ &\le \sum_{j=0}^{n-1} \frac{2}{\sqrt{n}} \int_{\mathbb{R}} \frac{|x|}{n} \tau(dx) = \frac{2}{\sqrt{n}} \int_{\mathbb{R}} |x| \tau(dx) \to 0\end{aligned}$$

as $n \to \infty$. This shows that

$$\sum_{j=0}^{n-1} \frac{2}{\sqrt{n}} \int_{\mathbb{R}} \frac{z_j}{x - \sqrt{n} z_j} \tau(dx) \to -2$$

and, together with (5.9), we conclude (5.8). □

It is worth noting that the monotone version of the central limit theorem yields a limiting distribution that differs significantly from the classical Gaussian distribution. Specifically, the arcsine law has compact support, meaning that the distribution of S_n converges to zero outside the interval $[-\sqrt{2}, \sqrt{2}]$.

In [**193**], a more general form of the monotone central limit theorem is established and subsequently applied to the ergodic theory of inner functions. Applications to quantum random variables are explored in [**93**, Sections 8.4–8.6], where the theorem is used in the spectral analysis of specific graphs, see also the book [**145**]. These applications include comb products of graphs (covered in Chapter 10) and so-called monotone trees, using monotone creation and annihilation operators acting on a Fock space. In [**116, 117**], Lenczewski investigates asymptotically monotonically independent random matrices.

Next, we consider semigroups of F-transforms, and prove results analogous to some of the equivalences in Theorem 2.63.

THEOREM 5.20. *Let $(\mu_t)_{t\geq 0} \subset \mathcal{P}(\mathbb{R})$ be a continuous $\triangleright$-semigroup, i.e.,*

- $\mu_0 = \delta_0$,
- $\mu_{s+t} = \mu_s \triangleright \mu_t$,
- *and $t \mapsto \mu_t$ is continuous.*

Then $(F_t)_{t\geq 0} := (F_{\mu_t})_{t\geq 0}$ is a continuous semigroup of holomorphic functions on $\mathbb{H}$ whose infinitesimal generator $G : \mathbb{H} \to \mathbb{H} \cup \mathbb{R}$ has the form

$$G(z) = a + \int_{\mathbb{R}} \frac{1 + xz}{x - z} \gamma(dx), \tag{5.10}$$

where $a \in \mathbb{R}$ and γ is a finite, non-negative measure. (Recall that $\frac{d}{dt}F_t = G(F_t)$.) Conversely, every generator of this form generates a semigroup of the form $(F_{\mu_t})_{t\geq 0}$ for a continuous $\triangleright$-semigroup $(\mu_t)_{t\geq 0}$.

PROOF. If $(\mu_t)_{t\geq 0}$ is a continuous $\triangleright$-semigroup, then $(F_{\mu_t})_{t\geq 0}$ is a continuous semigroup on $\mathbb{H}$. Lemma 3.30 implies that its generator G maps $\mathbb{H}$ into $\mathbb{H} \cup \mathbb{R}$. Theorem 3.27 (b) and Theorem 5.6 imply that G has a Nevanlinna triple of the form $(a, 0, \gamma)$, i.e., it is of the form (5.10).

Conversely, if $(F_t)_{t\geq 0}$ is a continuous semigroup generated by a G of this form, then Theorem 5.6 and Theorem 3.27 (b) imply that all F_t are F-transforms of probability measures $(\mu_t)_{t\geq 0}$. □

REMARK 5.21. The generators in (5.10) are exactly those holomorphic $f : \mathbb{H} \to \mathbb{H} \cup \mathbb{R}$ whose Nevanlinna triple has the form $(a, 0, \gamma)$. The pair (a, γ) can be regarded as the monotone analog of the Lévy triple, see Theorem 2.63. A more precise comparison will be given in Section 5.7. By Theorem 3.15, we see that, in particular, every $G(z) = \int_{\mathbb{R}} \frac{1}{x-z} \nu(dx)$, where $\nu \in \mathcal{P}(\mathbb{R})$, is a generator of the form (5.10).

THEOREM 5.22. *Let $\mu \in \mathcal{P}(\mathbb{R})$. The following statements are equivalent:*

(a) *There exists a continuous $\rhd$-semigroup $(\mu_t)_{t\geq 0} \subset \mathcal{P}(\mathbb{R})$ such that $\mu_1 = \mu$.*
(b) *μ is monotonically infinitely divisible.*

The semigroup in (a) is uniquely determined by μ.

PROOF. The measures appearing in continuous $\rhd$-semigroups are clearly monotonically infinitely divisible, for, if $\mu = \mu_1$ for a continuous $\rhd$-semigroup $(\mu_t)_{t\geq 0}$, then $\mu_1 = \mu_{1/n}^{\rhd n}$ for each $n \in \mathbb{N}$.

Conversely, if μ is monotonically infinitely divisible, then $\mu = \mu_{1/n}^{\rhd n}$ for some $\mu_n \in \mathcal{P}(\mathbb{R})$. One can show that $\mu_{1/n}$ is unique and that the definition $\mu_{m/n} := \mu_{1/n}^{\rhd m}$, $m, n \in \mathbb{N}$, and continuous extension to real numbers yield a $\rhd$-semigroup $(\mu_t)_{t\geq 0}$ with $\mu_1 = \mu$. This has been proven in [**139**, Proposition 5.4] for the case of compact support and the general case can be found in [**24**, Proposition 3.8]. We skip the rather technical proof, which involves solving the so-called Abel equation. (We note that the corresponding problem for the unit disk is easier to handle, see Theorem 7.13.) □

5.3. B-transform and Boolean convolution

DEFINITION 5.23. Let $\mu \in \mathcal{P}(\mathbb{R})$. The B-transform B_μ of μ is defined as

$$B_\mu(z) = z - \frac{1}{G_\mu(z)} = z - F_\mu(z).$$

EXAMPLE 5.24. Let $\mu = \delta_0$. Then $B_\mu(z) = 0$. A more important example is the distribution $\mu = \frac{1}{2}\delta_1 + \frac{1}{2}\delta_{-1}$, for which we obtain

$$B_\mu(z) = z - \frac{2}{1/(z-1) + 1/(z+1)} = z - \frac{z^2 - 1}{z} = \frac{1}{z}.$$

THEOREM 5.25. *Every B-transform maps $\mathbb{H}$ holomorphically into $\mathbb{H}^- \cup \mathbb{R}$. Conversely, if $B : \mathbb{H} \to \mathbb{H}^- \cup \mathbb{R}$ is holomorphic, then the following statements are equivalent:*

(a) *$B = B_\mu$ for a probability measure $\mu \in \mathcal{P}(\mathbb{R})$.*
(b) *There exist $a \in \mathbb{R}$ and a finite, non-negative measure γ such that*

$$B(z) = a + \int_{\mathbb{R}} \frac{1 + xz}{z - x} \gamma(dx).$$

PROOF. Consider the function $-B_\mu(z) = F_\mu(z) - z$. Theorem 5.6 implies that $-B_\mu$ maps $\mathbb{H}$ into $\mathbb{H} \cup \mathbb{R}$. Thus, if (a) holds, then $-B$ has a Nevanlinna triple $(a, 0, \gamma)$, which shows (b). If (b) holds, then Theorem 5.6 implies that $z - B(z)$ is the F-transform for some $\mu \in \mathcal{P}(\mathbb{R})$. Hence, $B = B_\mu$. □

THEOREM 5.26 (Boolean convolution). *Let X, Y be Boolean independent quantum random variables with distributions $\mu, \nu \in \mathcal{P}_c(\mathbb{R})$. Let α be the distribution of $X + Y$. Then*

$$B_\alpha(z) = B_\mu(z) + B_\nu(z) \text{ for all } z \in \mathbb{H}.$$

PROOF. Similarly to the monotone case, this relation is shown by analyzing the power series expansions of the involved B-transforms, see [**182**]. □

Theorem 5.25 shows that $B_\mu + B_\nu$ is again a B-transform for any $\mu, \nu \in \mathcal{P}(\mathbb{R})$.

DEFINITION 5.27. For $\mu, \nu \in \mathcal{P}(\mathbb{R})$, we define the *Boolean convolution* $\mu \uplus \nu \in \mathcal{P}(\mathbb{R})$ via

$$B_{\mu \uplus \nu} = B_\mu + B_\nu.$$

A probability measure $\mu \in \mathcal{P}(\mathbb{R})$ is called *Boolean infinitely divisible* if, for every $n \in \mathbb{N}$, there exists $\mu_n \in \mathcal{P}(\mathbb{R})$ such that $\mu = \mu_n \uplus \cdots \uplus \mu_n$ (n-fold convolution).

EXAMPLE 5.28. For $x \in \mathbb{R}$, we have

$$B_{\delta_x \uplus \mu}(z) = x + z - F_\mu(z), \quad \text{i.e.,} \quad F_{\delta_x \uplus \mu}(z) = F_\mu(z) - x.$$

Hence, $\delta_x \uplus \mu = \mu \uplus \delta_x = \delta_x \rhd \mu$.

LEMMA 5.29. *Let $\mu \in \mathcal{P}(\mathbb{R})$. For $\lambda > 0$, denote by μ_λ the distribution $\mu(\cdot/\lambda)$. Then*

$$B_{\mu_\lambda}(z) = \lambda B_\mu(z/\lambda) \quad \textit{for all } z \in \mathbb{H}.$$

PROOF. We have $G_{\mu_\lambda}(z) = G_{\mu_1}(z/\lambda)/\lambda$, and thus, $B_{\mu_\lambda}(z) = z - \lambda/G_{\mu_1}(z/\lambda) = \lambda B_{\mu_1}(z/\lambda)$. □

THEOREM 5.30 (Boolean central limit theorem). *Let $X_1, X_2, \ldots$ be a sequence of Boolean independent identically distributed random variables with mean c and variance $\sigma^2 > 0$. Let μ_n be the distribution of $S_n = (X_1 + \cdots + X_n - nc)/(\sigma\sqrt{n})$. As $n \to \infty$, μ_n converges weakly to the distribution*

$$\frac{1}{2}\delta_1 + \frac{1}{2}\delta_{-1}.$$

PROOF. Let $Y = (X_1 - c)/\sigma$ and denote its distribution by ν. Consider the power series expansion $B_\nu(z) = \sum_{k=1}^\infty \kappa_k / z^{k-1}$ at ∞. As $F_\nu(z) = z - \frac{1}{z} + \ldots$ at ∞, we have $\kappa_1 = 0$ and $\kappa_2 = 1$. By Lemma 5.29 and Theorem 5.26, we have

$$B_{\mu_n}(z) = \frac{1}{\sqrt{n}} B_{\nu^{\uplus n}}(\sqrt{n}z) = \frac{n}{\sqrt{n}} B_\nu(\sqrt{n}z) = 1/z + \sum_{k=3}^\infty \kappa_k n^{1-k/2}/z^{k-1} \to \frac{1}{z}$$

as $n \to \infty$ locally uniformly in $\mathbb{H}$. By Theorem 3.18 and Example 5.24, we have $\mu_n \to \frac{1}{2}\delta_1 + \frac{1}{2}\delta_{-1}$ as $n \to \infty$. □

THEOREM 5.31. *Every probability measure $\mu \in \mathcal{P}(\mathbb{R})$ is infinitely divisible with respect to Boolean convolution.*

PROOF. For $n \in \mathbb{N}$, let $B_n(z) = B_\mu(z)/n$. Then Theorem 5.25 implies that $B_n = B_{\mu_n}$ for some $\mu_n \in \mathcal{P}(\mathbb{R})$. Thus, $\mu = \mu_n^{\uplus n}$. □

5.4. R-transform and free convolution

Let $\mu \in \mathcal{P}_c(\mathbb{R})$. The Cauchy transform G_μ is analytic in a neighborhood of ∞ and $G'_\mu(\infty) \neq 0$. So we can define its compositional inverse G_μ^{-1} in a neighborhood U_μ of 0. For $z \in U_\mu$, we define the R-transform R_μ as

$$R_\mu(z) = G_\mu^{-1}(z) - \frac{1}{z}.$$

THEOREM 5.32 (Free convolution). *Let X, Y be freely independent quantum random variables with distributions $\mu, \nu \in \mathcal{P}_c(\mathbb{R})$. Let α be the distribution of $X + Y$. Then*

$$R_\alpha(z) = R_\mu(z) + R_\nu(z)$$

for all z in some neighborhood of 0.

PROOF. Similarly to the monotone case, this relation is shown by analyzing the power series expansions of the involved R-transforms, see [**189**] or [**132**, Section 2.4]. □

For an arbitrary $\mu \in \mathcal{P}(\mathbb{R})$, we can still define the inverse of F_μ within a domain of the form $\Gamma_{\alpha,\beta} = \{z \in \mathbb{H} \,|\, |\mathrm{Re}(z)| < \alpha \,\mathrm{Im}(z), |z| > \beta\}$ as in (5.3).

The set $\Delta_{\alpha,\beta} = \{1/z \,|\, z \in \Gamma_{\alpha,\beta}\}$ is then a wedge at 0 within the lower half-plane, symmetric with respect to the imaginary axis. We can now define R_μ as follows.

DEFINITION 5.33. For $\mu \in \mathcal{P}(\mathbb{R})$, let $\alpha, \beta > 0$ such that F_μ^{-1} exists on $\Gamma_{\alpha,\beta}$ as in (5.3). The *R-transform* R_μ is defined on $\Delta_{\alpha,\beta} = \{1/z \,|\, z \in \Gamma_{\alpha,\beta}\}$ as

$$R_\mu(z) := G_\mu^{-1}(z) - \frac{1}{z}.$$

As $\mathrm{Im}(F_\mu^{-1}(z)) \leq \mathrm{Im}(z)$, the function $F_\mu^{-1}(z) - z$ takes values in $\mathbb{H}^- \cup \mathbb{R}$. So $R_\mu(z) = F_\mu^{-1}(1/z) - 1/z$ maps $\Delta_{\alpha,\beta}$ into $\mathbb{H}^- \cup \mathbb{R}$.

REMARK 5.34. The transform $R_\mu(1/z) = F_\mu^{-1}(z) - z$ is also called *Voiculescu transform*.

EXAMPLE 5.35. Recall the semicircle distribution $W(0,1)$ given by the density

$$\frac{1}{2\pi}\sqrt{4 - x^2}, \quad x \in [-2, 2],$$

see Figure 5.6. We have $G_{W(0,1)}(z) = \frac{z - \sqrt{z^2-4}}{2}$, where the branch of the square root is chosen such that $\sqrt{\cdot}$ maps $\mathbb{C} \setminus [-4, \infty)$ into the upper half-plane. Solving the equation $\frac{z-\sqrt{z^2-4}}{2} = w$, gives $z = w + \frac{1}{w}$, and thus,

$$R_{W(0,1)}(w) = w.$$

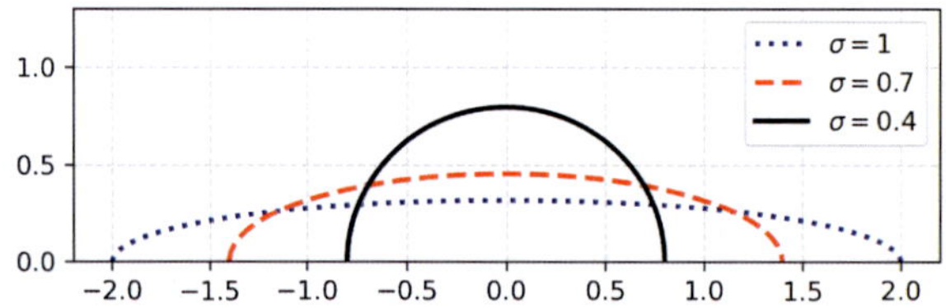

FIGURE 5.6. Densities of some centered semicircle distributions $W(0, \sigma^2)$.

For general $\mu, \nu \in \mathcal{P}(\mathbb{R})$, it is shown in [**34**, Corollary 5.8] that $R_\mu(z) + R_\nu(z)$, which is defined on a sufficiently small wedge $\Delta_{\alpha,\beta}$, is again an R-transform of a probability measure; see also [**132**, Section 3.6].

DEFINITION 5.36. For $\mu, \nu \in \mathcal{P}(\mathbb{R})$, the *free convolution* $\mu \boxplus \nu \in \mathcal{P}(\mathbb{R})$ is defined via

$$R_{\mu \boxplus \nu} = R_\mu + R_\nu.$$

A probability measure $\mu \in \mathcal{P}(\mathbb{R})$ is called *freely infinitely divisible* if, for every $n \in \mathbb{N}$, there exists $\mu_n \in \mathcal{P}(\mathbb{R})$ such that $\mu = \mu_n \boxplus \cdots \boxplus \mu_n$ (n-fold convolution).

EXAMPLE 5.37. For $x \in \mathbb{R}$ we have

$$R_{\delta_x \boxplus \mu}(w) = x + R_\mu(w), \quad \text{i.e.,} \quad G^{-1}_{\delta_x \boxplus \mu}(w) = x + G^{-1}_\mu(w).$$

So $G_\mu(G^{-1}_{\delta_x \boxplus \mu}(w) - x) = w$ or $G_\mu(z - x) = G_{\delta_x \boxplus \mu}(z)$, and thus, $\delta_x \boxplus \mu = \mu \boxplus \delta_x$ is simply μ shifted by x.

EXAMPLE 5.38. Let us take $\mu = \nu = \frac{1}{2}\delta_a + \frac{1}{2}\delta_b$, $a, b \in \mathbb{R}$. Then $G_\mu(z) = \frac{1}{2(z-a)} + \frac{1}{2(z-b)}$. From $R_{\mu \boxplus \mu} = 2R_\mu$, and from $G^{-1}_{\mu \boxplus \mu}(w) = 2R_\mu(w) + \frac{1}{w}$, we obtain

$$G_{\mu \boxplus \mu}(z) = \frac{1}{\sqrt{(z - (a+b))^2 - (a-b)^2}}.$$

Thus, $\mu \boxplus \mu$ is the arcsine distribution $Arc(a+b, (a-b)^2/2)$, see Example 5.3.

LEMMA 5.39. *Let X be a quantum random variable with distribution μ. For $\lambda > 0$, denote by μ_λ the distribution of λX. Then*

$$R_{\mu_\lambda}(z) = \lambda R_\mu(\lambda z) \quad \textit{for all } z \textit{ in a neighborhood of } 0.$$

PROOF. We have $G_{\mu_\lambda}(z) = G_\mu(z/\lambda)/\lambda =: G(z)$, and from

$$G_\mu(R_\mu(\lambda z) + 1/(\lambda z)) = \lambda z,$$

we see that $G(\lambda R_\mu(\lambda z) + 1/z) = z$. Hence, $R_{\mu_\lambda}(z) = \lambda R_\mu(\lambda z)$. □

THEOREM 5.40 (Free central limit theorem). *Let $X_1, X_2, \ldots$ be a sequence of freely independent identically distributed random variables with mean c and variance $\sigma^2 > 0$. Let μ_n be the distribution of $S_n = (X_1 + \cdots + X_n - nc)/(\sigma\sqrt{n})$. As $n \to \infty$, μ_n converges weakly to the semicircle law $W(0,1)$, given by the density*

$$\frac{1}{2\pi}\sqrt{4 - x^2}, \quad x \in [-2, 2].$$

PROOF. Let $Y = (X_1 - c)/\sigma$ and denote its distribution by ν. Consider the power series expansion $R_\nu(z) = \sum_{k=0}^\infty \kappa_{k+1} z^k$. Then $\kappa_1 = 0$ and $\kappa_2 = 1$. By Lemma 5.39 and Theorem 5.32, we have

$$\begin{aligned} R_{\mu_n}(z) &= \frac{1}{\sqrt{n}} R_{\nu^{\boxplus n}}(z/\sqrt{n}) = \frac{n}{\sqrt{n}} R_\nu(z/\sqrt{n}) \\ &= z + \sum_{k=2}^\infty \kappa_{k+1} z^k / n^{(k-1)/2} \to z \end{aligned}$$

as $n \to \infty$ locally uniformly in a neighborhood of 0. By Theorem 3.18 and Example 5.35, we have $\mu_n \to W(0,1)$ as $n \to \infty$. □

Freely infinitely divisible distributions can be related to $\boxplus$-semigroups, similarly to the case of monotone independence.

THEOREM 5.41 (Lemma 5.9 and Theorem 5.10 in [**34**]). *For a probability measure μ on $\mathbb{R}$, the following statements are equivalent:*

(1) *μ is $\boxplus$-infinitely divisible.*
(2) *$\mu = \mu_1$ for a $\boxplus$-semigroup $(\mu_t)_{t \geq 0}$ (i.e., $\mu_0 = \delta_0$, $\mu_{t+s} = \mu_t \boxplus \mu_s$ for all $s, t \geq 0$, and $t \mapsto \mu_t$ is continuous with respect to weak convergence).*
(3) *R_μ extends to a holomorphic function from $\mathbb{H}^-$ into $\mathbb{H}^- \cup \mathbb{R}$.*
(4) *There exist $a \in \mathbb{R}$ and a finite, non-negative measure γ on $\mathbb{R}$ such that*

$$R_\mu(1/z) = a + \int_\mathbb{R} \frac{1 + zx}{z - x}\gamma(dx), \qquad z \in \mathbb{H}. \tag{5.11}$$

Conversely, if $f : \mathbb{H} \to \mathbb{H}^- \cup \mathbb{R}$ *is holomorphic with* $\lim_{y\to\infty} f(iy)/y = 0$*, then* $f(z) = R_\mu(1/z)$ *for a* $\boxplus$*-infinitely divisible probability measure* μ.

Furthermore, a family $(\mu_t)_{t\geq 0}$ *of probability measures on* $\mathbb{R}$ *is a* $\boxplus$*-semigroup if and only if* $R_{\mu_t} = tR_{\mu_1}$ *for all* $t \geq 0$.

COROLLARY 5.42. *Let* $\mu \in \mathcal{P}(\mathbb{R})$ *be* $\boxplus$*-infinitely divisible. Then* F_μ *is univalent.*

PROOF. By definition of the R-transform, the function $g(z) = R_\mu(1/z) + z$ is a left inverse of F_μ on some $\Gamma_{\alpha,\beta}$, i.e., $(g \circ F_\mu)(z) = z$ for all $z \in \Gamma_{\alpha,\beta}$. As $R_\mu(1/z)$ extends analytically to $\mathbb{H}$, the identity theorem implies $(g \circ F_\mu)(z) = z$ for all $z \in \mathbb{H}$. So F_μ is univalent. □

EXAMPLE 5.43. The semicircle distribution $W(0, 1)$ is freely infinitely divisible, since $R_{W(0,1)}(z) = z$ extends analytically to the whole lower half-plane. Thus,

$$F^{-1}_{W(0,1)}(z) = z + \frac{1}{z}$$

extends analytically to the upper half-plane. While $F_{W(0,1)}(\mathbb{H})$ is the complement of a half-disk in $\mathbb{H}$, $F^{-1}_{W(0,1)}(\mathbb{H}) = \mathbb{C} \setminus ((-\infty, -2] \cup [2, \infty))$, see Figure 5.7.

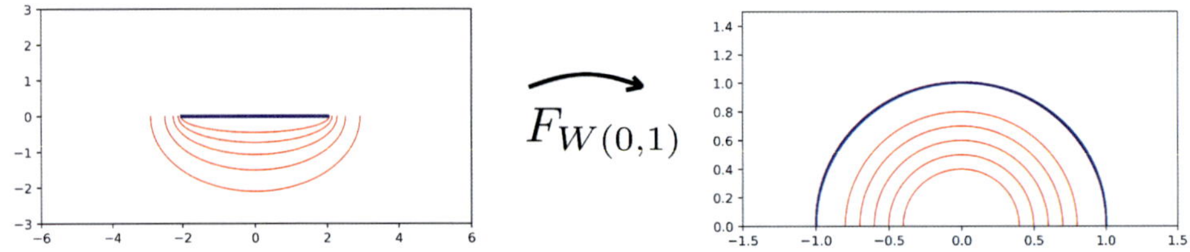

FIGURE 5.7. F-transform of the semicircle distribution $W(0, 1)$.

The normal distribution $\mathcal{N}(0, 1)$ was shown to be freely infinitely divisible in [**28**]. Further examples of freely infinitely divisible probability measures can be found in [**12, 13, 45, 86**].

5.5. Summary

(1) We have seen that the four non-classical convolutions can all be expressed by transforms derived from the Cauchy transform

$$G_\mu(z) = \int_{\mathbb{R}} \frac{1}{z - x}\mu(dx).$$

The central limit theorems can be summarized as follows:

Independence	**Central limit law**		**Transform**
Tensor	Normal distribution	$\frac{1}{\sqrt{2\pi}}e^{-x^2/2}dx$	characteristic function $\int_{\mathbb{R}} e^{itx}\mu(dx)$
Monotone	Arcsine distribution	$\frac{dx}{\pi\sqrt{2-x^2}}, x \in (-\sqrt{2}, \sqrt{2})$	F-transform $F_\mu(z) = 1/G_\mu(z)$
Boolean	Two point measure	$\frac{1}{2}\delta_{-1} + \frac{1}{2}\delta_1$	B-transform $B_\mu(z) = z - F_\mu(z)$
Free	Semicircle distribution	$\frac{1}{2\pi}\sqrt{4 - x^2}dx, x \in [-2, 2]$	R-transform $R_\mu(z) = G_\mu^{-1}(z) - 1/z$

(2) In the non-classical cases, the classes of infinitely divisible distributions can all be characterized by the set of holomorphic functions $f : \mathbb{H} \to \mathbb{H} \cup \mathbb{R}$ with
$$f(z) = a + \int_{\mathbb{R}} \frac{1+xz}{x-z} \gamma(dx),$$
where $a \in \mathbb{R}$ and γ is a finite, non-negative measure. If μ is $\rhd$- or $\boxplus$-infinitely divisible, then the F-transform of μ is a univalent function on $\mathbb{H}$. This relation will be further investigated in Chapter 6.

(3) The non-commutative shift: Let $x \in \mathbb{R}$. Apart from the shift by x, the ten convolutions $\delta_x \star \mu$, $\mu \star \delta_x$, with $\star \in \{*, \uplus, \boxplus, \rhd, \lhd\}$, yield only one new case $\delta_x \rhd \mu$ given by
$$F_{\delta_x \rhd \mu} = F_\mu - x.$$
Instead of shifting μ, we shift the image domain $F_\mu(\mathbb{H})$ by $-x$ in this case. This "non-commutative shift" will be used in Section 5.7 to define the class of $\rhd$-homogeneous Markov processes.

(4) The five convolutions can also be approached from a combinatorial point of view. Suppose μ has compact support. Then all moments $m_1, m_2, \ldots$ of μ exist. Furthermore, we can expand the R-transform as $R_\mu(z) = \sum_{n=0}^{\infty} \kappa_{n+1} z^n$, and the numbers $\kappa_1, \kappa_2, \ldots$ are called the *free cumulants* of μ. Similar to the classical case, see Remark 2.25, one can show that κ_n can be written as a polynomial in $m_1, \ldots, m_n$, and the free convolution of two probability measures $\mu, \nu \in \mathcal{P}_c(\mathbb{R})$ can now be expressed as
$$\kappa_n(\mu \boxplus \nu) = \kappa_n(\mu) + \kappa_n(\nu) \quad \text{for all } n \in \mathbb{N}.$$
We refer to [**144**] for the free cumulants and to [**182**] for the Boolean cumulants. The monotone case is studied in [**89**].

5.6. Additive processes and hemigroups

Let $(B(H), \varphi)$ be a quantum probability space. A quantum stochastic process is a collection $(X_t)_{t \in T}$ of random variables in $B(H)$. In this section, we define quantum stochastic processes with independent increments. First, we regard some further examples.

EXAMPLE 5.44. Any sequence $X_1, X_2, \dots \in \mathbb{C}^{N \times N}$ of (random) matrices can be regarded as a quantum stochastic process. Consider a multivariate time series $y_1, \ldots, y_n \in \mathbb{R}^N$, coming from N sensors of a certain machine. Fix some $m \in \mathbb{N}$ and let R_k be the sample covariance matrix calculated from the observations $x_k, \ldots, x_{k-m}$. If we model the time series by a random vector $Y \in \mathbb{R}^N$, we obtain a random matrix model for the R_k. The matrices R_k could be used to monitor the functioning of the machine by looking at the structure of their eigenvalues. We cite [**63**] as an example where free probability theory is applied to the sample covariance matrix of a certain model.

EXAMPLE 5.45. In quantum mechanics, the state of a particle is represented as a unit vector ψ_0 in some Hilbert space. An observable quantity is represented by a (possibly unbounded) self-adjoint operator A. One obtains a dynamical description of this particle by choosing a Hamiltonian H, which is again a self-adjoint operator, and now the state ψ_t depends on t (Schrödinger picture) and satisfies the Schrödinger equation

$$i\hbar\frac{\partial}{\partial t}\psi_t = H\psi_t.$$

In a mathematically equivalent way, one can leave the state ψ_0 constant and let the observable A_t depend on t (Heisenberg picture). Thus, the dynamics can be described by the quantum process $(A_t)_{t\geq 0}$, which satisfies the Heisenberg equation

$$\frac{d}{dt}A_t = \frac{i}{\hbar}[H, A_t] = \frac{i}{\hbar}(HA_t - A_tH).$$

Quantum mechanics motivated the development of a non-commutative probability theory, and conversely, results from this theory can be applied to quantum mechanical models. A mathematical introduction to quantum mechanics is provided in [**184**]. More information on quantum stochastic processes in quantum mechanics can be found in [**11, 14, 18, 131**]. See also [**105**] for an overview article on quantum random walks.

EXAMPLE 5.46. A linear mapping $\mathcal{T} : \mathbb{C}^{N\times N} \to \mathbb{C}^{N\times N}$ is called a *quantum channel* if

$$\mathcal{T}(X) = \sum_{j=1}^{M} E_j X E_j^*$$

for matrices $E_1, \ldots, E_M \in \mathbb{C}^{N\times N}$ with $\sum_{j=1}^{M} E_j^* E_j = I$. If $\rho \in \mathbb{C}^{N\times N}$ is a density matrix, then $\mathcal{T}(\rho)$ is also a density matrix since

$$\operatorname{Tr}(\mathcal{T}(\rho)) = \sum_j \operatorname{Tr}(E_j \rho E_j^*) = \sum_j \operatorname{Tr}(E_j^* E_j \rho) = \operatorname{Tr}((\sum_j E_j^* E_j)\rho) = \operatorname{Tr}(\rho) = 1,$$

and as each $E_j \rho E_j^*$ is non-negative, also the sum is non-negative.

A self-adjoint $X_0 \in \mathbb{C}^{N\times N}$ together with a family $(\rho_n)_{n\in\mathbb{N}_0} \subset \mathbb{C}^{N\times N}$ of density matrices is called a *(time-homogeneous) quantum Markov chain* if there exists a quantum channel $\mathcal{T}$ such that

$$\rho_{n+1} = \mathcal{T}(\rho_n) \quad \text{for all } n \in \mathbb{N}_0.$$

While X_0 encodes the "state space", the distribution at time n is given via the density matrix $\rho_n = \mathcal{T}^n(\rho_0)$ and the expectation $\varphi_n(X_0) = \operatorname{Tr}(X_0\rho_n)$. So the operator X_0 is fixed, but the state changes (Schrödinger picture). This concept generalizes classical time-homogeneous Markov chains. We refer to [**14, 82**].

We now define additive processes as quantum stochastic processes with independent increments. Similarly to Definition 2.57, we will further keep the initial value fixed at 0 and assume that the distributions induced by the process vary continuously.

DEFINITION 5.47 (Additive processes). Let $(B(H), \varphi)$ be a quantum probability space and $(X_t)_{t\geq 0} \subset B(H)$ a family of random variables. We call $(X_t)_{t\geq 0}$ a (tensor, Boolean, free, monotone, anti-monotone) *additive process* if the following conditions are satisfied:

(1) $X_0 = 0$.

(2) In case of tensor, Boolean or free independence: The random variables

$$X_{t_1}, X_{t_2} - X_{t_1}, \ldots, X_{t_n} - X_{t_{n-1}}$$

are independent for all $n \geq 2$ and all $t_1, \ldots, t_n \in \mathbb{R}$ with $0 \leq t_1 \leq t_2 \leq \cdots \leq t_n$.

In case of monotone or anti-monotone independence: The tuple

$$(X_{t_1}, X_{t_2} - X_{t_1}, \ldots, X_{t_n} - X_{t_{n-1}})$$

is independent for all $n \geq 2$ and all $t_1, \ldots, t_n \in \mathbb{R}$ with $0 \leq t_1 \leq t_2 \leq \cdots \leq t_n$.

(3) The mapping $(s,t) \mapsto \mu_{s,t}$ is continuous with respect to weak convergence, where $\mu_{s,t}$ denotes the distribution of the increment $X_t - X_s$, $0 \leq s \leq t$.

Such a process is called a (tensor, Boolean, free, monotone, anti-monotone) *Lévy process* if, in addition,

(4) the distribution of $X_{t+s} - X_s$ does not depend on s.

Consider an additive process $(X_t)_{t\geq 0}$ and let $\star \in \{*, \uplus, \boxplus, \rhd, \lhd\}$ be the convolution associated with the corresponding type of independence. By writing $X_u - X_s = (X_t - X_s) + (X_u - X_t)$ for $0 \leq s \leq t \leq u$, we see that

$$\mu_{s,u} = \mu_{s,t} \star \mu_{t,u}.$$

If the increments are also stationary, then we have $\mu_{s,t} = \mu_{0,t-s} =: \mu_{t-s}$, and

$$\mu_{s+t} = \mu_t \star \mu_s.$$

This motivates the following definition.

DEFINITION 5.48. Fix a convolution $\star \in \{*, \uplus, \boxplus, \rhd, \lhd\}$. A family $(\mu_{s,t})_{0\leq s\leq t} \subset \mathcal{P}(\mathbb{R})$ is called a (continuous) $\star$-*hemigroup* if the following conditions are satisfied:

(1) $\mu_{s,s} = \delta_0$ for all $s \geq 0$,
(2) $(s,t) \mapsto \mu_{s,t}$ is continuous with respect to weak convergence,
(3) $\mu_{s,u} = \mu_{s,t} \star \mu_{t,u}$ for all $0 \leq s \leq t \leq u$.

If, in addition, $\mu_{s,t} = \mu_{0,t-s}$ for all $0 \leq s \leq t$, then $(\mu_{0,t})_{t\geq 0}$ is called a (continuous) $\star$-*semigroup*.

The distributions of an additive process clearly form a hemigroup.

LEMMA 5.49. *Let $(X_t)_{t\geq 0}$ be an additive process and let $\mu_{s,t}$ be the distribution of $X_t - X_s$. Then $(\mu_{s,t})_{0\leq s\leq t} \subset \mathcal{P}(\mathbb{R})$ is a* $\star$-hemigroup, *where $\star \in \{*, \uplus, \boxplus, \rhd, \lhd\}$ is the convolution associated with the corresponding type of independence. If $(X_t)_{t\geq 0}$ is a Lévy process, then $(\mu_{0,t})_{t\geq 0}$ is a* $\star$-semigroup.

We have already seen that the $\star$-semigroup distributions correspond to the $\star$-infinitely divisible distributions, which can be encoded by Lévy triples/pairs. (In the Boolean case, every probability measure $\mu \in \mathcal{P}(\mathbb{R})$ is infinitely divisible and can be embedded into a semigroup, see the proof of Theorem 5.31.) The following natural question arises:

Which probability measures appear in $\star$-hemigroups?

THEOREM 5.50. *Let $\star \in \{*, \uplus, \boxplus\}$. If $(\mu_{s,t})_{0\leq s\leq t}$ is a $\star$-hemigroup, then each $\mu_{s,t}$ is $\star$-infinitely divisible.*

PROOF. In case $\star = \uplus$, every $\mu \in \mathcal{P}(\mathbb{R})$ is infinitely divisible, see Theorem 5.31.

Now fix $0 \leq s \leq t$ and let

$$\nu_{j,n} = \mu_{s+(j-1)\cdot(t-s)/n, s+j\cdot(t-s)/n}$$

for $n \in \mathbb{N}$, $j = 1, \dots, n$. Then $\nu_{1,n} \star \cdots \star \nu_{n,n} = \mu_{s,t}$. As $(\sigma, \tau) \mapsto \mu_{\sigma,\tau}$ is continuous, and as $S = \{(\sigma, \tau) \,|\, s \leq \sigma \leq \tau \leq t\}$ is a compact set, $(\sigma, \tau) \mapsto \mu_{\sigma,\tau}((-\varepsilon, \varepsilon))$ is uniformly continuous on S for any $\varepsilon > 0$. Hence, for $\delta \in (0, 1)$, we find $N \in \mathbb{N}$ such that $1 - \nu_{j,n}((-\varepsilon, \varepsilon)) < \delta$ for all $n \geq N$ and $j = 1, \dots, n$.

In the case $\star = *$, Theorem 2.63, direction (d) $\to$ (c), implies that $\mu_{s,t}$ is $*$-infinitely divisible. The same conclusion is possible in the case $\star = \boxplus$, see [**32**, Theorem 1]. $\square$

The $\rhd$-hemigroups will be handled in Chapter 6, where they lead us deeper into complex analysis. We will see that there are more hemigroup distributions than semigroup distributions in this case. Furthermore, the evolution of the measure-valued process $t \mapsto \mu_{0,t}$ turns out to have a nice geometric interpretation, namely a growth process in $\mathbb{H}$ described by univalent functions. Surprisingly, this geometric interpretation is also true for $\boxplus$-hemigroups, which we investigate in Chapter 9.

5.7. Monotonically space-homogeneous Markov processes

Recall that stochastic processes with classically independent increments have space-homogeneous transition kernels, i.e.,

$$k_{s,t}(x, B) = k_{s,t}(0, B - x), \text{ or } \delta_x * k_{s,t}(y, \cdot) = \delta_{x+y} * k_{s,t}(0, \cdot) = k_{s,t}(x + y, \cdot).$$

We have seen that the non-commutative convolutions of a probability measure μ with a point measure δ_x lead to only one new case that is different from the classical one, namely $\delta_x \rhd \mu$. This leads us to the definition of a new class of classical Markov processes.

DEFINITION 5.51. A probability kernel k on $\mathbb{R}$ is called $\rhd$-*homogeneous* if it satisfies

$$\delta_x \rhd k(y, \cdot) = k(x + y, \cdot)$$

for all $x, y \in \mathbb{R}$. A classical Markov process $(M_t)_{t\geq 0}$ on $\mathbb{R}$ is called a $\rhd$-*homogeneous Markov process* if its transition kernels $(k_{s,t})_{0\leq s\leq t}$ satisfy the following two conditions:

(a) The mapping $(s, t) \mapsto k_{s,t}(x, \cdot)$ is continuous with respect to weak convergence for all $x \in \mathbb{R}$.
(b) The kernel $k_{s,t}$ is $\rhd$-*homogeneous* for all $0 \leq s \leq t$.

$\rhd$-hemigroups correspond to $\rhd$-homogeneous Markov processes in a natural way.

THEOREM 5.52. *Let $(\mu_{s,t})_{0\leq s\leq t}$ be a $\rhd$-hemigroup. Then there exists a $\rhd$-homogeneous Markov process $(M_t)_{t\geq 0}$ with transition kernels $k_{s,t}$ with the property $k_{s,t}(x, \cdot) = \delta_x \rhd \mu_{s,t}(\cdot)$, i.e.,*

$$\int_{\mathbb{R}} \frac{1}{z - y} k_{s,t}(x, dy) = \frac{1}{F_{\mu_{s,t}}(z) - x}, \quad \text{for all } 0 \leq s \leq t,\ x \in \mathbb{R}. \tag{5.12}$$

Conversely, let $(M_t)_{t\geq 0}$ be a $\rhd$-homogeneous Markov process with transition kernels $k_{s,t}$. Then $(k_{s,t}(0, \cdot))_{0\leq s\leq t}$ is a $\rhd$-hemigroup.

PROOF. Let $(\mu_{s,t})_{0\leq s\leq t}$ be a $\triangleright$-hemigroup and define $k_{s,t}(x,\cdot) := \delta_x \triangleright \mu_{s,t}$ and $F_{s,t} := F_{\mu_{s,t}}$. We would like to apply Theorem 2.48, which gives the existence of the Markov process. We need to verify that $x \mapsto k_{s,t}(x,B)$ is a measurable function for all $B \in \mathcal{B}(\mathbb{R})$, $0 \leq s \leq t$, and that $k_{s,t}$ satisfies the Chapman–Kolmogorov equation.

By the inversion formulas (3.3) and (3.4), we have

$$k_{s,t}(x,\{\alpha\}) = \lim_{\varepsilon\searrow 0} \frac{i\varepsilon}{F_{s,t}(\alpha+i\varepsilon)-x},$$

$$\frac{1}{2}k_{s,t}(x,\{\alpha\}) + \frac{1}{2}k_{s,t}(x,\{\beta\}) + k_{s,t}(x,(\alpha,\beta)) = -\frac{1}{\pi}\lim_{\varepsilon\searrow 0}\int_\alpha^\beta \frac{1}{F_{s,t}(y+i\varepsilon)-x}\,dy,$$

which implies that $x \mapsto k_{s,t}(x,B)$ is measurable for open and closed intervals B. By using the monotone convergence theorem, we obtain the measurability for all Borel sets.

For $0 \leq s \leq t \leq u$, by using (5.12), we get

$$\begin{aligned}\int_{\mathbb{R}} \frac{1}{z-w}\int_{\mathbb{R}} k_{t,u}(y,dw)k_{s,t}(x,dy) &= \int_{\mathbb{R}} \frac{1}{F_{t,u}(z)-y}\,k_{s,t}(x,dy)\\ = \frac{1}{F_{s,t}(F_{t,u}(z))-x} &= \frac{1}{F_{s,u}(z)-x} = \int_{\mathbb{R}}\frac{1}{z-w}\,k_{s,u}(x,dw).\end{aligned}$$

We obtain the Chapman–Kolmogorov relation via the Stieltjes–Perron inversion formula.

The $\triangleright$-homogeneity of the transition kernels follows from

$$F_{\delta_x\triangleright k_{s,t}(y,\cdot)}(z) = F_{k_{s,t}(y,\cdot)}(z) - x = F_{s,t}(z)-(x+y) = F_{k_{s,t}(x+y,\cdot)}(z).$$

Conversely, let $(M_t)_{t\geq 0}$ be a $\triangleright$-homogeneous Markov process with transition kernels $k_{s,t}$, and let $\mu_{s,t} = k_{s,t}(0,\cdot)$. Then $k_{s,t}(x,\cdot) = \delta_x \triangleright \mu_{s,t}(\cdot)$. The Chapman–Kolmogorov relation and (5.12) imply that $F_{\mu_{s,u}} = F_{\mu_{s,t}} \circ F_{\mu_{t,u}}$. □

Let us consider the case of a $\triangleright$-semigroup. Then $(F_{\mu_t})_{t\geq 0}$ is a continuous semigroup whose generator $A : \mathbb{H} \to \mathbb{H}\cup\mathbb{R}$ has the form

$$A(z) = a + \int_{\mathbb{R}} \frac{1+xz}{x-z}\gamma(dx), \tag{5.13}$$

see Theorem 5.20. The corresponding Markov process is now time-homogeneous and the kernels satisfy $k_{s,t} = k_{0,t-s}$. We let $k_t := k_{0,t}$.

LEMMA 5.53. *Let $(M_t)_{t\geq 0}$ be a time-homogeneous, $\triangleright$-homogeneous Markov process with transition kernels $(k_t)_{t\geq 0}$. Let (a,γ) be the generating pair from* (5.13) *associated with the semigroup $(F_{k_t(0,\cdot)})_{t\geq 0}$. Then we have*

$$\frac{1}{t}\frac{(y-x)^2}{1+y^2}\,k_t(x,dy) \to \gamma(dy)$$

for all $x \in \mathbb{R}$ as $t \searrow 0$ with respect to weak convergence, and

$$\frac{1}{t}\int_{\mathbb{R}} \frac{(x-y)(1+xy)}{1+y^2}\,k_t(x,dy) \to a.$$

PROOF. Fix $x \in \mathbb{R}$, and let $G_{x,t}$ and $F_{x,t}$ be the Cauchy and F-transform, respectively, of the distribution $k_t(x,\cdot)$. Since the infinitesimal generator $A(z)$ in equation (5.13) is defined by $A(z) = \frac{d}{dt}F_{0,t}(z)\big|_{t=0}$, we have

$$\frac{d}{dt}G_{x,t}(z)\bigg|_{t=0} = \frac{d}{dt}\frac{1}{F_{0,t}(z)-x}\bigg|_{t=0} = -(z-x)^{-2}A(z),$$

and thus,

$$\begin{aligned}A(z) &= -\lim_{t\to 0}(z-x)^2\frac{1}{t}\left(\int_{\mathbb{R}}\frac{1}{z-y}\,k_t(x,dy) - \frac{1}{z-x}\right)\\ &= \lim_{t\to 0}\frac{1}{t}\int_{\mathbb{R}}\frac{(z-x)(x-y)}{z-y}\,k_t(x,dy).\end{aligned}$$

For $z = x+iv$, $v>0$, we get

$$\operatorname{Im}(A(x+iv)) = \lim_{t\to 0}\frac{1}{t}\int_{\mathbb{R}}\frac{(x-y)^2v}{v^2+(x-y)^2}\,k_t(x,dy).$$

There exists a constant $C>0$ not depending on y such that we have $1 \le C\frac{1+y^2}{1+(x-y)^2}$ for all $y\in\mathbb{R}$, and hence,

$$\frac{1}{t}\int_{\mathbb{R}}\frac{(x-y)^2}{1+y^2}\,k_t(x,dy) \le \frac{C}{t}\int_{\mathbb{R}}\frac{(x-y)^2}{1+(x-y)^2}\,k_t(x,dy) \to C\,\operatorname{Im}(A(x+i)) \quad \text{as } t\searrow 0.$$

This shows that the family $\{\rho_t := \frac{(x-y)^2}{t(1+y^2)}\,k_t(x,dy) : 0<t<t_0\}$ of measures is uniformly bounded for $t_0>0$ small enough.

By Helly's selection theorem, every subsequence ρ_{t_n}, with $t_n\to 0$, regarded as measures on $\hat{\mathbb{R}}$, has a converging subsequence. So assume that $\rho_{t_n}\to\rho$, $t_n\to 0$, as $n\to\infty$. Then

$$\begin{aligned}A(z) &= \lim_{n\to\infty}\frac{1}{t_n}\int_{\mathbb{R}}\frac{(z-x)(x-y)}{z-y}\,k_{t_n}(x,dy)\\ &= \lim_{n\to\infty}\frac{1}{t_n}\int_{\mathbb{R}}\left(x-y-\frac{(x-y)^2}{z-y}\right)k_{t_n}(x,dy)\\ &= \lim_{n\to\infty}\frac{1}{t_n}\int_{\mathbb{R}}\left(\frac{1+y^2}{y-z}-y\right)\frac{(x-y)^2}{1+y^2}k_{t_n}(x,dy)\\ &- \lim_{n\to\infty}\frac{1}{t_n}\int_{\mathbb{R}}\left(y-x-\frac{(x-y)^2y}{1+y^2}\right)k_{t_n}(x,dy)\\ &= \lim_{n\to\infty}\int_{\mathbb{R}}\frac{1+yz}{y-z}\rho_{t_n}(dy) + \lim_{n\to\infty}\frac{1}{t_n}\int_{\mathbb{R}}\left(x-y+\frac{(x-y)^2y}{1+y^2}\right)k_{t_n}(x,dy)\\ &= \int_{\mathbb{R}}\frac{1+yz}{y-z}\rho(dy) + \lim_{n\to\infty}\frac{1}{t_n}\int_{\mathbb{R}}\frac{(x-y)(1+xy)}{1+y^2}\,k_{t_n}(x,dy).\end{aligned}$$

By the uniqueness of the Nevanlinna representation, we conclude that $\rho=\gamma$, and thus, ρ_t converges as $t\to 0$ with

$$\rho_t\to\gamma \quad\text{and}\quad a = \lim_{t\to 0}\frac{1}{t}\int_{\mathbb{R}}\frac{(x-y)(1+xy)}{1+y^2}k_t(x,dy). \qquad \square$$

Let $B \in \mathcal{B}(\mathbb{R})$. Then $x \mapsto \int_{\mathbb{R}} \mathbf{1}_B(y) k_t(x, dy)$ is bounded and measurable. So denote by $B_b(\mathbb{R})$ the set of all bounded Borel measurable functions $f : \mathbb{R} \to \mathbb{C}$. By linear extension and by using the monotone convergence theorem, we see that

$$(T_t f)(x) = \int_{\mathbb{R}} f(y) k_t(x, dy), \qquad f \in B_b(\mathbb{R}), \tag{5.14}$$

defines a linear operator $T_t : B_b(\mathbb{R}) \to B_b(\mathbb{R})$. The family $(T_t)_{t \geq 0}$ are the *transition operators* induced by M_t. Using the Chapman–Kolmogorov relation yields

$$\begin{aligned}(T_t \circ T_s)(\mathbf{1}_B)(x) &= \int_{\mathbb{R}} \int_{\mathbb{R}} \mathbf{1}_B(y) k_s(z, dy) k_t(x, dz) = \int_{\mathbb{R}} k_s(z, B) k_t(x, dz)\\ &= k_{s+t}(x, B) = \int_{\mathbb{R}} \mathbf{1}_B(y) k_{s+t}(x, dy) = (T_{s+t} \mathbf{1}_B)(x).\end{aligned}$$

Again, by linear extension and monotone convergence, we have

$$T_{s+t} f = (T_t \circ T_s)(f) \qquad \text{for all } f \in B_b(\mathbb{R}).$$

Furthermore, T_0 is the identity. So we might be able to calculate an infinitesimal generator G of the semigroup $(T_t)_{t \geq 0}$, i.e., $Gf = \lim_{t \to 0}(T_t f - f)/t$.

For $f \in C^2(\mathbb{R})$, define

$$(\partial f)(x, y) = \begin{cases} \frac{f(x) - f(y)}{x - y}, & x \neq y, \\ f'(x), & x = y. \end{cases}$$

Then we have

$$\left(\frac{\partial}{\partial x} \partial f\right)(x, y) = \begin{cases} \frac{f(y) - f(x) - (y - x) f'(x)}{(y - x)^2}, & x \neq y, \\ \frac{1}{2} f''(x), & x = y. \end{cases}$$

THEOREM 5.54. *Let $(M_t)_{t \geq 0}$ be a time-homogeneous, $\triangleright$-homogeneous Markov process with transition kernels $(k_t)_{t \geq 0}$. Then the infinitesimal generator of the transition semigroup $(T_t)_{t \geq 0}$ is given by*

$$\begin{aligned}(Gf)(x) &:= \frac{d}{dt}(T_t f)(x)\Big|_{t=0}\\ &= -a f'(x) + \int_{\mathbb{R}} \left[(1 + y^2)\left(\frac{\partial}{\partial x} \partial f\right)(x, y) + y f'(x)\right] \gamma(dy)\end{aligned}$$

for all bounded $f \in C^2(\mathbb{R})$ and $x \in \mathbb{R}$, where (a, γ) is the pair in (5.13) associated with the semigroup $(F_{k_t(0, \cdot)})_{t \geq 0}$.

PROOF. For a bounded function $f \in C^2(\mathbb{R})$ we have

$$
\begin{aligned}
(T_t f)(x) - f(x) &= \int_{\mathbb{R}} f(y)\, k_t(x, dy) - f(x) = \int_{\mathbb{R}\setminus\{x\}} (f(y) - f(x))\, k_t(x, dy) \\
&= \int_{\mathbb{R}\setminus\{x\}} \left(\frac{(1+y^2)[f(y) - f(x) - (y-x)f'(x)]}{(y-x)^2} + yf'(x) \right) \frac{(y-x)^2}{1+y^2} k_t(x, dy) \\
&\quad + f'(x) \int_{\mathbb{R}\setminus\{x\}} \left(y - x - \frac{(y-x)^2 y}{1+y^2} \right) k_t(x, dy) \\
&= \int_{\mathbb{R}} \left[(1+y^2) \left(\frac{\partial}{\partial_x} \partial f \right)(x, y) + yf'(x) \right] \frac{(y-x)^2}{1+y^2} k_t(x, dy) \\
&\quad - f'(x) \int_{\mathbb{R}} \left(x - y + \frac{(y-x)^2 y}{1+y^2} \right) k_t(x, dy).
\end{aligned}
$$

Lemma 5.53 implies the desired formula. □

REMARK 5.55. A similar formula holds for the classical analog, i.e., for the case of Lévy processes, see [**168**, Theorem 31.5] or [**103**, Theorem 17.10]. Consider a Lévy process $(X_t)_{t\geq 0}$ with Lévy triple (a, σ, ν), i.e.,

$$
\mathbb{E}[e^{itX_u}] = \exp\left(u \left(iat - \frac{1}{2}\sigma^2 t^2 + \int_{\mathbb{R}} \left(e^{its} - 1 - its\mathbf{1}_{\{|s|<1\}}(s) \right) \nu(ds) \right) \right),
$$

$t \in \mathbb{R}$. Define the operators $(T_t)_{t\geq 0}$ as in (5.14). Then

$$
\begin{aligned}
(Gf)(x) &:= \left. \frac{d}{dt}(T_t f)(x) \right|_{t=0} \\
&= af'(x) + \frac{1}{2}\sigma^2 f''(x) + \int_{\mathbb{R}} \left[f(x+y) - f(x) - yf'(x)\mathbf{1}_{\{|y|<1\}}(y) \right] \nu(dy)
\end{aligned}
$$

for every $f \in C^2(\mathbb{R})$ with $\lim_{|x|\to\infty} f(x) = \lim_{|x|\to\infty} f'(x) = \lim_{|x|\to\infty} f''(x) = 0$.

REMARK 5.56. It is of interest to study the operators $(T_t)_{t\geq 0}$ from the perspective of functional analysis. For a $\triangleright$-homogeneous Markov process [**70**, Section 4.8] and for a classical Lévy process [**103**, Theorem 17.10], it can be shown that the semigroup $(T_t)_{t\geq 0}$ forms a *Feller semigroup*. This implies certain continuity properties of the Markov processes, such as the strong Markov property, see [**103**, Theorem 17.17].

5.8. Construction of monotone additive processes

In this section, we construct a monotone additive process from a given compactly supported hemigroup.

THEOREM 5.57. *Let $(M_t)_{t\geq 0}$ be a $\triangleright$-homogeneous Markov process such that $M_0 = 0$ and let $(\mathcal{F}_t)_{t\geq 0}$ be its natural filtration ($\mathcal{F}_t = \sigma((M_s)_{s\leq t})$). Assume that all distributions of M_t have compact support. Denote by $(\Omega, \mathcal{F}, \mathbb{P})$ the underlying probability space and by P_t the conditional expectation*

$$P_t = \mathbb{E}[\,\cdot\,|\mathcal{F}_t].$$

Define the operators $X_t \in B(L^2(\Omega, \mathcal{F}, \mathbb{P}))$, $t \geq 0$, by

$$X_t = P_t M_t, \tag{5.15}$$

where M_t acts by multiplication on $L^2(\Omega,\mathcal{F},\mathbb{P})$. Then $(X_t)_{t\geq 0}$ is a monotone additive process on $(B(L^2(\Omega,\mathcal{F},\mathbb{P})), \langle \mathbf{1}_\Omega, \cdot \mathbf{1}_\Omega\rangle)$, where $\mathbf{1}_\Omega$ is the constant function with value 1 on Ω. Let $\mu_{s,t} := k_{s,t}(0,\cdot)$, where $k_{s,t}$ are the transition kernels of $(M_t)_{t\geq 0}$. Then the distribution of $X_t - X_s$ is equal to $\mu_{s,t}$.

For the proof, we need a formula for the inverse of $z-(X_t - X_s)$, $z \in \mathbb{H}$, whose expectation will give us the Cauchy transform of the distribution of $X_t - X_s$. We write $F_{s,t} := F_{\mu_{s,t}}$.

LEMMA 5.58. *For $0 \leq s \leq t$ and $z \in \mathbb{H}$, the operator*

$$[R_{s,t}(z)](\psi) := \frac{\psi}{z} + \frac{M_t}{z(z-M_t)}P_t\psi - \frac{M_s\big(F_{s,t}(z)-M_s\big)}{F_{s,t}(z)(z-M_t)}P_s\frac{1}{z-M_t}\psi,$$

$\psi \in L^2(\Omega,\mathcal{F},\mathbb{P})$, satisfies

$$(z-(X_t-X_s))R_{s,t}(z) = R_{s,t}(z)(z-(X_t-X_s)) = I.$$

PROOF. Let $z \in \mathbb{H}$. The Markov property (2.8) and (5.12) imply

$$\mathbb{E}\left[\frac{1}{z-M_t}\middle|\mathcal{F}_s\right] = \frac{1}{F_{s,t}(z)-M_s} \ a.s.,$$

which can also be stated as

$$P_s\frac{1}{z-M_t}P_s = \frac{1}{F_{s,t}(z)-M_s}P_s. \tag{5.16}$$

For $\psi \in L^2(\Omega,\mathcal{F},\mathbb{P})$, expand the quantity

$$[z-(P_tM_t - P_sM_s)]\left(\frac{\psi}{z} + \frac{M_t}{z(z-M_t)}P_t\psi - \frac{M_s\left(F_{s,t}(z)-M_s\right)}{F_{s,t}(z)(z-M_t)}P_s\frac{1}{z-M_t}\psi\right)$$

$$= \psi + \underbrace{\frac{M_t}{z-M_t}P_t\psi}_{I_1} \underbrace{- \frac{zM_s(F_{s,t}(z)-M_s)}{F_{s,t}(z)(z-M_t)}P_s\frac{1}{z-M_t}\psi}_{I_2} \underbrace{+\frac{M_s}{z}P_s\psi}_{I_3} \underbrace{+\frac{M_s}{z}P_s\frac{M_t}{z-M_t}\psi}_{I_4}$$

$$\underbrace{-\frac{M_s^2(F_{s,t}(z)-M_s)}{F_{s,t}(z)}P_s\frac{1}{z-M_t}P_s\frac{1}{z-M_t}\psi}_{I_5} \underbrace{-\frac{M_t}{z}P_t\psi}_{I_6} \underbrace{-\frac{M_t^2}{z(z-M_t)}P_t\psi}_{I_7}$$

$$\underbrace{+\frac{M_sM_t(F_{s,t}(z)-M_s)}{F_{s,t}(z)(z-M_t)}P_s\frac{1}{z-M_t}\psi}_{I_8}.$$

Clearly, $I_1 + I_6 + I_7 = 0$. For I_5, property (5.16) shows that

$$P_s\frac{1}{z-M_t}P_s\frac{1}{z-M_t}\psi = \frac{1}{F_{s,t}(z)-M_s}P_s\frac{1}{z-M_t}\psi,$$

and we can simplify

$$I_2 + I_5 + I_8 = -M_sP_s\frac{1}{z-M_t}\psi.$$

Furthermore,

$$-M_sP_s\frac{1}{z-M_t}\psi + I_3 + I_4 = -M_sP_s\frac{1}{z-M_t}\psi + \frac{M_s}{z}P_s\psi + \frac{M_s}{z}P_s\frac{M_t}{z-M_t}\psi$$

$$= P_s\frac{-M_sz}{z(z-M_t)}\psi + P_s\frac{M_s(z-M_t)}{z(z-M_t)}\psi + P_s\frac{M_sM_t}{z(z-M_t)}\psi = 0.$$

Thus, we conclude that $(z-(X_t-X_s))R_{s,t}(z)=I$, and a similar computation shows that $R_{s,t}(z)(z-(X_t-X_s))=I$. □

PROOF OF THEOREM 5.57. As M_t has compact support, and P_t is a projection on $L^2(\Omega,\mathcal{F},\mathbb{P})$, X_t is a bounded operator on $L^2(\Omega,\mathcal{F},\mathbb{P})$. As $M_0=0$, we have $X_0=0$. To see that X_t is self-adjoint, let $v\in L^2(\Omega,\mathcal{F},\mathbb{P})$. By Lemma 2.39, we obtain

$$\begin{aligned}\langle P_tM_tv,v\rangle &= \mathbb{E}[\overline{P_tM_tv}v]=\mathbb{E}[\overline{\mathbb{E}[M_tv|\mathcal{F}_t]}v]=\mathbb{E}[\mathbb{E}[\overline{M_tv}|\mathcal{F}_t]v]=\mathbb{E}[\mathbb{E}[M_t\overline{v}|\mathcal{F}_t]v]=\\ &=\mathbb{E}[\mathbb{E}[\overline{v}|\mathcal{F}_t]M_tv]=\mathbb{E}[\overline{v}\mathbb{E}[M_tv|\mathcal{F}_t]]=\langle v,P_tM_tv\rangle .\end{aligned}$$

(In short: M_t is clearly self-adjoint and P_t is self-adjoint due to Lemma 2.39. P_t and M_t commute due to Theorem 2.36 (c), so P_tM_t is self-adjoint.)

For $0\le s\le t$, we have $\mu_{s,t}=k_{s,t}(0,\cdot)$, and next, we show that $\mu_{s,t}$ is the distribution of X_t-X_s by proving

$$P_s\left(z-(X_t-X_s)\right)^{-1}P_s=G_{\mu_{s,t}}(z)P_s \tag{5.17}$$

for $0\le s\le t$ and $z\in\mathbb{H}$. By applying both sides of (5.17) to the constant function $\mathbf{1}_\Omega$ and taking the expectation, we obtain with (5.4) that the distribution of X_t-X_s with respect to the expectation $\langle\mathbf{1}_\Omega,\cdot\mathbf{1}_\Omega\rangle$ is equal to $\mu_{s,t}$.

Let $\psi\in L^2(\Omega,\mathcal{F},\mathbb{P})$. By using (5.16) and Lemma 5.58, we obtain

$$\begin{aligned}&P_s\left(z-(X_t-X_s)\right)^{-1}P_s\psi=P_sR_{s,t}(z)P_s\psi=\\ &=\frac{1}{z}P_s\psi+P_s\frac{M_t}{z(z-M_t)}P_s\psi-P_s\frac{M_s\big(F_{s,t}(z)-M_s\big)}{F_{s,t}(z)(z-M_t)}P_s\frac{1}{z-M_t}P_s\psi\\ &=\frac{1}{z}P_s\psi+\frac{1}{z}\left(P_s\psi\right)\cdot\left(P_s\frac{M_t}{z-M_t}\mathbf{1}_\Omega\right)\\ &-\left(\frac{M_s(F_{s,t}(z)-M_s)}{F_{s,t}(z)}P_s\psi\right)\cdot\left(P_s\frac{1}{z-M_t}\mathbf{1}_\Omega\right)^2\\ &=\frac{1}{z}P_s\psi+\frac{1}{z}\left(P_s\psi\right)\cdot\left(-\mathbf{1}_\Omega+\frac{z}{F_{s,t}(z)-M_s}\mathbf{1}_\Omega\right)\\ &-\left(\frac{M_s(F_{s,t}(z)-M_s)}{F_{s,t}(z)}P_s\psi\right)\cdot\left(\frac{1}{F_{s,t}(z)-M_s}\mathbf{1}_\Omega\right)^2\\ &=\frac{1}{F_{s,t}(z)-M_s}P_s\psi-\frac{M_s}{F_{s,t}(z)(F_{s,t}(z)-M_s)}P_s\psi=\frac{1}{F_{s,t}(z)}P_s\psi=G_{\mu_{s,t}}(z)P_s\psi.\end{aligned}$$

The mapping $(s,t)\mapsto k_{s,t}(0,\cdot)$ is continuous by assumption. It remains to show that $(X_t)_{t\ge0}$ has monotonically independent increments.

We also have $P_r\left(z-(X_t-X_s)\right)^{-1}P_r\psi=G_{\mu_{s,t}}(z)P_r\psi$ for any $r\le s$, because $P_r=P_rP_s=P_sP_r$. Thus,

$$\begin{aligned}P_r\sum_{n=0}^{\infty}\frac{(X_t-X_s)^n}{z^{n+1}}P_r\psi&=\sum_{n=0}^{\infty}\frac{P_r(X_t-X_s)^nP_r\psi}{z^{n+1}}\\ &=\sum_{n=0}^{\infty}\langle\mathbf{1}_\Omega,(X_t-X_s)^n\mathbf{1}_\Omega\rangle\,P_r\psi/z^{n+1}\end{aligned}$$

for all $z \in \mathbb{H}$ large enough. Take any $w \in L^2(\Omega, \mathcal{F}, \mathbb{P})$. Then

$$\sum_{n=0}^{\infty} \frac{\langle w, P_r(X_t - X_s)^n P_r \psi \rangle}{z^{n+1}} = \sum_{n=0}^{\infty} \langle w, \langle \mathbf{1}_\Omega, (X_t - X_s)^n \mathbf{1}_\Omega \rangle P_r \psi \rangle / z^{n+1}.$$

The two power series (with values in $\mathbb{C}$) are the same, so

$$\langle w, P_r(X_t - X_s)^n P_r \psi \rangle = \langle w, \langle \mathbf{1}_\Omega, (X_t - X_s)^n \mathbf{1}_\Omega \rangle P_r \psi \rangle$$

for all $n \in \mathbb{N}_0$. We conclude that

$$P_r(X_t - X_s)^n P_r \psi = \langle \mathbf{1}_\Omega, (X_t - X_s)^n \mathbf{1}_\Omega \rangle P_r \psi$$

and $P_r(X_t - X_s)^n P_r = \langle \mathbf{1}_\Omega, (X_t - X_s)^n \mathbf{1}_\Omega \rangle P_r$ for all $n \in \mathbb{N}_0$.

Thus, for $0 \le r \le s \le t$ and any polynomial $p : \mathbb{R} \to \mathbb{R}$ with $p(0) = 0$, we have the relations

$$\begin{aligned} p(X_s - X_r)P_t = P_t p(X_s - X_r) &= p(X_s - X_r), \\ P_r p(X_t - X_s) P_r &= \langle \mathbf{1}_\Omega, p(X_t - X_s) \mathbf{1}_\Omega \rangle P_r. \end{aligned} \tag{5.18}$$

Step 1: Let $0 \le s' \le t' \le s \le t$ and $0 \le s'' \le t'' \le s \le t$. Let $p_1, p_2, p_3 : \mathbb{R} \to \mathbb{R}$ be polynomials with $p_j(0) = 0$ and set $X = p_1(X_{t'} - X_{s'})$, $Y = p_2(X_t - X_s)$, $Z = p_3(X_{t''} - X_{s''})$. From (5.18), we get

$$XYZ = XP_{t'}YP_{t''}Z = \langle \mathbf{1}_\Omega, Y\mathbf{1}_\Omega \rangle XZ.$$

This shows condition (i) from Remark 4.25.

Step 2: Let $t_1, \ldots, t_n, s_1, \ldots, s_n, t_1', \ldots, t_m', s_1', \ldots, s_m', t, s \ge 0$ be such that

$$t_1 \ge s_1 \ge t_2 \ge \cdots \ge t_n \ge s_n \ge t \ge s \le t \le s_m' \le t_m' \le \cdots \le t_2' \le s_1' \le t_1',$$

let $f_1, \ldots, f_n, g, h_1, \ldots, h_m : \mathbb{R} \to \mathbb{R}$ be polynomials fixing 0, and set

$$W_1 = f_1(X_{t_1} - X_{s_1}), \ldots, W_n = f_n(X_{t_n} - X_{s_n}),\ Y = g(X_t - X_s),$$
$$Z_1 = h_1(X_{t_1'} - X_{s_1'}), \ldots, Z_m = h_m(X_{t_m'} - X_{s_m'}).$$

From (5.18), we obtain that

$$\begin{aligned}
&\langle \mathbf{1}_\Omega, W_1 \cdots W_n Y Z_m \cdots Z_1 \mathbf{1}_\Omega \rangle \\
&= \langle P_{t_2}\mathbf{1}_\Omega, W_1(P_{t_2}W_2)W_3 \cdots W_n Y Z_m \cdots (Z_2 P_{t_2'}) Z_1 (P_{t_2'}\mathbf{1}_\Omega) \rangle \\
&= \langle \mathbf{1}_\Omega, (P_{t_2}W_1P_{t_2})W_2W_3 \cdots W_n Y Z_m \cdots Z_2(P_{t_2'}Z_1P_{t_2'})\mathbf{1}_\Omega \rangle \\
&= \langle \mathbf{1}_\Omega, W_1\mathbf{1}_\Omega \rangle \langle \mathbf{1}_\Omega, W_2W_3 \cdots W_n Y Z_m \cdots Z_2 \mathbf{1}_\Omega \rangle \langle \mathbf{1}_\Omega, Z_1 \mathbf{1}_\Omega \rangle \\
&= \cdots \\
&= \langle \mathbf{1}_\Omega, W_1\mathbf{1}_\Omega \rangle \cdots \langle \mathbf{1}_\Omega, W_n\mathbf{1}_\Omega \rangle \langle \mathbf{1}_\Omega, Y\mathbf{1}_\Omega \rangle \langle \mathbf{1}_\Omega Z_m \mathbf{1}_\Omega \rangle \cdots \langle \mathbf{1}_\Omega, Z_1\mathbf{1}_\Omega \rangle,
\end{aligned}$$

and we have shown condition (ii) from Remark 4.25. Hence, X_t has monotonically independent increments. □

By combining Theorems 5.52 and 5.57, we obtain the converse of Lemma 5.49, i.e., we can construct a monotone additive process whose distributions are prescribed by a given compactly supported $\rhd$-hemigroup.

REMARK 5.59. For $0 \leq s < t$, let $\mu_{s,t} = Arc(0, t-s)$ and define $\mu_{s,s} = \delta_0$. Then $\mu_{s,t}$ defines a $\triangleright$-hemigroup (see Example 5.15; $(\mu_{0,t})_{t\geq 0}$ is a $\triangleright$-semigroup) and due to the monotone central limit theorem, a corresponding monotone additive process is called a monotone Brownian motion. In [**138**], Muraki constructed a monotone Brownian motion on a monotone Fock space, see also [**59, 124**].

REMARK 5.60. Theorems 5.52 and 5.57 are due to U. Franz and T. Hasebe [**70**, Section 4], with preliminary work from [**69**]. Its proof also works for constructing additive processes consisting of unbounded operators. Monotone Lévy processes consisting of bounded self-adjoint operators were previously constructed in [**71**, Theorem 4.1], and monotone additive processes (bounded, but also for operator-valued expectation) in [**100**]. Earlier, connections between the theory of univalent functions and non-commutative probability theory were identified by P. Biane [**37**] and R. O. Bauer [**20–22**].

The classical case of the construction of (possibly unbounded) additive processes is handled by Theorem 2.48. For the free and Boolean case, we refer to [**18**, p. 112 and Section 4 on p. 229].

REMARK 5.61. A free additive process with distributions $\mu_{s,t} = W(0, t-s)$, $s < t$, is called a free Brownian motion. It can be constructed as follows. For a Hilbert space H and $\Omega \in H$, $\|\Omega\| = 1$, let $\mathcal{F}(H)$ be the full Fock space defined by

$$\mathcal{F}(H) = \mathbb{C}\Omega \oplus \bigoplus_{n=1}^{\infty} H^{\otimes n},$$

which becomes a Hilbert space via the product

$$\langle f_1 \otimes \cdots \otimes f_n, g_1 \otimes \cdots \otimes g_n \rangle = \langle f_1, g_1 \rangle \cdots \langle f_n, g_n \rangle,$$
$$\langle f_1 \otimes \cdots \otimes f_n, g_1 \otimes \cdots \otimes g_m \rangle = 0 \text{ if } n \neq m, \langle f_1 \otimes \cdots \otimes f_n, \Omega \rangle = 0, \langle \Omega, \Omega \rangle = 1.$$

For $f \in H$, let $a(f)$ (left annihilation operator) and $a^*(f)$ (left creation operator) be the elements from $B(\mathcal{F}(H))$ defined via

$$a(f)(f_1 \otimes \cdots \otimes f_{n+1}) = \langle f, f_1 \rangle f_2 \otimes \cdots \otimes f_{n+1}, \quad a(f)(f_1) = \langle f, f_1 \rangle \Omega, \quad a(f)(\Omega) = 0,$$
$$a^*(f)(f_1 \otimes \cdots \otimes f_n) = f \otimes f_1 \otimes f_2 \otimes \cdots \otimes f_n, \quad a^*(f)(\Omega) = f.$$

Now take $H = L^2([0, \infty), \mathbb{C})$. For $t \geq 0$, $\mathbf{1}_{[0,t]} \in H$. Then

$$t \mapsto a(\mathbf{1}_{[0,t]}) + a^*(\mathbf{1}_{[0,t]})$$

is a free Brownian motion within $(B(\mathcal{F}(H)), X \mapsto \langle \Omega, X\Omega \rangle)$, see [**179**].

5.9. Exercises

E 5.9.1. Let $\mu \in \mathcal{P}(\mathbb{R})$ and let $a \in \mathbb{R}$, $b > 0$. Show that $G_\mu(a + ib) \to 0$ as $a \to \pm\infty$.

E 5.9.2. Let $\mu_1, \mu_2 \in \mathcal{P}(\mathbb{R})$ and assume that μ_1 and $\mu_1 \rhd \mu_2$ have finite variances. Show that also μ_2 must have finite variance $\mathrm{Var}(\mu_1 \rhd \mu_2) - \mathrm{Var}(\mu_1)$.

E 5.9.3. Let F_μ and F_ν be F-transforms with parameters (c, γ) and (b, β), respectively, in representation (5.1). Furthermore, let γ' be the measure of $F_{\mu \rhd \nu}$ from (5.1). Show that there exists an example where $\gamma'(\mathbb{R}) < \gamma(\mathbb{R})$.

E 5.9.4. Let $\mu, \nu, \nu' \in \mathcal{P}(\mathbb{R})$ and assume that $\mu \rhd \nu = \mu \rhd \nu'$. Show that $\nu = \nu'$.

E 5.9.5. Let $\star \in \{\boxplus, \uplus\}$. Prove that if μ, ν are $\star$-infinitely divisible, then also $\mu \star \nu$ is $\star$-infinitely divisible.

E 5.9.6. Determine all $\mu \in \mathcal{P}(\mathbb{R})$ having a Boolean Lévy pair of the form $(0, b \cdot \delta_c)$ with $c \in \mathbb{R}$, $b > 0$, i.e., $B_\mu(z) = b\frac{1+cz}{z-c}$.

E 5.9.7. Determine F_1 from Theorem 5.20 when the generator G has the form

$$G(z) = \frac{1}{x - z},$$

$x \in \mathbb{R}$ (see Remark 5.21).

E 5.9.8 (Boolean Poisson limit theorem). For $n \in \mathbb{N}$, let $(X_{k,n})_{1 \le k \le n}$ be Boolean *iid* random variables with distribution $c_n\delta_1 + (1 - c_n)\delta_0$ and assume that $nc_n \to c > 0$ as $n \to \infty$. Determine the limiting distribution of $X_{1,n} + \cdots + X_{n,n}$.

E 5.9.9 (Free Poisson limit theorem). For $n \in \mathbb{N}$, let $(X_{k,n})_{1 \le k \le n}$ be freely *iid* random variables with distribution $c_n\delta_1 + (1 - c_n)\delta_0$ and assume that $nc_n \to c > 0$ as $n \to \infty$. Show that the limiting distribution μ of $X_{1,n} + \cdots + X_{n,n}$ exists and satisfies

$$G_\mu(z) = \frac{z + 1 - c - \sqrt{(z - (1 + c))^2 - 4c}}{2z}.$$

(μ is the Marchenko–Pastur distribution from Theorem 4.36 with $\sigma^2 = 1/\lambda = c$.)

The monotone Poisson limit theorem can be found in [**140**].

5.10. Solutions

S 5.10.1. Let $\varepsilon > 0$. We find a compact set $K \subset \mathbb{R}$ with $\mu(\mathbb{R} \setminus K) < \varepsilon$. Then

$$|G_\mu(a+ib)| \leq \int_K \frac{1}{|a+ib-x|}\mu(dx) + \int_{\mathbb{R}\setminus K} \frac{1}{|a+ib-x|}\mu(dx)$$
$$\leq \max_{x\in K} \frac{1}{|a+ib-x|} + \frac{1}{b}\mu(\mathbb{R}\setminus K) < \varepsilon + \frac{\varepsilon}{b}$$

for $|a|$ large enough. So $G_\mu(a+ib) \to 0$ as $a \to \pm\infty$.

S 5.10.2. Let τ and ρ be the measures from Theorem 5.7 corresponding to μ_1 and $\mu_1 \rhd \mu_2$. Then

$$z - \int_{\mathbb{R}} \frac{1}{z-x}\rho(dx) = F_{\mu_1}(F_{\mu_2}(z)) = F_{\mu_2}(z) - \int_{\mathbb{R}} \frac{1}{F_{\mu_2}(z)-x}\tau(dx),$$

so

$$F_{\mu_2}(z) = z - \int_{\mathbb{R}} \frac{1}{z-x}\rho(dx) + \int_{\mathbb{R}} \frac{1}{F_{\mu_2}(z)-x}\tau(dx) =: z - H(z).$$

The function H must map $\mathbb{H}$ into $\mathbb{H}^- \cup \mathbb{R}$ as $\operatorname{Im}(F_{\mu_2}(z)) \geq \operatorname{Im}(z)$. Note that H is the difference of two Cauchy transforms: $H(z) = \int_{\mathbb{R}} \frac{1}{z-x}\rho(dx) - \tau(\mathbb{R}) \int_{\mathbb{R}} \frac{1}{z-x}(\tau/\tau(\mathbb{R}) \rhd \mu_2)(dx)$ (if $\tau(\mathbb{R}) > 0$). By Theorem 3.15, we conclude that H is the Cauchy transform of some non-negative measure with mass $\rho(\mathbb{R}) - \tau(\mathbb{R}) = \operatorname{Var}(\mu_1 \rhd \mu_2) - \operatorname{Var}(\mu_1)$, and thus, μ_2 has finite variance $\operatorname{Var}(\mu_1 \rhd \mu_2) - \operatorname{Var}(\mu_1)$ by Theorem 5.7.

S 5.10.3. Let $(c,\gamma) = (0,\delta_0)$, i.e., $\gamma(\mathbb{R}) = 1$ and $F_\mu(z) = z - \frac{1}{z}$. Furthermore, let $F_\nu(z) = z + b$. Then $\beta = 0$ and
$F_{\mu\rhd\nu}(i) = F_\mu(F_\nu(i)) = i + b - \frac{1}{i+b} = i + b - \frac{b-i}{1+b^2}$. By Remark 3.14, we have

$$\gamma'(\mathbb{R}) = \frac{1}{1+b^2} < 1 = \gamma(\mathbb{R})$$

if $b \neq 0$.

S 5.10.4. Let $\mu, \nu, \nu' \in \mathcal{P}(\mathbb{R})$ with $\mu \rhd \nu = \mu \rhd \nu'$.

We have $F_\mu \circ F_\nu = F_\mu \circ F_{\nu'}$ and by applying F_μ^{-1} as in (5.3), we obtain $F_\nu = F_{\nu'}$ on a sufficiently small $\Gamma_{\alpha,\beta}$. By the identity theorem, $F_\nu = F_{\nu'}$ on $\mathbb{H}$, and thus, $\nu = \nu'$.

S 5.10.5. Let $\star = \boxplus$. Then R_μ and R_ν both extend to holomorphic functions on $\mathbb{H}^-$ by Theorem 5.41. Thus $R_{\mu\boxplus\nu} = R_\mu + R_\nu$ extends to a holomorphic function on $\mathbb{H}^-$ as well, which shows that $\mu \boxplus \nu$ is $\boxplus$-infinitely divisible.

If $\star = \uplus$, there is nothing to show due to Theorem 5.31.

S 5.10.6. We have $b\frac{1+cz}{z-c} = z - \frac{1}{G_\mu(z)}$, and thus,

$$G_\mu(z) = \frac{1}{z - b\frac{1+cz}{z-c}} = \frac{z-c}{z^2 - zc(1+b) - b}.$$

The two zeros z_1, z_2, $z_1 < z_2$, of $z^2 - zc(1+b) - b$ satisfy $z_1 < c < z_2$, and thus, $\mu = w\delta_{z_1} + (1-w)\delta_{z_2}$ for some $w \in [0,1]$. By (3.3), we have

$$\begin{aligned} w = \mu(\{z_1\}) &= \lim_{y \searrow 0} iyG_\mu(z_1 + iy) = \lim_{y \searrow 0} iy \frac{(z_1+iy) - c}{((z_1+iy)-z_1)((z_1+iy)-z_2)} \\ &= \lim_{y \searrow 0} \frac{(z_1+iy)-c}{(z_1+iy)-z_2} = \frac{z_1 - c}{z_1 - z_2}. \end{aligned}$$

S 5.10.7. For $x = 0$, we have to solve the equation $\frac{d}{dt}y(t) = -\frac{1}{y(t)}, y(0) = z$. The solution is given by $y(t) = \sqrt{z^2 - 2t}$, and thus, $F_\mu(z) = F_1(z) = \sqrt{z^2-2}$, i.e., $\mu = Arc(0,1)$, see Example 5.3.

Let $\mu_{x,t} = \delta_{-x} \rhd Arc(0,t) \rhd \delta_x$. We obtain $F_{\mu_{x,t}}(z) = \sqrt{(z-x)^2 - 2t} + x$ and differentiation yields

$$\frac{d}{dt} F_{\mu_{x,t}}(z) = \frac{-2}{2\sqrt{(z-x)^2 - 2t}} = \frac{1}{x - F_{\mu_{x,t}}(z)}.$$

Thus $F_{\mu_{x,1}}(z) = \sqrt{(z-x)^2 - 2} + x$.

S 5.10.8. Let ν_n be the distribution of $X_{k,n}$. Then $G_{\nu_n}(z) = \frac{c_n}{z-1} + \frac{1-c_n}{z}$ and

$$B_{\nu_n}(z) = z - \frac{z(z-1)}{z-1+c_n} = \frac{c_n z}{z - 1 + c_n}.$$

Let μ_n be the distribution of $X_{1,n} + \cdots + X_{n,n}$. By Theorem 5.26, we get

$$B_{\mu_n}(z) = nB_{\nu_n}(z) = \frac{nc_n z}{z-1+c_n} \to \frac{cz}{z-1}.$$

Thus, $\mu_n \to \mu$, and μ is given by $G_\mu(z) = \frac{1}{z - \frac{cz}{z-1}} = \frac{z-1}{z(z-(1+c))}$, which gives $\mu = \frac{1}{1+c}\delta_0 + \frac{c}{1+c}\delta_{1+c}$.

S 5.10.9. Let ν_n be the distribution of $X_{k,n}$. Then $G_{\nu_n}(z) = \frac{c_n}{z-1} + \frac{1-c_n}{z}$ and

$$R_{\nu_n}(z) = \frac{z - 1 - \sqrt{1 - 2z + 4c_n z + z^2}}{2z}.$$

Let μ_n be the distribution of $X_{1,n} + \cdots + X_{n,n}$. Due to Theorem 5.32, $R_{\mu_n} = nR_{\nu_n}$, and we obtain

$$\begin{aligned} R_{\mu_n}(z) &= n\frac{(z-1)^2 - (1 - 2z + 4c_n z + z^2)}{2z(z - 1 + \sqrt{1 - 2z + 4c_n z + z^2})} \\ &= \frac{-4nc_n z}{2z(z - 1 + \sqrt{1 - 2z + 4c_n z + z^2}))} \to \frac{-cz}{z(z-1)}. \end{aligned}$$

So $G_\mu^{-1}(z) = \frac{z-1-cz}{z(z-1)}$ and

$$G_\mu(z) = \frac{z + 1 - c - \sqrt{(z-(1+c))^2 - 4c}}{2z}.$$

CHAPTER 6

Univalent Functions

In this chapter, we will prove that $\rhd$-hemigroups consist of probability measures with univalent F-transforms. More precisely, we have the following theorem.

THEOREM 6.1.

(a) *Let* $(\mu_{s,t})_{0\leq s\leq t}$ *be a* $\rhd$*-hemigroup. Then* $F_{\mu_{s,t}}$ *is univalent for all* $0 \leq s \leq t$.

(b) *Let* $\mu \in \mathcal{P}(\mathbb{R})$ *such that* F_μ *is univalent. Then there exists a* $\rhd$*-hemigroup* $(\mu_{s,t})_{0\leq s\leq t}$ *such that* $\mu_{0,1} = \mu$.

Statement (b) will be proven in the next chapter.

Together with Theorems 5.52 and 5.57 (and Remark 5.60), it follows that a probability distribution appears in a monotone additive process if and only if its Cauchy transform is univalent, which is Theorem 1.2 from the introduction.

$\rhd$-hemigroups are in one-to-one correspondence with certain flows of conformal mappings, so-called Loewner chains. We will first focus on specific Loewner chains that enable us to prove Theorem 6.1 (a) in Section 6.3.

In Section 6.5, we regard Loewner chains corresponding to $\rhd$-hemigroups that consist of probability measures with mean 0 and finite variance.

6.1. Loewner chains

Let $D \subsetneq \mathbb{C}$ be a simply connected domain. Loewner chains generalize the notion of a continuous semigroup on D from Section 3.6. Recall that a semigroup $(f_t)_{t\geq 0}$ on D satisfies $f_t = f_s \circ f_{t-s}$ whenever $0 \leq s \leq t$. We generalize this relation by replacing f_{t-s} with a function $f_{s,t}$, which depends on both s and t. While a semigroup $(f_t)_{t\geq 0}$ satisfies an autonomous differential equation, $\frac{d}{dt} f_t = G(f_t)$, we will show that, under certain conditions, Loewner chains satisfy a non-autonomous version of this equation.

DEFINITION 6.2.

(1) Let $(f_{s,t})_{0\leq s\leq t}$ be a family of non-constant holomorphic self-mappings $f_{s,t} : D \to D$ satisfying:
  (a) $f_{s,s}(z) = z$ for all $z \in D$ and $s \geq 0$,
  (b) $f_{s,u} = f_{s,t} \circ f_{t,u}$ for all $0 \leq s \leq t \leq u$,
  (c) $(s, t) \mapsto f_{s,t}$ is continuous with respect to locally uniform convergence.

  The family $(f_t)_{t\geq 0} := (f_{0,t})_{t\geq 0}$ is called a *(decreasing) Loewner chain on* D. We will call the mappings $f_{s,t}$ the *transition mappings* of the Loewner chain.

(2) We call a Loewner chain $(f_t)_{t\geq 0}$ an *additive Loewner chain* if $D = \mathbb{H}$ and $\lim_{y\to\infty} f_{s,t}(iy)/(iy) = 1$, or equivalently

$$f_{s,t} = F_{\mu_{s,t}}$$

for all $0 \leq s \leq t$, where each $\mu_{s,t}$ is a probability measure on $\mathbb{R}$.

Due to property (b), the domains $f_t(D)$ are decreasing, i.e., $f_t(D) \subseteq f_s(D)$ whenever $s \leq t$; see Figure 6.1.

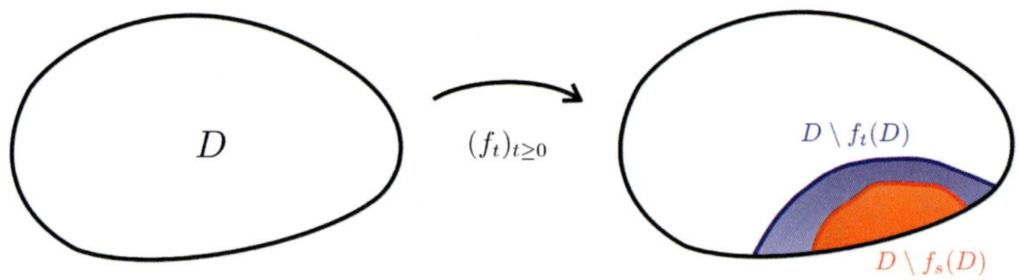

FIGURE 6.1. The complements $D \setminus f_t(D)$ of a Loewner chain $(f_t)_{t\geq 0}$ are growing subsets of D.

A few remarks and examples are in order.

REMARK 6.3. The literature usually treats *increasing* Loewner chains. This notion is due to Charles Loewner (Karel Löwner in Czech, Karl Löwner in German), who derived a differential equation for the flow of such chains of univalent functions in 1923 in order to tackle the Bieberbach conjecture, see Section 8.6.

EXAMPLE 6.4. Let $D = \mathbb{D}$ and $f_{s,t}(z) = e^{s-t}z$. Each $f_{s,t}$ is non-constant and properties (a) and (c) are clearly satisfied. For all $0 \leq s \leq t \leq u$, we have $(f_{s,t} \circ f_{t,u})(z) = e^{s-t}(e^{t-u}z) = e^{s-u}z = f_{s,u}(z)$, and we conclude that $(e^{-t}z)_{t\geq 0}$ is a Loewner chain on $\mathbb{D}$.

EXAMPLE 6.5. In the literature, a Loewner chain $(f_t)_{t\geq 0}$ is often assumed to consist of univalent functions, and then, $f_{s,t}$ is given by

$$f_{s,t} = f_s^{-1} \circ f_t.$$

Our definition is in accordance with $\triangleright$-hemigroups, see Lemma 6.7 below, and we will prove that Loewner chains consist of univalent functions in Theorem 6.17.

Let $D = \mathbb{D}$ and let $f_t(z) = (1-t)z$ for $t < 1$, and $f_t(z) = 0$ for $t \geq 1$. As f_t is constant for $t \geq 1$, $(f_t)_{t\geq 0}$ is not a Loewner chain. For $0 \leq s \leq t < 1$, we still have uniquely determined transition mappings defined by $f_{s,t}(z) = \frac{1-t}{1-s}z$. However, this family cannot be extended continuously to the case $s = t = 1$. Fix some $C \geq 1$, and consider the sequence $f_{1-\frac{C}{n},1-\frac{1}{n}}$ for $n \geq C$. Then

$$f_{1-\frac{C}{n},1-\frac{1}{n}}(z) = \frac{z}{C}.$$

Hence, as $n \to \infty$, the limit depends on C, and we cannot extend $f_{s,t}$ continuously to $(s,t) = (1,1)$. For $1 \leq s \leq t$, we can still define transition mappings, but the definition is arbitrary and thus meaningless in regard to the mappings $(f_t)_{t\geq 0}$.

REMARK 6.6. In Loewner theory, a family $(\phi_{s,t})_{0\leq s\leq t}$ of non-constant holomorphic mappings $\phi_{s,t} : D \to D$ is called an *evolution family on* D (modulo slight variations) if

(a) $\phi_{s,s}(z) = z$ for all $z \in D$ and all $s \geq 0$,
(b) $\phi_{s,u} = \phi_{t,u} \circ \phi_{s,t}$ whenever $0 \leq s \leq t \leq u$,
(c) $(s,t) \mapsto \phi_{s,t}$ is continuous with respect to locally uniform convergence.

If (b) is replaced by $\phi_{s,u} = \phi_{s,t} \circ \phi_{t,u}$, then the family is usually called a *reverse evolution family.* Thus, the transition mappings of a decreasing Loewner chain form a reverse evolution family.

We have chosen the name *additive* for the special Loewner chains in part (2) of Definition 6.2 since they are in one-to-one correspondence with $\rhd$-hemigroups.

LEMMA 6.7. *Let $(\mu_{s,t})_{0\leq s\leq t}$ be a $\rhd$-hemigroup. Then $(F_{\mu_{0,t}})_{t\geq 0}$ is an additive Loewner chain with transition mappings $F_{\mu_{s,t}}$. Conversely, if $(f_t)_{t\geq 0}$ is an additive Loewner chain with transition mappings $f_{s,t}$, then $f_{s,t} = F_{\mu_{s,t}}$ for a $\rhd$-hemigroup $(\mu_{s,t})_{0\leq s\leq t}$.*

PROOF. By Theorem 3.18, condition (c) is equivalent to the continuity of $(s,t) \mapsto \mu_{s,t}$ with respect to weak convergence. Note that F-transforms are always non-constant functions on $\mathbb{H}$. □

EXAMPLE 6.8. The measures $\mu_{s,t} = Arc(0, t-s)$ for $s < t$ and $\mu_{s,s} = \delta_0$ form a $\rhd$-hemigroup with $F_{\mu_{s,t}}(z) = \sqrt{z^2 - 2(t-s)}$, see Example 5.15. $F_t := F_{\mu_{0,t}}$ maps $\mathbb{H}$ conformally onto $\mathbb{H}$ minus the vertical line segment from 0 to $\sqrt{2t}i$. In fact, $t \mapsto F_t$ is a semigroup and its infinitesimal generator is $-1/z$ because

$$\frac{d}{dt}F_t(z) = \frac{-1}{\sqrt{z^2 - 2t}} = \frac{-1}{F_t(z)}.$$

6.2. Radial Loewner chains

DEFINITION 6.9.

(1) A Loewner chain $(f_t)_{t\geq 0}$ on $\mathbb{D}$ with $f_{s,t}(0) = 0$ for all $0 \leq s \leq t$ will be called *radial.*
(2) A *normalized radial Loewner chain* is a radial Loewner chain with $f'_{s,t}(0) = e^{s-t}$ for all $0 \leq s \leq t$.

Radial Loewner chains, like additive Loewner chains, also have a probabilistic interpretation. We can define $\rhd$-hemigroups also for distributions on the unit circle $\partial\mathbb{D}$, which arise by replacing self-adjoint random variables with unitary random variables, see Section 7.1. (Accordingly, we could call radial Loewner chains also *multiplicative Loewner chains*, but we will stick to the more common name.)

For the rest of this section, we fix a normalized radial Loewner chain $(f_t)_{t\geq 0}$. We will see that $t \mapsto f_t$ is differentiable almost everywhere, and that each $f_{s,t} : \mathbb{D} \to \mathbb{D}$ is a univalent function. We follow [**151**, Section 6.1] with slight modifications.

Recall the class $\mathcal{P}$ defined as

$$\mathcal{P} = \{p : \mathbb{D} \to \mathbb{C} \,|\, p \text{ holomorphic, } \operatorname{Re}(p) > 0,\ p(0) = 1\}.$$

By the Herglotz representation formula, Theorem 3.11, every $p \in \mathcal{P}$ can be written as

$$p(z) = \int_{\partial\mathbb{D}} \frac{u+z}{u-z}\rho(du) \quad \text{for a probability measure } \rho \text{ on } \partial\mathbb{D}.$$

LEMMA 6.10. *Let $p \in \mathcal{P}$. Then*

$$|p(z)| \leq \frac{1+|z|}{1-|z|} \quad \textit{and} \quad |p'(z)| \leq \frac{2}{(1-|z|)^2} \quad \textit{for all } z \in \mathbb{D}.$$

PROOF. The Herglotz representation formula yields

$$|p(z)| \leq \int_{\partial\mathbb{D}} \left|\frac{1+z/u}{1-z/u}\right| \rho(du) \leq \int_{\partial\mathbb{D}} \frac{1+|z/u|}{1-|z/u|}\rho(du) = \int_{\partial\mathbb{D}} \frac{1+|z|}{1-|z|}\rho(du) = \frac{1+|z|}{1-|z|}.$$

Furthermore,

$$\begin{aligned} |p'(z)| &= \left|\int_{\partial\mathbb{D}} \frac{2u}{(u-z)^2}\rho(du)\right| \leq \int_{\partial\mathbb{D}} \left|\frac{2u}{(u-z)^2}\right| \rho(du) \\ &\leq \int_{\partial\mathbb{D}} \frac{2}{(1-|z|)^2}\rho(du) = \frac{2}{(1-|z|)^2}. \end{aligned}$$

□

The set of all holomorphic $f : \mathbb{D} \to \mathbb{D}$ forms a normal family, and hence, also the set of its derivatives f' is a normal family. In the following, we let $m(r)$, $r \in [0,1)$, be a non-decreasing function such that

$$|f'(z)| \leq m(|z|)$$

for all $z \in \mathbb{D}$ and all holomorphic $f : \mathbb{D} \to \mathbb{D}$ with $f(0) = 0$. One can derive an explicit bound m. Dieudonné has shown that one can choose $m(r) = 1$ for $r \in [0, \sqrt{2}-1]$ and $m(r) = \frac{(1+r^2)^2}{4r(1-r^2)}$ for $r \in [\sqrt{2}-1, 1)$, see [**61**, p. 352].

LEMMA 6.11. *Let $s \geq 0$. Then*

$$|f_{s,u}(z) - f_{s,t}(z)| \leq 2|z|\frac{1+|z|}{1-|z|}m(|z|)|e^t - e^u| \tag{6.1}$$

for all $t, u \geq s$ and all $z \in \mathbb{D}$.

PROOF. We have $f_t = f_s \circ f_{s,t}$ and $|f_{s,t}(z)| \leq |z|$ due to the Schwarz lemma. Hence, $f_{s,t}(z)/z$ is an analytic function in $\mathbb{D}$ with modulus ≤ 1. Define

$$p(z,s,t) = \frac{1+e^{s-t}}{1-e^{s-t}}\frac{1-f_{s,t}(z)/z}{1+f_{s,t}(z)/z}. \tag{6.2}$$

Then $p(\cdot, s, t)$ belongs to the class $\mathcal{P}$. Since $|p(z,s,t)| \leq \frac{1+|z|}{1-|z|}$ by Lemma 6.10, (6.2) gives us

$$|z - f_{s,t}(z)| \leq 2|z|(1-e^{s-t})\frac{1+|z|}{1-|z|}.$$

Assume that $t \leq u$. Then

$$|f_{s,t}(z) - f_{s,u}(z)| = \left| \int_{f_{t,u}(z)}^{z} f'_{s,t}(w)dw \right| \leq |z - f_{t,u}(z)| m(|z|)$$
$$\leq 2|z| \frac{e^u - e^t}{e^u} \frac{1+|z|}{1-|z|} m(|z|) \leq 2|z| \frac{1+|z|}{1-|z|} m(|z|)(e^u - e^t). \quad \square$$

Now we are able to derive Loewner's partial differential equation. We recall the fundamental theorem of calculus for Lebesgue integration: a function $f : [a, b] \to \mathbb{C}$ is absolutely continuous if and only if its derivative f' exists almost everywhere and f' is Lebesgue integrable with

$$f(x) = f(a) + \int_a^x f'(y)dy$$

for all $x \in [a, b]$.

THEOREM 6.12. *The function $t \mapsto f_t(z)$ is locally absolutely continuous for every $z \in \mathbb{D}$ and there exist a function $p : \mathbb{D} \times [0, \infty) \to \mathbb{C}$, such that $p(\cdot, t) \in \mathcal{P}$ for almost all $t \geq 0$ and $t \mapsto p(z, t)$ is measurable for all $z \in \mathbb{D}$, and a set $\mathcal{N} \subset [0, \infty)$ of Lebesgue measure 0 such that*

$$(6.3) \quad \frac{\partial}{\partial t} f_t(z) = -z \frac{\partial}{\partial z} f_t(z) p(z, t) \quad \text{for all } t \in [0, \infty) \setminus \mathcal{N} \text{ and all } z \in \mathbb{D}, f_0(z) = z.$$

PROOF. The estimate (6.1) implies that $t \mapsto f_t(z)$ is locally absolutely continuous for all $z \in \mathbb{D}$. Thus, for each $z \in Q := \{z \in \mathbb{D} \,|\, \mathrm{Re}(z), \mathrm{Im}(z) \in \mathbb{Q}\}$, there exists a set $N_z \subset [0, \infty)$ of Lebesgue measure 0 such that $t \mapsto f_t(z)$ is differentiable for all $t \in [0, \infty) \setminus N_z$. Therefore, $t \mapsto f_t(z)$ is differentiable for all $z \in Q$ and all $t \in [0, \infty) \setminus \mathcal{N}$, where $\mathcal{N} = \cup_{z \in Q} N_z$ has Lebesgue measure 0 since Q is countable.

Let $t \in [0, \infty) \setminus \mathcal{N}$, and let h_n be a null-sequence such that $F_n(z) := (f_{t+h_n}(z) - f_t(z))/h_n$ is defined. Then the sequence $(F_n)_n$ is locally uniformly bounded due to (6.1), and thus, it is a normal family. For each $z \in Q$, F_n converges as $n \to \infty$. Each locally uniformly converging subsequence of F_n converges on Q to the same limit. Thus, each locally uniformly converging subsequence of F_n converges on $\mathbb{D}$ to the same limit, which proves that F_n is locally uniformly converging. Consequently, $t \mapsto f_t(z)$ is differentiable for all $t \in [0, \infty) \setminus \mathcal{N}$ and all $z \in \mathbb{D}$.

Let $t \notin \mathcal{N}$ and let $u > t$. By using (6.2), we have

$$\frac{f_t(z) - f_u(z)}{t - u} = \frac{f_t(z) - f_u(z)}{z - f_{t,u}(z)} \frac{z - f_{t,u}(z)}{t - u}$$
$$= \frac{f_t(z) - f_t(f_{t,u}(z))}{z - f_{t,u}(z)} \frac{1 - e^{t-u}}{t - u} \frac{z + f_{t,u}(z)}{1 + e^{t-u}} p(z, t, u).$$

As $u \to t$, we obtain

$$\frac{\partial}{\partial t} f_t(z) = -z \frac{\partial}{\partial z} f_t(z) p(z, t),$$

for some $p(\cdot, t) \in \mathcal{P}$, as $\mathcal{P}$ is closed with respect to locally uniform convergence. As $\frac{\partial}{\partial t} f_t(z)$ and $\frac{\partial}{\partial z} f_t(z)$ are both measurable in t, also $p(z, t)$ is measurable in t. We can extend p to $\mathbb{D} \times [0, \infty)$ such that $p(z, t)$ is measurable in t. □

REMARK 6.13. Fix $s \geq 0$, let $t \in [s, \infty) \setminus \mathcal{N}$, and $u > t$. Then

$$\begin{aligned}\frac{f_{s,t}(z) - f_{s,u}(z)}{t - u} &= \frac{f_{s,t}(z) - f_{s,u}(z)}{z - f_{t,u}(z)} \frac{z - f_{t,u}(z)}{t - u} \\ &= \frac{f_{s,t}(z) - f_{s,t}(f_{t,u}(z))}{z - f_{t,u}(z)} \frac{1 - e^{t-u}}{t - u} \frac{z + f_{t,u}(z)}{1 + e^{t-u}} p(z, t, u).\end{aligned}$$

We know that the right side converges as $u \to t$. Thus, $\frac{\partial}{\partial t} f_{s,t}(z)$ exists with

$$\frac{\partial}{\partial t} f_{s,t}(z) = -z \frac{\partial}{\partial z} f_{s,t}(z) p(z, t) \quad \text{for almost every (a.e.) } t \geq s \text{ and all } z \in \mathbb{D}.$$

In order to study Loewner's partial differential equation, it is helpful to consider a related ordinary differential equation.

THEOREM 6.14. *Let $p : \mathbb{D} \times [0, \infty) \to \mathbb{C}$ be such that $p(\cdot, t) \in \mathcal{P}$ for almost all $t \geq 0$ and $t \mapsto p(z, t)$ is measurable for all $z \in \mathbb{D}$. There exists a unique solution of holomorphic functions $g_{s,t} : \mathbb{D} \to \mathbb{D}$, such that $t \mapsto g_{s,t}(z)$ is locally absolutely continuous for every $z \in \mathbb{D}$, to the initial value problem*

$$\frac{\partial}{\partial t} g_{s,t}(z) = -g_{s,t}(z) p(g_{s,t}(z), t), \quad \textit{for a.e. } t \geq s, g_{s,s}(z) = z. \tag{6.4}$$

Furthermore, each $g_{s,t} : \mathbb{D} \to \mathbb{D}$ is univalent and $g_{s,u} = g_{t,u} \circ g_{s,t}$ for all $0 \leq s \leq t \leq u$.

PROOF. Let $s \geq 0$ and $r \in (0, 1)$. For $t \geq s$ and $|z| \leq r$, we define $g^0_{s,t}(z) \equiv 0$ and for $n \in \mathbb{N}$, let

$$g^n_{s,t}(z) := z \exp\left(-\int_s^t p(g^{n-1}_{s,\tau}(z), \tau) d\tau\right).$$

As $\operatorname{Re} p(\cdot, t) > 0$ in $\mathbb{D}$ for almost all t, we see by induction that $|g^n_{s,t}(z)| \leq r$, which justifies the definition. Since $|e^{-a} - e^{-b}| \leq |a - b|$ for complex numbers a, b with $\operatorname{Re}(a), \operatorname{Re}(b) \geq 0$, it follows, together with the second estimate from Lemma 6.10, that

$$\begin{aligned}|g^{n+1}_{s,t}(z) - g^n_{s,t}(z)| &\leq \int_s^t |p(g^n_{s,\tau_n}(z), \tau_n) - p(g^{n-1}_{s,\tau_n}(z), \tau_n)| d\tau_n \\ &\leq \frac{2}{(1-r)^2} \int_s^t |g^n_{s,\tau_n}(z) - g^{n-1}_{s,\tau_n}(z)| d\tau_n.\end{aligned}$$

Inductively, we obtain

$$|g_{s,t}^{n+1}(z) - g_{s,t}^{n}(z)| \leq \frac{2^n}{(1-r)^{2n}} \int_s^t \int_s^{\tau_n} \cdots \int_s^{\tau_2} |g_{s,\tau_1}^1(z) - g_{s,\tau_1}^0(z)| d\tau_1 \dots d\tau_n$$

$$= \frac{2^n}{(1-r)^{2n}} \int_s^t \int_s^{\tau_n} \cdots \int_s^{\tau_2} |g_{s,\tau_1}^1(z)| d\tau_1 \dots d\tau_n$$

$$\leq \frac{2^n}{(1-r)^{2n}} \int_s^t \int_s^{\tau_n} \cdots \int_s^{\tau_2} 1 d\tau_1 \dots d\tau_n = \frac{2^n (t-s)^n}{(1-r)^{2n} n!}$$

for all $n \geq 0$. Using this estimate, we see that $g_{s,t}^n(z)$ is a Cauchy sequence, and thus, $\lim_{n\to\infty} g_{s,t}^n(z) =: g_{s,t}(z)$ exists, uniformly for $|z| \leq r$, $s \leq t \leq T$, for every $T > s$. Thus, $g_{s,t}(z)$ is defined for all $t \geq s$ and all $z \in \mathbb{D}$, and $z \mapsto g_{s,t}(z)$ is holomorphic with $g_{s,t}(\mathbb{D}) \subset \mathbb{D}$. Furthermore,

$$g_{s,t}(z) = z \exp\left(-\int_s^t p(g_{s,\tau}(z), \tau) d\tau\right),$$

which shows that $t \mapsto g_{s,t}(z)$ is locally absolutely continuous for every $z \in \mathbb{D}$ and that $g_{s,t}$ solves (6.4).

Next we show that $g_{s,t}$ is univalent. Suppose that $g_{s,T}(z) = g_{s,T}(w)$ for some $z, w \in \mathbb{D}$ with $z \neq w$ and some $T > s$. Then, for a.e. t with $s \leq t \leq T$, we have

$$\frac{d}{dt}(g_{s,t}(z) - g_{s,t}(w)) = g_{s,t}(w) p(g_{s,t}(w), t) - g_{s,t}(z) p(g_{s,t}(z), t)$$
$$= g_{s,t}(w)(p(g_{s,t}(w), t) - p(g_{s,t}(z), t)) + p(g_{s,t}(z), t)(g_{s,t}(w) - g_{s,t}(z)).$$

Thus, with Lemma 6.10,

$$\left|\frac{d}{dt}|g_{s,t}(z) - g_{s,t}(w)|\right| \leq \left|\frac{d}{dt}(g_{s,t}(z) - g_{s,t}(w))\right| \leq C|g_{s,t}(z) - g_{s,t}(w)|$$

for some constant $C > 0$. In particular,

$$\frac{d}{dt}|g_{s,t}(z) - g_{s,t}(w)| \geq -C|g_{s,t}(z) - g_{s,t}(w)|,$$

which implies $\frac{d}{dt}(e^{Ct}|g_{s,t}(z) - g_{s,t}(w)|) \geq 0$. Integration from s to T yields

$$e^{CT}|g_{s,T}(z) - g_{s,T}(w)| - e^{Cs}|g_{s,s}(z) - g_{s,s}(w)| = -e^{Cs}|z - w| \geq 0.$$

Thus $z = w$, a contradiction.

Now, let $h_{s,t}$ be another solution to (6.4) such that $t \mapsto h_{s,t}(z)$ is locally absolutely continuous for every $z \in \mathbb{D}$. Fix $z \in \mathbb{D}$. Then, for a.e. t with $s \leq t$,

$$\frac{d}{dt}|h_{s,t}(z) - g_{s,t}(z)| \leq \left|\frac{d}{dt}(h_{s,t}(z) - g_{s,t}(z))\right| =$$
$$|h_{s,t}(z) p(h_{s,t}(z), t) - g_{s,t}(z) p(g_{s,t}(z), t)| =$$
$$|h_{s,t}(z) p(h_{s,t}(z), t) - g_{s,t}(z) p(h_{s,t}(z), t) + g_{s,t}(z) p(h_{s,t}(z), t) - g_{s,t}(z) p(g_{s,t}(z), t)|$$
$$\leq C|h_{s,t}(z) - g_{s,t}(z)|$$

for some $C > 0$, again by Lemma 6.10. We obtain $\frac{d}{dt}(e^{-Ct}|g_{s,t}(z) - h_{s,t}(z)|) \leq 0$, and thus, $e^{-Ct}|g_{s,t}(z) - h_{s,t}(z)| \leq 0$ as $h_{s,s}(z) = z = g_{s,s}(z)$. So $h_{s,t}(z) = g_{s,t}(z)$ for all $t \geq s$.

Let $0 \leq s \leq \tau$. Consider the function $t \mapsto g_{\tau,t} \circ g_{s,\tau}$ for $t \geq \tau$. It coincides with $g_{s,t}$ for $t = \tau$ and it satisfies the differential equation (6.4). Hence, $g_{s,t} = g_{\tau,t} \circ g_{s,\tau}$ for all $0 \leq s \leq \tau \leq t$. □

COROLLARY 6.15. *Let $p : \mathbb{D} \times [0, \infty) \to \mathbb{C}$ be as in Theorem 6.12. Then the initial value problem (6.3) has exactly one solution $(h_t)_{t \geq 0}$ of holomorphic mappings $h_t : \mathbb{D} \to \mathbb{D}$, such that $t \mapsto h_t(z)$ is locally absolutely continuous for every $z \in \mathbb{D}$. The family $(h_t)_{t \geq 0}$ is a normalized radial Loewner chain and each h_t is univalent.*

PROOF. Fix some $T > 0$ and let $g_{s,t}$ be the solution to (6.4) for $p(\cdot, T - t)$.

Define $h_t = g_{T-t,T}$, $0 \leq t \leq T$. We have $h_t = g_{T-t,T} = g_{T-s,T} \circ g_{T-t,T-s} = h_s \circ g_{T-t,T-s}$ whenever $s \leq t$. Hence, $(h_t)_{0 \leq t \leq T}$ is (a part of) a decreasing Loewner chain consisting of univalent functions. From $g_{0,T} = h_t \circ g_{0,T-t}$, we have (for $z \neq 0$)

$$0 = \frac{(h_{t+h} \circ g_{0,T-t-h})(z) - (h_t \circ g_{0,T-t})(z)}{h} = \frac{(h_{t+h} \circ g_{0,T-t})(z) - (h_t \circ g_{0,T-t})(z)}{h}$$
$$+ \frac{(h_{t+h} \circ g_{0,T-t-h})(z) - (h_{t+h} \circ g_{0,T-t})(z)}{h} = \frac{(h_{t+h} \circ g_{0,T-t})(z) - (h_t \circ g_{0,T-t})(z)}{h}$$
$$+ \frac{(h_{t+h} \circ g_{0,T-t-h})(z) - (h_{t+h} \circ g_{0,T-t})(z)}{g_{0,T-t-h}(z) - g_{0,T-t}(z)} \cdot \frac{g_{0,T-t-h}(z) - g_{0,T-t}(z)}{h}$$

and we see that $u \mapsto h_u(g_{0,T-t}(z))$ is differentiable at $u = t$ provided $u \mapsto g_{0,T-u}(z)$ is differentiable at t. In that case

$$\begin{aligned} \frac{\partial}{\partial u} h_u(g_{0,T-t}(z))\bigg|_{u=t} &= -h_t'(g_{0,T-t}(z)) \frac{\partial}{\partial t} g_{0,T-t}(z) \\ &= -h_t'(g_{0,T-t}(z)) g_{0,T-t}(z) p(g_{0,T-t}(z), t). \end{aligned}$$

As $g_{0,T-t}(\mathbb{D})$ is open, the limit of $(h_{t+h}(z) - h_t(z))/h$ exists for all $z \in \mathbb{D}$ by Vitali's theorem. Hence,

$$\frac{\partial}{\partial t} h_t(z) = -z \frac{\partial}{\partial z} h_t(z) \cdot p(z, t) \quad \text{for a.e. } 0 \leq t \leq T, \quad h_0(z) = z \in \mathbb{D}.$$

By choosing another $\hat{T} > T$, we obtain a family $(\hat{g}_{s,t})_{0 \leq s \leq t \leq \hat{T}}$ with

$$\hat{g}_{s,\hat{T}} = g_{s+T-\hat{T},\hat{T}+T-\hat{T}} = g_{s+T-\hat{T},T}$$

for all $s \in [\hat{T} - T, \hat{T}]$. Hence, $\hat{h}_t := \hat{g}_{\hat{T}-t,\hat{T}} = g_{T-t,T} = h_t$ for all $t \in [0, T]$.

As we can choose $T > 0$ arbitrarily large, we conclude that there exits a decreasing Loewner chain $(h_t)_{t\geq 0}$ of univalent functions satisfying (6.3). From $g_{s,t}(0) = 0$ and $g'_{s,t}(0) = e^{s-t}$, we see that $(h_t)_{t\geq 0}$ is a normalized radial Loewner chain.

Conversely, let $(J_t)_{t\geq 0}$ be a family of holomorphic mappings, such that $t \mapsto J_t(z)$ is locally absolutely continuous for every $z \in \mathbb{D}$, satisfying (6.3). For $T > 0$, we have

$$\frac{\partial}{\partial t}[(J_t \circ g_{0,T-t})(z)] = -g_{0,T-t}(z)J_t'(g_{0,T-t}(z))p(g_{0,T-t}(z),t) \\ + J_t'(g_{0,T-t}(z)) \cdot (g_{0,T-t}(z)p(g_{0,T-t}(z),t)) = 0$$

for a.e. $t \leq T$, and thus, $J_t(g_{0,T-t}(z)) = J_0(g_{0,T}(z)) = g_{0,T}(z)$ for all $z \in \mathbb{D}$ and $0 \leq t \leq T$. This implies $J_t = g_{0,T} \circ g_{0,T-t}^{-1} = g_{T-t,T}$ on $g_{0,T-t}(\mathbb{D})$. As $g_{0,T-t}(\mathbb{D})$ is an open set, the identity theorem implies $J_t = g_{T-t,T} = h_t$ on $\mathbb{D}$. We can choose T arbitrarily large, so $J_t = h_t$ for all $t \geq 0$. □

REMARK 6.16. The uniqueness of the solution to (6.3) can also be shown as follows. Consider the power series expansions $p(z,t) = \sum_{n=0}^{\infty} c_n(t)z^n$ and $f_t(z) = \sum_{n=1}^{\infty} a_n(t)z^n$, where $f_t(z)$ is a solution to (6.3), locally absolutely continuous in t for every z. Then

$$\frac{d}{dt}a_1(t)z + \frac{d}{dt}a_2(t)z^2 + \dots \\ = -z(a_1(t) + 2a_2(t)z + 3a_3(t)z^2 + \dots)(1 + c_1(t)z + c_2(t)z^2 + \dots).$$

By comparing coefficients of both sides, we obtain

$$\frac{d}{dt}a_1(t) = -a_1(t), \quad a_1(0) = 1, \quad \text{and}$$

$$\frac{d}{dt}a_n(t) = -\sum_{k=1}^{n} ka_k(t)c_{n-k}(t), \quad a_n(0) = 0, n \geq 2.$$

In order to show the uniqueness of these initial value problems, we cannot apply the Picard–Lindelöf uniqueness theorem, as the equations only hold almost everywhere. However, a similar result also holds in this situation, see [**83**, Theorem 5.3]. We can use Exercise 3.9.6 here, which shows that $|c_n(t)| \leq 2$ for almost all t and all $n \geq 1$. Thus, each function $t \mapsto a_n(t)$ is uniquely determined, and there exists at most one solution of holomorphic mappings, locally absolutely continuous in t for every z, to (6.3).

6.3. Univalent F-transforms

THEOREM 6.17. *Let $(f_t)_{t\geq 0}$ be a Loewner chain. Then every transition mapping $f_{s,t}$, in particular every $f_t = f_{0,t}$, is a univalent function.*

It follows that each transition function $f_{s,t}$ can be expressed as

$$f_{s,t} = f_s^{-1} \circ f_t.$$

PROOF. Because of $f_t = f_s \circ f_{s,t}$, it is sufficient to prove that f_t is univalent for all $t \geq 0$.

Step 1: Assume that $(f_t)_{t\geq 0}$ is a Loewner chain on $\mathbb{D}$ with $f_{s,t}(0) = 0$ for all $0 \leq s \leq t$. Let $a_t := f_t'(0)$. Due to the Schwarz lemma, we have $|a_t| \leq 1$ and, as $f_t = f_s \circ f_{s,t}$, $t \mapsto |a_t|$ is non-increasing. Furthermore, as $t \mapsto f_t$ is continuous, also $t \mapsto a_t$ is continuous, and we conclude that $a_t \neq 0$ for all $t \in [0, \varepsilon]$ and some $\varepsilon > 0$.

First, assume that $a_t \neq 0$ for all $t \geq 0$. Then we have $0 < |a_t| \leq 1$ for all $t \geq 0$ and there exists a uniquely determined continuous function $C : [0, \infty) \to \{z \in \mathbb{C} \,|\, \mathrm{Re}(z) \leq 0\}$ with $C(0) = 0$ such that $a_t = e^{C(t)}$. It is easy to see that

$$g_t(z) := f_t(e^{-i\,\mathrm{Im}(C(t))}z)$$

is also a radial Loewner chain with

$$g_t'(0) = e^{\mathrm{Re}(C(t))}.$$

The function $t \mapsto \mathrm{Re}(C(t))$ is non-increasing and continuous. We note that if $\mathrm{Re}(C(t)) = \mathrm{Re}(C(s))$, $s \leq t$, then $g_t = g_s$, as $g_t = g_s \circ g_{s,t}$ with $g_{s,t} : \mathbb{D} \to \mathbb{D}$, $g_{s,t}(0) = 0$, $g_{s,t}'(0) = 1$, i.e., $g_{s,t}$ is the identity by the Schwarz lemma.

If $\mathrm{Re}(C(t)) = 0$ for all $t \geq 0$, then all f_t are rotations, which are univalent functions. So assume that $\mathrm{Re}(C(t))$ is negative for some t.

We can reparametrize g_t to $h_s := g_{\tau(s)}$ such that $h_s'(0) = e^{-s}$ for all $s \in [0, S)$ for some $0 < S \leq \infty$, where $\tau(s)$ is defined by

$$\tau(s) = \inf\{t \geq 0 : \mathrm{Re}(C(t)) = -s\},$$

which is a strictly increasing, possibly discontinuous function. The reparametrization $(h_s)_{s\in[0,S)}$ is (part of) a normalized radial Loewner chain and Theorem 6.12 plus Corollary 6.15 imply that each h_s is univalent, which implies that each f_t is univalent. (We refer to Section 7.2 for an alternative proof.)

Now assume that $a_\tau = 0$ for some $\tau > 0$ and $a_t \neq 0$ for $t < \tau$. The previous case implies that f_t is univalent for all $t < \tau$. Hence, $f_\tau = \lim_{t\nearrow\tau} f_t$ is the limit of univalent self-mappings of $\mathbb{D}$ fixing 0. It follows that $f_\tau(z) \equiv 0$. This is a contradiction as all elements of a Loewner chain are non-constant by definition.

Step 2: Now we consider the general case. We can use a conformal mapping $I : D \to \mathbb{D}$ to transfer the problem to the unit disk, i.e., we define $F_{s,t} := I \circ f_{s,t} \circ I^{-1}$, which gives transition mappings of a Loewner chain on $\mathbb{D}$. Next, we use an idea from Proposition 2.9 in [**52**]. Fix some $T > 0$. We define

$$a(t) := F_{t,T}(0), \quad h_t(z) := \frac{z + a(t)}{1 + \overline{a(t)}z},$$

for $0 \leq t \leq T$, $z \in \mathbb{D}$. Note that h_t is an automorphism of $\mathbb{D}$ mapping 0 onto $a(t)$. Define

$$(G_{s,t})_{0\leq s\leq t\leq T} := (h_s^{-1} \circ F_{s,t} \circ h_t)_{0\leq s\leq t\leq T}$$

and $G_t = G_{0,t}$, $F_t = F_{0,t}$. Then $(G_t)_{0\leq t\leq T}$ is (part of) a Loewner chain on $\mathbb{D}$ with transition mappings $(G_{s,t})_{0\leq s\leq t\leq T}$. We have

$$G_{s,t}(0) = (h_s^{-1} \circ F_{s,t} \circ h_t)(0) = (h_s^{-1} \circ F_{s,t})(F_{t,T}(0)) = h_s^{-1}(F_{s,T}(0)) = 0.$$

Hence, $(G_t)_t$ is a radial Loewner chain and step 1 implies that every G_t, $t \in [0, T]$, is univalent. As $T > 0$ can be chosen arbitrarily large, we conclude that every F_t is univalent. □

By Lemma 6.7, we conclude the following corollary, which is Theorem 6.1 (a).

COROLLARY 6.18. *Let $(\mu_{s,t})_{0\leq s\leq t}$ be a $\rhd$-hemigroup. Then $F_{\mu_{s,t}}$ is univalent for all $0 \leq s \leq t$.*

Which subdomains of $\mathbb{H}$ take the form $F_\mu(\mathbb{H})$ for a univalent F_μ? Since the Denjoy–Wolff point of F_μ is at ∞, these subdomains should include all simply connected subdomains of $\mathbb{H}$ that have ∞ as a boundary point in a certain sense. However, a geometric characterization is not known to the author. We refer the reader to [**152**, Chapters 4 and 11]. A partial result can be found in [**70**, Theorem 3.17]. Let us consider the case $\mu \in \mathcal{P}_c(\mathbb{R})$, which is relatively straightforward to handle.

THEOREM 6.19. *Let $\Omega \subseteq \mathbb{H}$ be a simply connected domain such that $\mathbb{H} \setminus \Omega$ is a bounded set. Then there exists a unique probability measure μ on $\mathbb{R}$ with mean 0 and compact support such that $F_\mu(\mathbb{H}) = \Omega$.*

PROOF. By the Riemann mapping theorem, we find a conformal mapping $f : \Omega \to \mathbb{H}$. Consider the complement $B = \mathbb{H} \setminus \Omega$. As B is bounded, we find a disk $R \cdot \mathbb{D}$ such that $\overline{B} \subset R \cdot \mathbb{D}$. Then f maps the curve $\partial(R \cdot \mathbb{D}) \cap \mathbb{H}$ onto a simple curve in $\mathbb{H}$ with two endpoints $a, b \in \mathbb{R}$, $a < b$, due to [**152**, Prop. 2.14]. Thus, f maps the Jordan domain $\mathbb{H} \setminus R \cdot \overline{\mathbb{D}}$ (as a domain in $\hat{\mathbb{C}}$) onto another Jordan domain, and Theorem B.4 implies that f extends to a homeomorphism between the closures of these domains. We can postcompose f with an automorphism of $\mathbb{H}$ to get that $f(\infty) = \infty$. As $f(x) \in \mathbb{R}$ whenever $x \in (-\infty, -R) \cup (R, \infty)$, we can extend f to a holomorphic function on $\hat{\mathbb{C}} \setminus R \cdot \overline{\mathbb{D}}$ with power series expansion $f(z) = az + b + \sum_{n=1}^{\infty} c_n/z^n$, $a \neq 0$, at ∞, as f has a simple pole at ∞. Furthermore, all coefficients are real numbers as $f(x) \in \mathbb{R}$ for $x \in \mathbb{R}$ and $|x|$ large enough. Because $\mathrm{Im}(f(iy))/y \to a$ as $y \to \infty$, we see that $a > 0$. Thus, we can postcompose f with the automorphism $(z - b)/a$ of $\mathbb{H}$ to obtain a conformal mapping $f : \Omega \to \mathbb{H}$ such that $f^{-1}(z) = z - \frac{c}{z} + \ldots$ with $c \in \mathbb{R}$. Due to Corollary 3.16, $f^{-1}(z) - z$ maps $\mathbb{H}$ into $\mathbb{H} \cup \mathbb{R}$, which implies that $c \geq 0$.

Remark 5.8 implies that $f^{-1} = F_\mu$ for a probability measure $\mu \in \mathcal{P}_c(\mathbb{R})$ with mean 0 and variance c.

Any other conformal mapping $G : \mathbb{H} \to \Omega$ has the form $F_\mu \circ \alpha$, for some automorphism α of $\mathbb{H}$, which is an F-transform of some $\nu \in \mathcal{P}(\mathbb{R})$ only if $\alpha(z) = z - d$, $d \in \mathbb{R}$. However, for $d \neq 0$, ν does not have mean 0. □

REMARK 6.20. The number c in the proof is equal to the variance of μ. This value is also called the *half-plane capacity* of the "hull" $\mathbb{H} \setminus \Omega$, see [**115**, Section 3.4]. It has a more or less geometric interpretation, see [**113**]. An explicit probabilistic formula is given in [**115**, Proposition 3.41].

REMARK 6.21. In general, if there are two univalent F-transforms with $F_\mu(\mathbb{H}) = F_{\mu'}(\mathbb{H})$, then $\alpha = F_\mu^{-1} \circ F_{\mu'}$ is an automorphism of $\mathbb{H}$ with $\alpha(\infty) = \infty$ and $\alpha'(\infty) = 1$, which implies $\alpha(z) = z + d$ for some $d \in \mathbb{R}$. Hence, μ' is a translation of μ. Thus, we can determine a unique μ by requiring its first moment to be 0, provided it exists. If we drop univalence, then we find also further probability measures μ'. Let H be the F-transform of $\frac{1}{2}\delta_{-1} + \frac{1}{2}\delta_1$. Then

$$H(z) = \frac{1}{\frac{1/2}{z+1} + \frac{1/2}{z-1}} = \frac{z^2 - 1}{z} = z - \frac{1}{z}$$

and $H : \mathbb{H} \to \mathbb{H}$ is surjective, as a rational function of degree 2, mapping $\mathbb{R} \cup \{\infty\}$ onto itself, but not injective ($H(i/2 \pm \sqrt{3}/2) = i$). Consequently, $F_\mu \circ H$ is a non-injective F-transform with $(F_\mu \circ H)(\mathbb{H}) = F_\mu(\mathbb{H})$.

From the probabilistic point of view, we would like to have conditions on μ that guarantee the univalence of F_μ. There are at least the following three strategies:

- We know that $\rhd$-infinitely divisible distributions (e.g., the arcsine law) yield univalent Cauchy transforms. A class of such measures (monotonically stable distributions) is studied in [**85, 91**]. We can, of course, generate such distributions indirectly by choosing an infinitesimal generator on $\mathbb{H}$ in Theorem 5.20.
- We also know that $\boxplus$-infinitely divisible distributions have a univalent Cauchy transform. Examples include the semicircle law $W(0,1)$, the normal distribution $\mathcal{N}(0,1)$, the Marchenko–Pastur law from Exercise 5.9.9 (its R-transform is shown to be a rational function that extends analytically to $\mathbb{H}^-$ in the solution).
- The literature knows many results on (analytic) univalence criteria, which might be applied to the Cauchy- or F-transform of μ.

An important example for such a univalence criterion is the following result due to Aksent'ev [**8**] (see also the survey paper [**16**, Theorem 39]).

THEOREM 6.22. *Let $a, b \in \mathbb{R}$ with $a < b$ and assume that $u : [a, b] \to [0, \infty)$ is a function that is non-decreasing on $[a, c]$ and non-increasing on $[c, b]$, for some $c \in (a, b)$. Then $\int_a^b \frac{1}{z-x} u(x)dx$ is either constant or univalent on $\mathbb{C} \setminus [a, b]$.*

EXAMPLE 6.23. Let $a < b$ and let μ be the uniform distribution on $[a, b]$. Then

$$G_\mu(z) = \int_a^b \frac{1}{z-x}\mu(dx) = \frac{1}{b-a}\log\left(\frac{z-a}{z-b}\right)$$

is univalent.

DEFINITION 6.24. A probability measure μ on $\mathbb{R}$ is called *unimodal* with mode $c \in \mathbb{R}$ if there exist a non-decreasing function $f : (-\infty, c) \mapsto [0, \infty)$, a non-increasing function $g : (c, \infty) \mapsto [0, \infty)$, and $\lambda \geq 0$ such that

$$\mu(dx) = f(x)\mathbf{1}_{(-\infty,c)}(x)\, dx + \lambda\delta_c + g(x)\mathbf{1}_{(c,\infty)}(x)\, dx.$$

COROLLARY 6.25. *Let μ be a unimodal probability measure on $\mathbb{R}$. Then G_μ is univalent.*

PROOF. Let c, f, g, λ as in the definition. Assume that $\lambda = 0$ and that f and g are bounded. For $n \in \mathbb{N}$, let $I_n = [c-n, c+n]$, and let $u_n : I_n \to [0, \infty)$ with $u_n(x) = f(x)$ for $x < c$, $u_n(x) = g(x)$ for $x > c$, $u_n(c) = \max(\lim_{x\nearrow c} f(x), \lim_{x\searrow c} g(x))$. Then $f_n(z) := \int_{I_n} \frac{1}{z-x} u_n(x)dx$ is univalent on $\mathbb{H}$ and as the measure $u_n(x)dx$ converges weakly to μ, f_n converges locally uniformly on $\mathbb{H}$ to G_μ, which is thus also univalent on $\mathbb{H}$. The general case also follows by approximation. □

This result gives another proof for the univalence of the Cauchy transform of $\mathcal{N}(0, 1)$, and we see that many other well-known distributions, e.g., the Cauchy distribution, have a univalent Cauchy transform.

REMARK 6.26. It can be shown that if μ is unimodal, then $G_\mu(\mathbb{H})$ is horizontally convex, i.e., if $z_1, z_2 \in G_\mu(\mathbb{H})$ with the same imaginary part, then $(1-t)z_1 + tz_2 \in G_\mu(\mathbb{H})$ for any $t \in (0, 1)$, see [**70**, Theorem 6.23].

The measure $\delta_x \rhd Arc(0, \sigma^2)$ has a univalent F-transform and an isolated atom when $x \neq 0$, see Example 5.17. If μ has at least two isolated atoms, then G_μ can be extended meromorphically to neighborhoods of these atoms, and G_μ cannot be univalent. If we assume that μ has an isolated atom, then the univalence of F_μ has strong implications for the regularity of μ ([**70**, Proposition 6.32]).

THEOREM 6.27. *Let $\mu \in \mathcal{P}(\mathbb{R})$ such that F_μ is univalent. Suppose that μ has an isolated atom at $a \in \mathbb{R}$, i.e., $\mu(\{a\}) > 0$ and $\mu(\{a\}) = \mu((a-\varepsilon, a+\varepsilon))$ for some $\varepsilon > 0$. Then $\mu|_{\mathbb{R}\setminus\{a\}}$ is Lebesgue absolutely continuous with L^∞-density.*

PROOF. Denote by $B_\varepsilon(a)$ the disk with center a and radius ε. Then G_μ extends to a holomorphic function on $B_\varepsilon(a) \setminus \{a\}$ with a simple pole at a. Thus, $G_\mu(B_\varepsilon(a))$

is a neighborhood of ∞, and the univalence of G_μ implies

$$G_\mu(\mathbb{H} \setminus B_\varepsilon(a)) \text{ is a bounded subset of } \mathbb{H}^-. \tag{6.5}$$

The singular part of μ is supported on the set

$$\left\{x \in \mathbb{R} \,|\, \lim_{h \searrow 0} \frac{\mu((x-h, x+h))}{2h} = \infty\right\},$$

see [**165**, Theorem 7.15]. For $x \in \mathbb{R}$, $h > 0$, we have the estimate

$$\frac{\mu((x-h,x+h))}{2h} \leq \int_{|u-x|<h} \frac{h}{h^2 + (u-x)^2}\,\mu(du) \leq -\operatorname{Im}(G_\mu(x+ih)), \tag{6.6}$$

and it follows that the singular part of μ is supported on the set

$$\{x \in \mathbb{R} \,|\, \lim_{y \searrow 0} \operatorname{Im}(G_\mu(x+iy)) = -\infty\}. \tag{6.7}$$

Obviously, (6.5) implies that the set (6.7) is $\{a\}$, and hence, $\mu|_{\mathbb{R}\setminus\{a\}}$ is Lebesgue absolutely continuous. The density is essentially bounded since, due to (6.5), the right hand side of (6.6) is bounded by a uniform constant independent of h and $x \in \mathbb{R} \setminus (a-\varepsilon, a+\varepsilon)$, and by the Stieltjes–Perron inversion formula. □

QUESTION 6.28. If F_μ is univalent, can μ have a non-zero singular part that is not a one-point measure?

REMARK 6.29. If μ is $\boxplus$-infinitely divisible, then μ has at most one atom [**34**, Proposition 5.12] and its singular continuous part is zero [**26**, Theorem 3.4]. See also [**25**] for the regularity of measures of the form $\mu \boxplus \nu$ for any $\mu, \nu \in \mathcal{P}(\mathbb{R})$.

6.4. Starlike Cauchy transforms

DEFINITION 6.30. We call a Cauchy transform $G_\mu : \mathbb{H} \to \mathbb{H}^-$ *starlike* if G_μ is univalent and $r \cdot G_\mu(\mathbb{H}) \subset G_\mu(\mathbb{H})$ for every $r \in (0,1)$.

This definition is equivalent to: F_μ is univalent and starlike with respect to ∞ in the sense that $c \cdot F_\mu(\mathbb{H}) \subset F_\mu(\mathbb{H})$ for all $c \in (1, \infty)$.

THEOREM 6.31. *Let $\mu \in \mathcal{P}(\mathbb{R})$. Then G_μ is starlike if and only if*

$$\operatorname{Im}(G'_\mu(z)/G_\mu(z)) > 0 \quad \textit{for all } z \in \mathbb{H}.$$

This is equivalent to $\operatorname{Im}(F'_\mu(z)/F_\mu(z)) < 0$ *for all* $z \in \mathbb{H}$.

PROOF. Assume that F_μ is starlike. Then $iy \in F_\mu(\mathbb{H})$ for all $y \geq M$, with some $M > 0$, and the curve $F_\mu^{-1}(iy)$ belongs to a sector $\{z \in \mathbb{H} \,|\, |\operatorname{Re}(z)| \leq C \operatorname{Im}(z)\}$ for some $C > 0$, see Lemma 5.10.

Let $y > 0$. First, we show that $F_\mu(z + iy)$ is also starlike. Let $c > 1$. From $cF_\mu(\mathbb{H}) \subset F_\mu(\mathbb{H})$, we see that we can write $cF_\mu(z) = F_\mu(\omega(z))$ for some $\omega : \mathbb{H} \to \mathbb{H}$. Let $z_n = F_\mu^{-1}(in)$ for all n large enough. As z_n and $F_\mu^{-1}(cin)$ approach ∞ non-tangentially, Theorem 3.13 yields

$$\begin{aligned}
\lim_{n\to\infty} \frac{\omega(z_n)}{z_n} &= \lim_{n\to\infty} \frac{F_\mu^{-1}(cF_\mu(z_n))}{z_n} = \lim_{n\to\infty} \frac{F_\mu^{-1}(cF_\mu(z_n))}{cF_\mu(z_n)} \frac{cF_\mu(z_n)}{z_n} \\
&= \lim_{n\to\infty} \frac{F_\mu^{-1}(cin)}{cin} \frac{cF_\mu(z_n)}{z_n} = \lim_{n\to\infty} \frac{F_\mu^{-1}(cin)}{F_\mu(F_\mu^{-1}(cin))} \frac{cF_\mu(z_n)}{z_n} = 1 \cdot c = c.
\end{aligned}$$

Corollary 3.16 implies that $\operatorname{Im}(\omega(z)) \geq \operatorname{Im}(z)$ for all $z \in \mathbb{H}$. Hence,

$$cF_\mu(\mathbb{H} + iy) = (F_\mu \circ \omega)(\mathbb{H} + iy) \subset F_\mu(\mathbb{H} + iy).$$

Similar to the proof of Theorem 3.42, we have, for any $y > 0$, that

$$\frac{d}{dx} \arg F_\mu(x + iy) = \operatorname{Im}\left(\frac{d}{dz} \log F_\mu(z)\right) = \operatorname{Im}(F_\mu'(z)/F_\mu(z)) \geq 0 \text{ or } \leq 0,$$

where $z = x + iy$. As $F_\mu'(z)/F_\mu(z) = 1/z + o(1/z)$ as $z \to \infty$ non-tangentially, see Theorem 3.13, there exists at least one $y_1 > 0$ such that $\frac{d}{dx} \arg F_\mu(x + iy_1) \leq 0$. Assume there exists $y_2 > 0$ with $\frac{d}{dx} \arg F_\mu(x + iy_2) \geq 0$. Then, due to continuity, we find y_3 between y_1 and y_2 satisfying $\frac{d}{dx} \arg F_\mu(x + iy_3) = 0$. So F_μ maps $\{x + iy_3 \,|\, x \in \mathbb{R}\}$ onto a straight line segment (with constant argument) and as $|F_\mu(x + iy_3)| \to \infty$ as $x \to \pm\infty$, see Exercise 5.9.1, we get a contradiction to the univalence of F_μ. Hence, $\operatorname{Im}(F_\mu'(z)/F_\mu(z)) \leq 0$ for all $z \in \mathbb{H}$. The maximum principle implies $\operatorname{Im}(F_\mu'(z)/F_\mu(z)) < 0$ for all $z \in \mathbb{H}$.

Conversely, assume that $\operatorname{Im}(F_\mu'(z)/F_\mu(z)) = \operatorname{Im}(\frac{d}{dz} \log F_\mu(z)) < 0$. For $z_1, z_2 \in \mathbb{H}$ with $z_1 \neq z_2$ we have

$$\begin{aligned} |(\log F_\mu)(z_2) - (\log F_\mu)(z_1)| &= |(z_2 - z_1) \int_0^1 (\log F_\mu)'(z_1 + t(z_2 - z_1))dt| \\ &= |z_2 - z_1| \cdot \left| \int_0^1 (\log F_\mu)'(z_1 + t(z_2 - z_1))dt \right|, \end{aligned}$$

and the complex number inside the second absolute value belongs to $\mathbb{H}^-$. Hence, $\log F_\mu$ is univalent, and thus, F_μ is univalent. Furthermore, we see that $F_\mu(\mathbb{H} + iy)$ is starlike with respect to ∞ for any $y > 0$, and thus, F_μ is starlike with respect to ∞. □

Let $\mu \in \mathcal{P}(\mathbb{R})$ and $c > 0$. We let $D_c\mu$ be the scaled measure $D_c\mu(A) = \mu(A/c)$. We also define $D_0\mu := \delta_0$.

DEFINITION 6.32. A probability measure $\mu \in \mathcal{P}(\mathbb{R})$ is called $\rhd$-*self-decomposable* if, for any $c \in [0, 1]$, there exists a probability measure ν_c such that $\mu = (D_c\mu) \rhd \nu_c$.

Due to Exercise 5.9.4, ν_c is uniquely determined. We now have the following relation [**70**, Theorem 6.49].

THEOREM 6.33. *A probability measure $\mu \in \mathcal{P}(\mathbb{R})$ is $\rhd$-self-decomposable if and only if G_μ is starlike.*

PROOF. Let μ be $\rhd$-self-decomposable and $c \in (0, 1]$. The relation $\mu = (D_c\mu) \rhd \nu_c$ gives us $G_\mu(z) = c^{-1} G_\mu(F_{\nu_c}(z)/c)$, see Lemma 5.18. So $cG_\mu(\mathbb{H}) \subset G_\mu(\mathbb{H})$. It remains to show that G_μ is univalent.

For $t \in [0, 1]$, let $F_t = F_{D_t\mu}$. Then $t \mapsto F_t$ is continuous and $F_0(z) = z$. Let $0 < s \leq t \leq 1$. From $F_\mu = \frac{s}{t} F_\mu(F_{\nu_{s/t}}(z)/(s/t))$ and $F_{D_t\mu}(z) = tF_\mu(z/t)$, we obtain

$$F_{D_t\mu}(z) = tF_\mu(z/t) = t\frac{s}{t} F_\mu(F_{\nu_{s/t}}(z/t)/(s/t)) = sF_\mu(tF_{\nu_{s/t}}(z/t)/s),$$

i.e., $D_t\mu = D_s\mu \rhd D_t\nu_{s/t}$. We can let $s \to 0$ and this relation still holds with $\nu_0 = \mu$. For $0 \leq s \leq t \leq 1$, let $F_{s,t} = F_{D_t\nu_{s/t}}$. Then F_t is (a part of) a Loewner chain with transition mappings $F_{s,t}$. Due to Theorem 6.17, each $F_{s,t}$ is univalent, and in particular, $F_1 = F_\mu$ is univalent.

Now assume that G_μ is starlike and let $c \in (0,1)$. We consider F_μ. As $F_{D_c\mu}(z) = cF_\mu(z/c)$, see Lemma 5.18, we have $F_\mu(\mathbb{H}) \subset F_{D_c\mu}(\mathbb{H})$. As $F_{D_c\mu}$ is univalent, we can define $H_c = F_{D_c\mu}^{-1} \circ F_\mu : \mathbb{H} \to \mathbb{H}$. Also H_c is an F-transform (which we show as in the proof of Theorem 6.31). So $H_c = F_{\nu_c}$ for some $\nu_c \in \mathcal{P}(\mathbb{R})$ and $(D_c\mu) \triangleright \nu_c = \mu$. □

6.5. Normalized additive Loewner chains

We have seen that normalized radial Loewner chains are differentiable almost everywhere. The normalization $f_t'(0) = e^{-t}$ guarantees locally absolute continuity of $t \mapsto f_t(z)$ for every $z \in \mathbb{D}$. An arbitrary radial Loewner chain can be normalized by applying rotations such that $f_t'(0) \in (0,1]$, followed by a reparametrization of $t \mapsto f_t$.

For additive Loewner chains, the situation is a little different. If $(F_t)_{t\geq 0}$ is an additive Loewner chain and $t \mapsto c_t \in \mathbb{R}$ a continuous function with $c_0 = 0$, then also

$$(F_t(z + c_t))_{t\geq 0}$$

is an additive Loewner chain. While the image domains remain unaffected, the probability measures are now shifted by $-c_t$. Assuming that the first moment m_t of each μ_t exists, we can eliminate this degree of freedom by centralizing the measures with $c_t = m_t$. If the variances of $(\mu_t)_{t\geq 0}$ also exist, then the function $t \mapsto \mathrm{Var}(\mu_t)$ is continuous and non-decreasing (due to Theorem 5.13), thereby serving as a time parameter, similar to how $t \mapsto -\log f_t'(0)$ functions in the radial case.

Therefore, we regard the class

$$\mathcal{P}^0(\mathbb{R}) = \left\{ \mu \in \mathcal{P}(\mathbb{R}) \,|\, \int_{\mathbb{R}} x^2 \mu(dx) < \infty, \int_{\mathbb{R}} x\mu(dx) = 0 \right\}.$$

We recall that Theorem 5.7 characterizes the F-transforms of $\mathcal{P}^0(\mathbb{R})$ as

$$F_\mu(z) = z - \int_{\mathbb{R}} \frac{1}{z-x} \tau(dx), \tag{6.8}$$

where τ is a non-negative measure on $\mathbb{R}$ with $\tau(\mathbb{R}) = \mathrm{Var}(\mu) = \int_{\mathbb{R}} x^2 \mu(dx)$. In the context of Loewner theory, this class was investigated in [**75**]. We will proceed similarly to the proofs for the radial case.

LEMMA 6.34. *Let $T \geq 0$. Then the family $\{F_\mu \,|\, \mu \in \mathcal{P}^0(\mathbb{R}), \mathrm{Var}(\mu) \leq T\}$ is sequentially compact.*

PROOF. Let $(\mu_n)_n \subset \mathcal{P}^0(\mathbb{R})$ be a sequence with $\mathrm{Var}(\mu_n) \leq T$, and let τ_n be the corresponding non-negative measures from (6.8). Then $\tau_n(\mathbb{R}) \leq T$ and by Helly's selection theorem, $(\tau_n)_n$ has a converging subsequence $(\tau_{n_k})_k$ on $\hat{\mathbb{R}}$ with limit τ, $\tau(\hat{\mathbb{R}}) \leq T$. Then $F_{\mu_{n_k}}$ converges locally uniformly to $z - \int_{\mathbb{R}} \frac{1}{z-x}\tau(dx) = z - \int_{\mathbb{R}} \frac{1}{z-x}\tau|_{\mathbb{R}}(dx)$, which is again an F-transform of a probability measure $\mu \in \mathcal{P}^0(\mathbb{R})$ with $\mathrm{Var}(\mu) \leq T$. □

THEOREM 6.35. *Let $(f_t)_{t\geq 0} = (F_{\mu_t})_{t\geq 0}$ be an additive Loewner chain such that $\mu_t \in \mathcal{P}^0(\mathbb{R})$ for all $t \geq 0$ and $\mathrm{Var}(\mu_t) = t$. Then $t \mapsto f_t(z)$ is locally absolutely continuous for every $z \in \mathbb{H}$, and there exists a set $\mathcal{N} \subset [0, \infty)$ of Lebesgue measure 0 such that f_t satisfies the partial differential equation*

$$\frac{\partial}{\partial t} f_t(z) = \frac{\partial}{\partial z} f_t(z) \cdot M(z,t) \quad \text{for all } t \in [0,\infty)\setminus\mathcal{N} \text{ and all } z \in \mathbb{H}, f_0(z) = z, \tag{6.9}$$

where M is of the form

$$M(z,t) = \int_{\mathbb{R}} \frac{1}{x - z}\,\rho_t(dx)$$

with ρ_t being a probability measure on $\mathbb{R}$ for a.e. $t \geq 0$, and $t \mapsto M(z,t)$ is measurable for every $z \in \mathbb{H}$.

Conversely, for any such M, there exists a unique solution $(f_t)_{t\geq 0}$ to (6.9), locally absolutely continuous in t for every $z \in \mathbb{H}$. This solution is an additive Loewner chain of probability measures $(\mu_t)_{t\geq 0}$, where each μ_t has mean 0 and variance t.

PROOF. We begin with proving the converse statement for the more general case, where ρ_t is a non-negative measure with $\rho_t(\mathbb{R}) \leq 1$ for a.e. $t \geq 0$. The existence and uniqueness of the solution to (6.9) can be shown similarly to Corollary 6.15. We first consider the initial value problem

$$\frac{\partial}{\partial t} g_{s,t}(z) = M(g_{s,t}(z), t), \quad \text{for a.e. } t \geq s, g_{s,s}(z) = z. \tag{6.10}$$

Let $g^0_{s,t}(z) = z$ and for $n \in \mathbb{N}$ let

$$g^n_{s,t}(z) := z + \int_s^t M(g^{n-1}_{s,\sigma}(z), \sigma)d\sigma =: z - B_{n,s,t}(z).$$

We show inductively that $g^n_{s,t}$ is an F-transform of a probability measure with mean 0 and variance $\int_s^t \rho_\sigma(\mathbb{R})d\sigma$ for each $n \in \mathbb{N}$, which also shows that the definition of $g^n_{s,t}$ is well-defined. Assume that this is true for some $n \in \mathbb{N}$. Then, for almost every $\sigma \geq 0$, $-M(g^n_{s,\sigma}(z), \sigma)$ is the Cauchy transform of a non-negative measure with mass $\rho_\sigma(\mathbb{R})$. So $\lim_{y\to\infty} -iyM(g^n_{s,\sigma}(iy), \sigma) = \rho_\sigma(\mathbb{R})$. We need to show that $B_{n+1,s,t}$ is the Cauchy transform of a non-negative measure with mass $\int_s^t \rho_\sigma(\mathbb{R})d\sigma$. Clearly, $B_{n+1,s,t}$ maps $\mathbb{H}$ holomorphically into $\mathbb{H}^- \cup \mathbb{R}$. Due to the estimate $\left|\frac{iy}{g^n_{s,\sigma}(iy)-x}\right| \leq y/\,\mathrm{Im}(g^n_{s,\sigma}(iy)) \leq 1$ for a.e. $\sigma \geq s$, we can use the dominated convergence theorem to get

$$\lim_{y\to\infty} iyB_{n+1,s,t}(iy) = \lim_{y\to\infty} \int_s^t \int_{\mathbb{R}} \frac{iy}{g^n_{s,\sigma}(iy) - x}\rho_\sigma(dx)d\sigma$$

$$= \int_s^t \lim_{y\to\infty} \int_{\mathbb{R}} \frac{iy}{g^n_{s,\sigma}(iy) - x}\rho_\sigma(dx)d\sigma = \int_s^t \rho_\sigma(\mathbb{R})d\sigma.$$

(In the same way, we show that the statement is true for the initial case $n = 1$.)

Now let $0 \leq s < T$, and $K \subset \mathbb{H}$ be a compact subset. The set of all $g^n_{s,t}$, $n \in \mathbb{N}$, $s \leq t \leq T$, forms a normal family. So we find a second compact set $L \subset \mathbb{H}$, $K \subset L$, such that $g^n_{s,t}(z) \in L$ for all $z \in K$. Furthermore, as the set of all Cauchy transforms is a normal family on $\mathbb{H}$, there exists a constant $C > 0$ such that $|M(z,\sigma) - M(w,\sigma)| \leq C|z-w|$ and $|M(z,\sigma)| \leq C$ for all $z, w \in L$ and a.e. $\sigma \geq 0$. Let $z \in K$. Similarly to Theorem 6.14, we get

$$\begin{aligned}
|g^{n+1}_{s,t}(z) - g^n_{s,t}(z)| &\leq \int_s^t |M(g^n_{s,\sigma_n}(z), \sigma_n) - M(g^{n-1}_{s,\sigma_n}(z), \sigma_n)| d\sigma_n \\
&\leq C \int_s^t |g^n_{s,\sigma_n}(z) - g^{n-1}_{s,\sigma_n}(z)| d\sigma_n \leq \dots \\
&\leq C^n \int_s^t \cdots \int_s^{\sigma_2} |g^1_{s,\sigma_1}(z) - g^0_{s,\sigma_1}(z)| d\sigma_1 \dots d\sigma_n \\
&\leq C^n \int_s^t \cdots \int_s^{\sigma_1} |M(z,\sigma_0)| d\sigma_0 d\sigma_1 \dots d\sigma_n \\
&\leq C^{n+1} \int_s^t \cdots \int_s^{\sigma_1} 1 d\sigma_0 d\sigma_1 \dots d\sigma_n = \frac{C^{n+1}(t-s)^{n+1}}{(n+1)!}.
\end{aligned}$$

We conclude that $g^n_{s,t}$ converges locally uniformly to a holomorphic function $g_{s,t}$ with

$$g_{s,t}(z) = z + \int_s^t M(g_{s,\tau}(z), \tau) d\tau,$$

which shows that $t \mapsto g_{s,t}(z)$ is locally absolutely continuous for every $z \in \mathbb{H}$ and that $g_{s,t}$ solves (6.10). As the limit of $g^n_{s,t}$, each $g_{s,t}$ is the F-transform of a probability measure with mean 0 and variance $\leq \int_s^t \rho_\sigma(\mathbb{R}) d\sigma$, by Lemma 6.34. The same calculation as for $g^n_{s,t}$ shows that the variance is equal to $\int_s^t \rho_\sigma(\mathbb{R}) d\sigma$.

Univalence of $g_{s,t}$ and uniqueness of the solution to (6.10) can be shown just as in Theorem 6.14. A time reversal as in Corollary 6.15 yields the unique solution to (6.9), i.e., we consider

$$\frac{\partial}{\partial t} g_{s,t}(z) = M(g_{s,t}(z), T-t) \quad \text{for a.e. } t \in [s,T], g_{s,s}(z) = z,$$

and then $f_t = g_{T-t,T}$ yields the unique solution to (6.9).

Now let $(f_t)_{t\geq 0}$ be an additive Loewner chain as in the statement of the theorem. Let $f_{s,t} = F_{\mu_{s,t}}$ be the transition mappings. Then also $\mu_{s,t}$ has finite variance $\mathrm{Var}(\mu_{s,t}) = t - s$, see Exercise 5.9.2, and we find non-negative measures $\tau_{s,t}$ with $\tau_{s,t}(\mathbb{R}) = t - s$ such that $f_{s,t}(z) = z + \int_{\mathbb{R}} \frac{1}{x-z} \tau_{s,t}(dx)$. Then

$$|f_{s,t}(z) - z| = \left| \int_{\mathbb{R}} \frac{1}{x-z} \tau_{s,t}(dx) \right| \leq \frac{t-s}{\mathrm{Im}(z)}. \tag{6.11}$$

Furthermore, $|f_s'(z)| \leq 1 + \int_{\mathbb{R}} \frac{1}{|x-z|^2} \tau_s(dx) \leq 1 + \frac{s}{\mathrm{Im}(z)^2}$. For $s \leq t$, we have $f_t = f_s \circ f_{s,t}$, and thus,

$$|f_t(z) - f_s(z)| = \left| \int_z^{f_{s,t}(z)} f_s'(w) dw \right| \leq \frac{t-s}{\mathrm{Im}(z)} \cdot \left(1 + \frac{s}{\mathrm{Im}(z)^2}\right). \tag{6.12}$$

Thus, $t \mapsto f_t(z)$ is locally absolutely continuous for each $z \in \mathbb{H}$. As in the proof of Theorem 6.12, we see that $\frac{\partial}{\partial t} f_t(z)$ exists for all $t \in [0,\infty) \setminus \mathcal{N}$ for a set $\mathcal{N}$ of Lebesgue measure 0 and all $z \in \mathbb{H}$. For $s < t$ we have

$$\begin{aligned}\frac{f_t(z) - f_s(z)}{t-s} &= \frac{f_s(f_{s,t}(z)) - f_s(z)}{f_{s,t}(z) - z} \frac{f_{s,t}(z) - z}{t-s} \\ &= \frac{f_s(f_{s,t}(z)) - f_s(z)}{f_{s,t}(z) - z} \int_{\mathbb{R}} \frac{1}{x-z} \frac{\tau_{s,t}}{t-s}(dx).\end{aligned}$$

Hence,

$$\frac{\partial}{\partial t} f_t(z) = \frac{\partial}{\partial z} f_t(z) \int_{\mathbb{R}} \frac{1}{x-z} \rho_t(dx),$$

where ρ_t is a non-negative measure with $\rho_t(\mathbb{R}) \leq 1$ for a.e. $t \geq 0$. As $\frac{\partial}{\partial t} f_t(z)$ and $\frac{\partial}{\partial z} f_t(z)$ are both measurable in t, also $M(z,t) = \int_{\mathbb{R}} \frac{1}{x-z} \rho_t(dx)$ is measurable in t for every $z \in \mathbb{H}$. We can extend $M(z,t)$ to $\mathbb{H} \times [0,\infty)$ such that $M(z,t)$ is measurable in t. The converse statement from above implies that $t = \int_0^t \rho_s(\mathbb{R}) ds$, so $\rho_t(\mathbb{R}) = 1$ for almost every $t \geq 0$. □

REMARK 6.36. From (6.9), we obtain for the Cauchy transforms $G_{\mu_t} = 1/f_t$ that

$$\frac{\partial}{\partial t} G_{\mu_t}(z) = \frac{\partial}{\partial z} G_{\mu_t}(z) \cdot M(z,t) = -\frac{\partial}{\partial z} G_{\mu_t}(z) \cdot G_{\rho_t}(z).$$

Assume that all μ_t and ρ_t have compact support and denote by $m_n(t)$ and $k_n(t)$ the nth moment of μ_t and ρ_t, respectively. Then $m_1(t) = 0$ and $m_2(t) = t$. Let $m_0(t) = k_0(t) = 1$ for all $t \geq 0$. The power series expansions at ∞ yield

$$\begin{aligned}&\frac{\frac{d}{dt} m_1(t)}{z^2} + \frac{\frac{d}{dt} m_2(t)}{z^3} + \cdots = \\ &\left(\frac{m_0(t)}{z^2} + \frac{2m_1(t)}{z^3} + \frac{3m_2(t)}{z^4} + \ldots \right) \cdot \left(\frac{k_0(t)}{z} + \frac{k_1(t)}{z^2} + \ldots \right)\end{aligned}$$

for a.e. $t \geq 0$. Comparing coefficients yields

$$\frac{d}{dt} m_n(t) = \sum_{j=1}^{n-1} j m_{j-1}(t) k_{n-j-1}(t), \quad m_n(0) = 0, \quad n \in \mathbb{N}. \tag{6.13}$$

For practical as well as for theoretical purposes, it is useful to express the solution of the partial differential equation (6.9) via an ordinary differential equation. Besides (6.13) for the case of compact supports, there are two ways. Fix $T > 0$. We can consider the ordinary differential equation
(6.14)

$$\frac{\partial}{\partial t} g_{s,t}(z) = M(g_{s,t}(z), T-t) \quad \text{for a.e. } t \in [s,T] \quad (\textbf{Backward flow equation})$$

with $g_{s,s}(z) = z$. Then $f_t = g_{T-t,T}$ for $0 \leq t \leq T$ and, in particular, $f_T = g_{0,T}$.

The second possibility is to consider the inverse mappings f_t^{-1}. Let $z \in f_T(\mathbb{H})$. Then $z \in f_t(\mathbb{H})$ for all $t \in [0,T]$ and from $(f_t \circ f_t^{-1})(z) = z$, we obtain, by using the chain rule (as in the proof of Corollary 6.15),

$$\frac{\partial}{\partial t} f_t^{-1}(z) = -M(f_t^{-1}(z), t) \quad \text{for a.e. } t \in [0,T]. \quad (\textbf{Inverse equation}) \tag{6.15}$$

Equation (6.9) has the following scaling behavior.

LEMMA 6.37. *Let $c, d > 0$ and let $f_t = F_{\mu_t}$ be the solution to (6.9). Consider the scaled measures $\nu_t(B) = \mu_{d\cdot t}(c \cdot B)$. Let $h_t = F_{\nu_t}$. Then h_t solves*

$$\frac{\partial}{\partial t}h_t(z) = \frac{\partial}{\partial z}h_t(z) \cdot \int_{\mathbb{R}} \frac{d/c^2}{x-z}\rho_{d\cdot t}(c\cdot dx) \quad \textit{for a.e. } t \geq 0, h_0(z) = z \in \mathbb{H}.$$

PROOF. We have

$$\begin{aligned} h_t(z) &= \left(\int_{\mathbb{R}} \frac{1}{z-x}\mu_{d\cdot t}(c\cdot dx)\right)^{-1} = \left(\int_{\mathbb{R}} \frac{1}{z-x/c}\mu_{d\cdot t}(dx)\right)^{-1} \\ &= \left(\int_{\mathbb{R}} \frac{c}{cz-x}\mu_{d\cdot t}(dx)\right)^{-1} = f_{d\cdot t}(cz)/c, \end{aligned}$$

which leads to

$$\begin{aligned} \frac{\partial}{\partial t}h_t(z) &= \frac{d}{c}\frac{\partial}{\partial z}h_t(z) \cdot \int_{\mathbb{R}} \frac{1}{x-cz}\rho_{d\cdot t}(dx) \\ &= \frac{\partial}{\partial z}h_t(z) \cdot \int_{\mathbb{R}} \frac{d/c^2}{x/c-z}\rho_{d\cdot t}(dx) = \frac{\partial}{\partial z}h_t(z) \cdot \int_{\mathbb{R}} \frac{d/c^2}{x-z}\rho_{d\cdot t}(c\cdot dx). \end{aligned}$$ □

Finally, we will show the following embedding theorem, which implies part (b) of Theorem 6.1 for the case of measures with finite second moment. Its proof will be based on a similar embedding theorem for radial Loewner chains and is given in Section 7.8.

THEOREM 6.38. *Let μ be a probability measure on $\mathbb{R}$ with finite first and second moment such that F_μ is univalent.*

(a) *Then there exists an additive Loewner chain $(f_t)_{t\geq 0}$ such that $f_1 = F_\mu$.*
(b) *If*

$$\int_{\mathbb{R}} x\,\mu(dx) = 0, \qquad \int_{\mathbb{R}} x^2\,\mu(dx) =: T > 0,$$

then there exists an additive Loewner chain $(f_t)_{t\geq 0}$ having the normalization from Theorem 6.35 for $t \in [0, T]$ such that $f_T = F_\mu$.

In the case of finite variances, there is a simpler proof that the F-transforms of $\triangleright$-hemigroups are always univalent.

LEMMA 6.39. *Let $\mu \in \mathcal{P}(\mathbb{R})$ and assume that the first and second moment of μ exist. Then F_μ is univalent in $\{z \in \mathbb{H} \,|\, \mathrm{Im}(z) > \sqrt{\mathrm{Var}(\mu)}\}$.*

PROOF. Let $z, w \in \mathbb{H}$, $z \neq w$, with $\mathrm{Im}(z), \mathrm{Im}(w) > \sqrt{\mathrm{Var}(\mu)}$. By representation (5.2), we have

$$\begin{aligned} &|F_\mu(z) - F_\mu(w)| = \left| z - w + \int_{\mathbb{R}} \frac{z-w}{(x-z)(x-w)}\,\tau(dx)\right| \\ &\geq |z-w|\left(1 - \int_{\mathbb{R}} \frac{1}{\mathrm{Im}(z)\,\mathrm{Im}(w)}\,\tau(x)\right) = |z-w|\left(1 - \frac{\mathrm{Var}(\mu)}{\mathrm{Im}(z)\,\mathrm{Im}(w)}\right) > 0. \end{aligned}$$ □

THEOREM 6.40. *Let $(\mu_{s,t})_{s\leq t}$ be a $\triangleright$-hemigroup such that all $\mu_{s,t}$ have finite variances. Then each $F_{\mu_{s,t}}$ is univalent.*

PROOF. It is sufficient to consider only $\mu_{0,T}$, $T > 0$. Fix $n \in \mathbb{N}$. Then

$$\mu_{0,T} = \mu_{0,T/n} \rhd \mu_{T/n,2T/n} \rhd \cdots \rhd \mu_{T-T/n,T}.$$

As $(s,t) \mapsto \mathrm{Var}(\mu_{s,t})$ is continuous, it is uniformly continuous on the set of all (s,t) with $0 \le s \le t \le T$. So, for each $\varepsilon > 0$, we find $n \in \mathbb{N}$ such that $\mathrm{Var}(\mu_{kT/n,(k+1)T/n}) \le \varepsilon$ for all $k = 0, \ldots, n-1$. Each $F_{\mu_{kT/n,(k+1)T/n}}$ is univalent in $\{z \in \mathbb{H} \mid \mathrm{Im}(z) > \sqrt{\varepsilon}\}$ by the previous lemma. Hence, $F_{\mu_{0,T}}$ is univalent in $\{z \in \mathbb{H} \mid \mathrm{Im}(z) > \sqrt{\varepsilon}\}$ for every $\varepsilon > 0$, and thus, it is univalent in $\mathbb{H}$. □

6.6. Value ranges

We can apply our results on F-transforms to compute some simple value ranges. Fix some $z_0 \in \mathbb{H}$. How does the set

$$V(z_0) := \{F_\mu(z_0) \mid \mu \in \mathcal{P}(\mathbb{R})\}$$

look like? As $\mathrm{Im}(F_\mu(z_0)) \ge \mathrm{Im}(z_0)$, $V(z_0) \subset \{z \in \mathbb{H} \mid \mathrm{Im}(z) \ge \mathrm{Im}(z_0)\}$. By using the arcsine law with mean $\mathrm{Re}(z_0)$ and a convolution by δ_{-a} from the left, we obtain the F-transform

$$F(z) = a + \sqrt{(z - \mathrm{Re}(z_0))^2 - 2\sigma^2}$$

with $\sigma \ge 0$ and $a \in \mathbb{R}$. We have $F(z_0) = a + i\sqrt{\mathrm{Im}(z_0)^2 + 2\sigma^2}$. Thus

$$V(z_0) = \{z \in \mathbb{H} \mid \mathrm{Im}(z) \ge \mathrm{Im}(z_0)\}.$$

Note that this value range does not change if we only allow for univalent F-transforms. As observed in [**163**, Theorem 2.4], this set does not change much if we only take univalent F-transforms with $\mu \in \mathcal{P}^0(\mathbb{R})$.

THEOREM 6.41. *Let $z_0 \in \mathbb{H}$ and define $V^0(z_0) := \{F_\mu(z_0) \mid \mu \in \mathcal{P}^0(\mathbb{R})\}$. Then*

$$V^0(z_0) = \{F_\mu(z_0) \mid \mu \in \mathcal{P}^0(\mathbb{R}),\ F_\mu \text{ univalent}\} = \{z \in \mathbb{H} \mid \mathrm{Im}(z) > \mathrm{Im}(z_0)\} \cup \{z_0\}.$$

PROOF. First, note that $F_\mu(z_0) = z_0 + a$ for some $\mu \in \mathcal{P}(\mathbb{R})$ and some $a \in \mathbb{R}$, if and only if $F_\mu(z) = z + a$ (Corollary 3.16), i.e., $\mu = \delta_{-a}$, which belongs to $\mathcal{P}^0(\mathbb{R})$ only if $a = 0$. So $V^0(z_0) \subset \{z \in \mathbb{H} \mid \mathrm{Im}(z) > \mathrm{Im}(z_0)\} \cup \{z_0\}$.

In order to show equality, we use the backward flow equation (6.14) with $M(z, T-t) = \frac{1}{U(t)-z}$ for a continuous function $U : [0,\infty) \to \mathbb{R}$. (We can ignore the dependence on T, as T can be chosen arbitrarily large.) Taking the initial value z_0 leads to

$$\frac{d}{dt} z(t) = \frac{1}{U(t) - z(t)}, \quad z(0) = z_0.$$

By writing $z(t) = x(t) + iy(t)$, we obtain

$$\frac{d}{dt}x(t) = \frac{U(t) - x(t)}{(U(t) - x(t))^2 + y(t)^2}, \quad x(0) = \mathrm{Re}(z_0),$$
$$\frac{d}{dt}y(t) = \frac{y(t)}{(U(t) - x(t))^2 + y(t)^2}, \quad y(0) = \mathrm{Im}(z_0).$$

Let $z_1 \in \mathbb{H}$ with $\mathrm{Im}(z_1) > \mathrm{Im}(z_0)$ and $c = \frac{\mathrm{Re}(z_1 - z_0)}{\mathrm{Im}(z_1 - z_0)}$. Assuming that we can choose U such that $U(t) - x(t) = cy(t)$ yields

$$\frac{d}{dt}x(t) = \frac{c}{(1+c^2)y(t)}, \quad x(0) = \mathrm{Re}(z_0), \quad \frac{d}{dt}y(t) = \frac{1}{(1+c^2)y(t)}, \quad y(0) = \mathrm{Im}(z_0).$$

The last equation can be solved explicitly and leads to $y(t) = \sqrt{y_0^2 + 2t/(1+c^2)}$ and $x(t) = cy(t) + x_0 - cy_0$. From $U(t) = cy(t) + x(t)$, we now see that such U does exist, namely,

$$U(t) = 2c\sqrt{y_0^2 + 2t/(1+c^2)} + x_0 - cy_0.$$

In this case, $z(t) = x(t) + iy(t)$ describes a straight half-line through z_0 and z_1. We conclude that there exists a univalent F-transform F_μ with $\mu \in \mathcal{P}^0(\mathbb{R})$ and $F_\mu(z_0) = z_1$. □

Next we consider $\mu \in \mathcal{P}^0(\mathbb{R})$ with fixed variance.

THEOREM 6.42. *Let $z_0 \in \mathbb{H}$ and $T > 0$. Define*

$$V_T^0(z_0) := \{F_\mu(z_0) \mid \mu \in \mathcal{P}^0(\mathbb{R}), \mathrm{Var}(\mu) = T\}.$$

Let $D(z_0, T)$ be the closed disk that is symmetric with respect to the imaginary axis such that $\partial D(z_0, T)$ passes through 0 and $Ti/\mathrm{Im}(z_0)$. Then

$$V_T^0(z_0) = (z_0 + D(z_0, T)) \setminus \{z_0\}.$$

PROOF. We have $\mu \in \mathcal{P}^0(\mathbb{R})$ with $\mathrm{Var}(\mu) = T$ if and only if

$$F_\mu(z) = z - \int_{\mathbb{R}} \frac{1}{z - x}\tau(dx),$$

where τ is a non-negative measure on $\mathbb{R}$ with $\tau(\mathbb{R}) = T$. As $T > 0$, $z_0 \notin V_T^0(z_0)$.

Consider $\tau = T\delta_x$. Then $F_\mu(z_0) = z_0 + \frac{T}{x - z_0}$. The set

$$\left\{\frac{T}{x - z_0} \mid x \in \mathbb{R}\right\} \cup \{0\} = \left\{\frac{T}{x - i\,\mathrm{Im}(z_0)} \mid x \in \mathbb{R}\right\} \cup \{0\}$$

describes the circle C through 0 and $Ti/\mathrm{Im}(z_0)$ that is symmetric with respect to the imaginary axis. For any other τ, $\int_{\mathbb{R}} \frac{1}{x - z_0}\tau(dx)$ is a convex combination of points from C. Thus, $V_T^0(z_0) \subset (z_0 + D(z_0, T)) \setminus \{z_0\}$.

Conversely, any w from the interior of $D(z_0, T)$ can be written as a convex combination $w = s\frac{T}{x_1 - z_0} + (1-s)\frac{T}{x_2 - z_0}$ for some $x_1, x_2 \in \mathbb{R}$ and $s \in (0,1)$. Hence,

$$z_0 + w = F_\mu(z_0) \quad \text{for} \quad \mu = s(T\delta_{x_1}) + (1-s)(T\delta_{x_2}).$$

□

REMARK 6.43. As $V_S^0(z_0) \subset V_T^0(z_0)$ whenever $S \leq T$, it follows that

$$\{F_\mu(z_0) \,|\, \mu \in \mathcal{P}^0(\mathbb{R}), 0 \leq \mathrm{Var}(\mu) \leq T\} = z_0 + D(z_0, T).$$

A probability measure $\mu \in \mathcal{P}(\mathbb{R})$ is symmetric in the sense $\mu(-B) = \mu(B)$ if and only if $-\overline{G_\mu(-\overline{z})} = -G_\mu(-z) = -\int_{\mathbb{R}} \frac{1}{-z-x}\mu(dx) = \int_{\mathbb{R}} \frac{1}{z+x}\mu(dx) = \int_{\mathbb{R}} \frac{1}{z-x}\mu(dx) = G_\mu(z)$, i.e., G_μ is symmetric with respect to reflection across the imaginary axis.

Let $z_0 \in \mathbb{H}$ and $T \geq 0$. Define

$$W_T^0(z_0) := \{F_\mu(z_0) \,|\, \mu \in \mathcal{P}^0(\mathbb{R}), \mathrm{Var}(\mu) = T, \mu \text{ symmetric}\}.$$

THEOREM 6.44. *Let $T > 0$ and denote by $C(z_0, T)$ the circular arc*

$$\left\{ z_0 + \frac{Tz_0}{y - z_0^2} \,|\, y \in [0, \infty) \right\}.$$

($C(z_0, T)$ is a straight line segment if $z_0 \in i\mathbb{R}$.) Then $W_T^0(z_0)$ is the convex hull of $C(z_0, T)$ (which is a disk segment if $z_0 \notin i\mathbb{R}$), see Figure 6.2.

PROOF. We have $\mu \in \mathcal{P}^0(\mathbb{R})$ with $\mathrm{Var}(\mu) = T$, μ symmetric, if and only if

$$F_\mu(z) = z - \int_{\mathbb{R}} \frac{1}{z - x}\tau(dx),$$

for a symmetric non-negative measure τ on $\mathbb{R}$ with $\tau(\mathbb{R}) = T$. So let ρ be a finite non-negative measure supported on $[0, \infty)$ such that $\tau(\cdot) = \rho(\cdot) + \rho(-\cdot)$. Then

$$F_\mu(z) = z - \int_{\mathbb{R}} \frac{1}{z-x} + \frac{1}{z+x}\rho(dx) = z - \int_{\mathbb{R}} \frac{2z}{z^2 - x^2}\rho(dx).$$

Consider $\rho = T/2 \cdot \delta_x$, $x \geq 0$. Then $F_\mu(z_0) = z_0 + \frac{Tz_0}{x^2 - z_0^2}$ and

$$\left\{ z_0 + \frac{Tz_0}{x^2 - z_0^2} \,|\, x \in [0, \infty) \right\} = \left\{ z_0 + \frac{Tz_0}{y - z_0^2} \,|\, y \in [0, \infty) \right\} = C(z_0, T) \subset W_T^0(z_0).$$

As in the previous theorem, we conclude that $W_T^0(z_0)$ is the convex hull of $C(z_0, T)$.

□

Note that the convex hull of $C(z_0, T)$ is not closed. If $z_0 \in i\mathbb{R}$, it does not contain z_0, and if $z_0 \notin i\mathbb{R}$, it does not contain the line segment $[z_0, z_0 - \frac{T}{z_0})$.

Assume that $\mathrm{Re}(z_0) > 0$. Let us consider $C(z_0, T)$ in more detail. First, $\mathrm{Re}(z_0 - \frac{T}{z_0}) < \mathrm{Re}(z_0)$ and $\mathrm{Im}(z_0 - \frac{T}{z_0}) > \mathrm{Im}(z_0)$. Let $L_1(z_0)$ be the half-line

$$L_1(z_0) = \left\{ z_0 - \frac{T}{z_0} \,|\, T \geq 0 \right\}.$$

Next, we compute the tangent vector of the curve $z_0 + \frac{Tz_0}{y - z_0^2}$ at the point z_0. We are only interested in its argument, so let us reparametrize to

$$z(y) = z_0 + \frac{Tz_0}{1/y - z_0^2} = z_0 + \frac{Tz_0 y}{1 - z_0^2 y}.$$

We have

$$\frac{d}{dy}z(y) = \frac{Tz_0/y^2}{(1/y - z_0^2)^2} = \frac{Tz_0}{(1 - z_0^2 y)^2},$$

which is equal to Tz_0 at $y = 0$. Define the half-line

$$L_2(z_0) = \{z_0 + Tz_0 \,|\, T \geq 0\}.$$

The union

$$W^0(z_0) := \{F_\mu(z_0) \,|\, \mu \in \mathcal{P}^0(\mathbb{R}), \mu \,\text{symmetric}\} = \bigcup_{T \geq 0} W_T^0(z_0)$$

lies within the sector given by $L_1(z_0)$ and $L_2(z_0)$. As $z_0 - \frac{T}{z_0} \in W_T^0(z_0)$, $L_1(z_0)$ belongs to $W^0(z_0)$, while $L_2(z_0) \cap W^0(z_0) = \{z_0\}$.

COROLLARY 6.45. *If* $\mathrm{Re}(z_0) > 0$, *then*

$$W^0(z_0) = \{z_0\} \cup \{w \in \mathbb{H} \,|\, \mathrm{Im}(w) > \mathrm{Im}(z_0), \arg(z_0) < \arg(w - z_0) \leq \arg(-1/z_0)\}.$$

The case $\mathrm{Re}(z_0) < 0$ follows by reflection and if $z_0 \in i\mathbb{R}$, then $W^0(z_0) = \{z_0 + it \,|\, t \geq 0\}$.

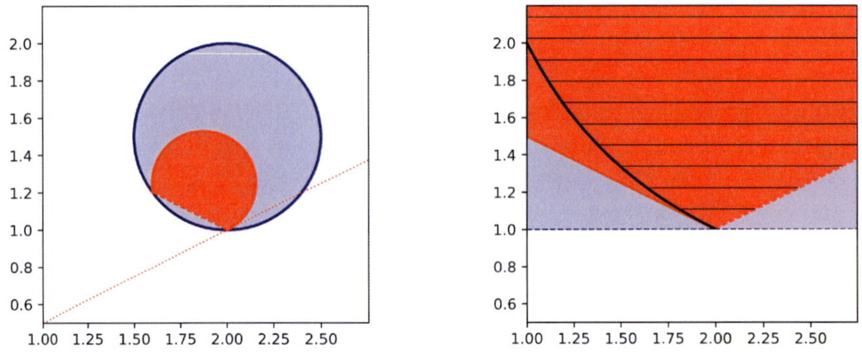

FIGURE 6.2. Left: The value ranges $V_T^0(z_0)$ and $W_T^0(z_0)$ for $z = 2 + i$ and $T = 1$. Right: $W_u^0(z_0) \subset W^0(z_0) \subset V^0(z_0)$.

Finally, we consider

$$W_u^0(z_0) := \{F_\mu(z_0) \,|\, \mu \in \mathcal{P}^0(\mathbb{R}), \mu \,\text{symmetric}, \, F_\mu \text{ univalent}\}.$$

Let $E(z_0)$ be the curve $\left\{\sqrt{z_0^2 - 2t} \,|\, t \geq 0\right\}$ and for $\mathrm{Re}(z_0) \neq 0$ let $J(z_0)$ be the Jordan domain whose boundary is given by $E(z_0) \cup L_2(z_0)$ (and lies within $\{w \in \mathbb{H} \,|\, \mathrm{Im}(w) > \mathrm{Im}(z_0)\}$).

THEOREM 6.46 (Theorem 2.2 in [**108**]). *If* $\mathrm{Re}(z_0) = 0$, *then* $W_u^0(z_0) = \{z_0 + it \,|\, t \geq 0\}$, *and if* $\mathrm{Re}(z_0) \neq 0$, *then*

$$W_u^0(z_0) = E(z_0) \cup J(z_0).$$

PROOF. We can embed any univalent F_μ with symmetric $\mu \in \mathcal{P}^0(\mathbb{R})$ into an additive Loewner chain F_{μ_t}, where each $\mu_t \in \mathcal{P}^0(\mathbb{R})$ is symmetric as well, which will be shown in Theorem 7.37. The function $M(z,t)$ from Loewner's differential equation (6.9) is then symmetric as well. Let us consider this equation only with a symmetric two point measure, i.e.,

$$\frac{\partial}{\partial t}F_{\mu_t}(z) = \frac{\partial}{\partial z}F_{\mu_t}(z)\left(\frac{1/2}{U(t)-z} + \frac{1/2}{-U(t)-z}\right),$$

where U is continuous. The solutions to this equation are dense in the set of all univalent F_μ with $\mu \in \mathcal{P}^0(\mathbb{R})$ symmetric. This statement will be proven in Lemma 8.26 for the general equation, and the same arguments hold for the symmetric case.

We consider the corresponding backward equation:

$$\frac{d}{dt}z(t) = \frac{1}{2}\frac{1}{U(t)-z(t)} + \frac{1}{2}\frac{1}{-U(t)-z(t)} = \frac{z(t)}{U(t)^2 - z(t)^2} = \frac{z(t)}{V(t)-z(t)^2},$$

$z(0) = z_0 = x_0 + iy_0$, where we put $V(t) := U(t)^2$. Denote the reachable set of the solutions to this initial value problem for all such U by $R(z_0)$. Then $R(z_0) \subset W_u^0(z_0) \subset \overline{R(z_0)}$.

We have

$$\frac{d}{dt}x(t) = \frac{x(t)(V(t) - |z(t)|^2)}{|V(t)-z(t)^2|^2}, \quad \frac{d}{dt}y(t) = \frac{y(t)(V(t)+|z(t)|^2)}{|V(t)-z(t)^2|^2}.$$

As $t \mapsto y(t)$ is strictly increasing, we can parametrize x and V by y, which yields

$$\frac{dx}{dy} = \frac{x(y)(V(y) - |z(y)|^2)}{y(V(y)+|z(y)|^2)}. \tag{6.16}$$

Let us assume that $\mathrm{Re}(z_0) = x_0 > 0$. Due to the symmetry and the univalence of the generated F-transforms, $z(t)$ must have positive real part for all $t \geq 0$, i.e., $x(y) > 0$. As $V \geq 0$, $|V(y) - |z(y)|^2| \leq V(y) + |z(y)|^2$, so

$$-\frac{x(y)}{y} \leq \frac{dx}{dy} \leq \frac{x(y)}{y}.$$

Solving the differential equations $\frac{dx}{dy} = x/y$ and $\frac{dx}{dy} = -x/y$ gives

$$\frac{x_0 y_0}{y} \leq x(y) \leq \frac{x_0}{y_0} \cdot y.$$

We have equality for the left side when $V(t) \equiv 0$. In this case, the solution is given by $z(t) = \sqrt{z_0^2 - 2t}$. Hence, $E(z_0) \subset R(z_0)$. Equality in the right side simply leads to the line $L_2(z_0)$ and we know that $L_2(z_0) \cap W_u^0(z_0) = \{z_0\}$. Let $p \in [-1, 1)$ and consider the equation

$$\frac{dx}{dy} = p \cdot \frac{x}{y}, \quad \text{i.e.,} \quad x(t) = \frac{x_0}{y_0^p} \cdot y(t)^p.$$

If this equation is of the form (6.16), then $p = \frac{V(y) - x(y)^2 - y^2}{V(y) + x(y)^2 + y^2}$, which leads to

$$V(y) = \frac{1+p}{1-p}(x(y)^2 + y^2) = \frac{1+p}{1-p}\left(\frac{y_0^{2p} y^2 + x_0^2 y^{2p}}{y_0^{2p}}\right).$$

Indeed, V is continuous and $V \geq 0$. The solutions $x(t) = \frac{x_0}{y_0^p} \cdot y(t)^p$ deform the curve $E(z_0)$ into $L_2(z_0)$ as p runs from -1 to 1, and they cover $E(z_0) \cup J(z_0)$. We conclude that $E(z_0) \cup J(z_0) = R(z_0)$. As $W_u^0(z_0) \subset \overline{R(z_0)} = E(z_0) \cup J(z_0) \cup L_2(z_0)$, $R(z_0) \subset W_u^0(z_0)$, and $L_2(z_0) \cap W_u^0(z_0) = \{z_0\}$, we have $W_u^0(z_0) = E(z_0) \cup J(z_0)$.

If $\text{Re}(z_0) = 0$, then the two curves $E(z_0)$ and $L_2(z_0)$ coincide and we simply have $W_u^0(z_0) = \{z_0 + it \mid t \geq 0\}$. □

The set

$$\{F_\mu(z_0) \mid \mu \in \mathcal{P}^0(\mathbb{R}), \text{Var}(\mu) = T, F_\mu \text{ univalent}\} \subset V_T^0(z_0)$$

corresponds to the reachable set of the backward flow equation, i.e., to the set of all $z(T)$ with $\frac{d}{dt}z(t) = M(z(t), t)$ and $z(0) = z_0$ for a corresponding vector field M. Computing this set can be handled by using Pontryagin's maximum principle from control theory, see [**156**]. For further results in this direction, we refer to the survey paper [**157**].

6.7. Loewner's differential equation for general Loewner chains

The considerations of Sections 6.2 and 6.5 can be generalized to arbitrary Loewner chains, by endowing them with a stronger regularity property.

Definition 6.47. Let $d \in [1, \infty]$ and let $D \subsetneq \mathbb{C}$ be a simply connected domain. A (decreasing) Loewner chain $(f_t)_{t \geq 0}$ on D is called a *Loewner chain of order d* if it satisfies the condition

(*) for any compact set $K \subset D$ and any $T > 0$, there exists a non-negative function $k_{K,T} \in L^d([0, T], \mathbb{R})$ such that

$$|f_t(z) - f_s(z)| \leq \int_s^t k_{K,T}(x)dx$$

for all $0 \leq s \leq t \leq T$ and all $z \in K$.

Example 6.48. Let $D = \mathbb{H}$ and $f_t(z) = z - C(t)$, where $C : [0, \infty) \to \mathbb{R}$ is continuous but not locally absolutely continuous. Then $(f_t)_{t \geq 0}$ is an additive Loewner chain with $\mu_t = \delta_{C(t)}$, and we have

$$|f_t(z) - f_s(z)| = |C(t) - C(s)|.$$

Hence, $(f_t)_{t\geq 0}$ is not a Loewner chain of any order d.

Property (∗) ensures that $t \mapsto f_t(z)$ is locally absolutely continuous for any $z \in D$ and thus differentiable almost everywhere. For a precise statement, we also need the following notion.

DEFINITION 6.49. A *Herglotz vector field of order* $d \in [1, \infty]$ *on* D is a function $M : D \times [0, \infty) \to \mathbb{C}$ with the following properties:

(i) The function $t \mapsto M(z, t)$ is measurable for every $z \in D$.
(ii) The function $z \mapsto M(z, t)$ is an infinitesimal generator on D for a.e. $t \geq 0$.
(iii) For any compact set $K \subset D$ and any $T > 0$ there exists a non-negative function $k_{K,T} \in L^d([0, T], \mathbb{R})$ such that $|M(z, t)| \leq k_{K,T}(t)$ for all $z \in K$ and for a.e. $t \in [0, T]$.

A Herglotz vector field of order 1 will simply be called a *Herglotz vector field.*

One can now prove the following one-to-one correspondence.

THEOREM 6.50. *A Loewner chain $(f_t)_{t\geq 0}$ of order d satisfies the Loewner partial differential equation*

$$\frac{\partial}{\partial t} f_t(z) = \frac{\partial}{\partial z} f_t(z) \cdot M(z, t) \quad \textit{for a.e. } t \geq 0 \textit{ and all } z \in D, f_0(z) = z, \tag{6.17}$$

for a Herglotz vector field M of order d.

Conversely, for a given Herglotz vector field M of order d, the unique solution of holomorphic functions $f_t : D \to D$ to (6.17)*, such that $t \mapsto f_t(z)$ is locally absolutely continuous for every $z \in D$, is always a Loewner chain of order d.*

PROOF. We refer the reader to [**53**, Theorem 3.2]. One can also obtain (6.17) from [**47**, Theorem 1.1] by considering the family $(f_{T-t,T-s})_{0\leq s\leq t\leq T}$ for some fixed $T > 0$. It can be verified that it forms an evolution family (see Remark 6.6). Let $g_{s,t} = f_{T-t,T-s}$. Then $g_{s,t}$ satisfies the backward equation

$$\frac{\partial}{\partial t} g_{s,t}(z) = M(g_{s,t}(z), T - t) \quad \text{for a.e. } t \in [s, T]. \tag{6.18}$$

□

REMARK 6.51. From the relation $f_t = f_s \circ f_{s,t}$, we obtain

$$\frac{\partial}{\partial t} f_{s,t}(z) = \frac{\partial}{\partial z} f_{s,t}(z) \cdot M(z, t) \quad \text{for a.e. } t \geq s, f_{s,s}(z) = z \in D.$$

Furthermore, we can also differentiate $f_{s,t}$ with respect to s and obtain

$$\frac{\partial}{\partial s} f_{s,t}(z) = -M(f_{s,t}(z), s) \quad \text{for a.e. } s \leq t, f_{t,t}(z) = z \in D. \tag{6.19}$$

Conversely, this equation has a unique solution, which gives the transition mappings of a decreasing Loewner chain of order d, see again [**53**, Theorems 1.11 and 3.2].

General additive Loewner chains now satisfy the following relationship.

THEOREM 6.52. *Let $(f_t)_{t\geq 0}$ be an additive Loewner chain satisfying condition* (∗) *from Definition* 6.47 *for $d = 1$. Then f_t satisfies* (6.17) *with a Herglotz vector field M having the form*

$$M(z, t) = a_t + \int_{\mathbb{R}} \frac{1 + xz}{x - z} \rho_t(dx), \tag{6.20}$$

where $a_t \in \mathbb{R}$ and ρ_t is a finite, non-negative Borel measure on $\mathbb{R}$ for a.e. $t \geq 0$.

Conversely, let M be a Herglotz vector field of the above form. Then the solution $(f_t)_{t\geq 0}$ to (6.17) is an additive Loewner chain satisfying condition $(*)$ for $d = 1$.

PROOF. We sketch how Theorem 6.52 can be obtained by adjusting the proof of Theorem 6.35.

Let $M(z,t)$ be a Herglotz vector field on $\mathbb{H}$ such that $M(\cdot,t)$ maps $\mathbb{H}$ into $\mathbb{H}\cup\mathbb{R}$ for a.e. $t \geq 0$. Then we can write $M(z,t) = \alpha_t + \beta_t z + \int_{\mathbb{R}} \frac{1+xz}{x-z}\gamma_t(dx)$ for a.e. $t \geq 0$. We have $M(i,t) = \alpha_t + i(\beta_t + \gamma_t(\mathbb{R}))$ (Remark 3.14) and property (iii) implies that $|a_t|, \beta_t, \gamma_t(\mathbb{R})$ are bounded by an L^1-function on any interval $[0,T]$.

We can now show that the initial value problem

$$\frac{\partial}{\partial t} g_{s,t}(z) = M(g_{s,t}(z),t), \quad \text{for a.e. } t \geq s, g_{s,s}(z) = z, \tag{6.21}$$

has a unique solution of univalent functions $g_{s,t} : \mathbb{H} \to \mathbb{H}\cup\mathbb{R}$, $g_{s,t}(z) = a_{s,t} + b_{s,t}z + \int_{\mathbb{R}} \frac{1+xz}{x-z}\tau_{s,t}(dx)$, and that $b_{s,t} = 1$ for all $t \geq s$ if and only if $\beta_t = 0$ for a.e. $t \geq s$. In this case, each $g_{s,t}$ is an F-transform. Now we consider the solution $g_{s,t}$ for $M(z, T-t)$, and $f_t = g_{T-t,T}$ yields an additive Loewner chain satisfying condition $(*)$ for $d = 1$.

Conversely, let $(f_t)_{t\geq 0} = (F_{\mu_t})_{t\geq 0}$ be an additive Loewner chain with transition mappings $f_{s,t}$, satisfying condition $(*)$ for $d = 1$. Property $(*)$ yields that $t \mapsto f_t(z)$ is locally absolutely continuous. By arguing as in the proof of Theorem 6.12, $\frac{\partial}{\partial t} f_t(z)$ exists for a.e. $t \geq 0$ and all $z \in \mathbb{H}$.

Write $f_{s,t}(z) = z + a_{s,t} + \int_{\mathbb{R}} \frac{1+xz}{x-z}\tau_{s,t}(dx)$. For $0 \leq s < t$ we have

$$\begin{aligned}\frac{f_t(z) - f_s(z)}{t-s} &= \frac{f_s(f_{s,t}(z)) - f_s(z)}{f_{s,t}(z) - z}\frac{f_{s,t}(z) - z}{t-s} \\ &= \frac{f_s(f_{s,t}(z)) - f_s(z)}{f_{s,t}(z) - z}\left(\frac{a_{s,t}}{t-s} + \int_{\mathbb{R}} \frac{1+xz}{x-z}\frac{\tau_{s,t}}{t-s}(dx)\right),\end{aligned}$$

provided that $f_{s,t}(z) \neq z$. If $t \in [0,\infty)$ is such that the limit exists as $s \to t$, then the function in parentheses converges locally uniformly to a holomorphic function $z \mapsto M(z,t)$ with $M(\mathbb{H},t) \subset \mathbb{H}\cup\mathbb{R}$ and we can write $M(z,t) = \alpha_t + \beta_t z + \int_{\mathbb{R}} \frac{1+xz}{x-z}\gamma_t(dx)$.

Now we can extend M to a function on $\mathbb{H} \times [0,\infty)$ such that M is a Herglotz vector field. Since each f_t is an F-transform, $\beta_t = 0$ for a.e. $t \geq 0$. □

REMARK 6.53. Theorem 6.52 can be derived from a more general result in [**48**] as follows.

"$\Longrightarrow$": Theorem 6.50 yields a Herglotz vector field M. We have

$$\operatorname{Im}(f_{s,u}(z)) = \operatorname{Im}(f_{s,t}(f_{t,u}(z))) \geq \operatorname{Im}(f_{t,u}(z))$$

for all $0 \leq s \leq t \leq u$. So $s \mapsto \operatorname{Im}(f_{s,t}(z))$ is non-increasing for every $z \in \mathbb{H}$. From (6.19), we see that $\operatorname{Im}(M(z,t)) \geq 0$ for almost every $t \geq 0$ and every $z \in \mathbb{H}$. Hence, $M(\cdot,t)$ maps $\mathbb{H}$ into $\mathbb{H}\cup\mathbb{R}$ for a.e. $t \geq 0$. (See also [**47**, Thm. 8.1].)

Assume that $\lim_{y\to\infty} M(iy,t)/(iy) > 0$ for a set $I \subset [0,T]$ of positive Lebesgue measure. Then, by [**48**, Thm. 1.1], we obtain that $\lim_{y\to\infty} f_T(iy)/(iy) > 1$, a contradiction. This proves that $\lim_{y\to\infty} M(iy,t)/(iy) = 0$ for a.e. $t \geq 0$, and with the Nevanlinna representation formula we see that $M(z,t)$ has the form (6.20).

"$\Longleftarrow$": We have to show that every f_t can be written as $f_t = F_{\mu_t}$ for a probability measure μ_t. Consider the Nevanlinna representation of f_t:

$$f_t(z) = a_t + b_t z + \int_{\mathbb{R}} \frac{1+xz}{x-z} \tau_t(dx).$$

As $M(\cdot, t)$ has the form (6.20) for a.e. $t \geq 0$, $M(\cdot, t)$ has a "boundary regular null point" at ∞ with dilation 0 for a.e. $t \geq 0$, see [**48**, Def. 2.6] which handles the unit disk case. By [**48**, Thm. 1.1], the "spectral function" of f_t at ∞ is equal to 0, which translates in our setting to

$$b_t = 1$$

for every $t \geq 0$. Hence, $f_t = F_{\mu_t}$ for a probability measure μ_t for every $t \geq 0$.

6.8. Exercises

E 6.8.1. We consider the Loewner equation (6.3), i.e.,

$$(*) \qquad \frac{\partial}{\partial t} f_t(z) = -z \frac{\partial}{\partial z} f_t(z) p(z,t), \quad f_0(z) = z \in \mathbb{D}.$$

(a) Let $(f_t)_{t \geq 0}$ be a normalized radial Loewner chain satisfying $(*)$. For $\alpha \in \mathbb{R}$, let $g_t(z) := e^{-i\alpha} f_t(e^{i\alpha} z)$. Which differential equation does g_t satisfy?
(b) Let $p : \mathbb{D} \times [0, \infty) \to \mathbb{C}$ be given by $p(z,t) = 1 + z$. Obviously, $p(\cdot, t) \in \mathcal{P}$ for every $t \geq 0$. Find the unique solution to the Loewner equation $(*)$.
(c) Now let

$$p(z,t) = 1 + z \quad \text{for} \quad t \in [0,1) \cup [2,3) \cup \ldots \quad \text{and}$$
$$p(z,t) = 1 - z \quad \text{for} \quad t \in [1,2) \cup [3,4) \cup \ldots$$

Give a (qualitative) description of the solution of the Loewner equation $(*)$.

E 6.8.2. By Theorem 6.50 and the Berkson–Porta formula (3.15) with

$$\tau = 1 \quad \text{and} \quad p(z) = \frac{1+z}{1-z},$$

the following Loewner equation has a unique solution:

$$\frac{\partial}{\partial t} f_t(z) = \frac{\partial}{\partial z} f_t(z) \cdot (1-z)^2 \cdot \frac{1+z}{1-z} = \frac{\partial}{\partial z} f_t(z) \cdot (1-z^2), \quad f_0(z) = z \in \mathbb{D}.$$

Determine the solution f_t and describe the images $f_t(\mathbb{D})$.

E 6.8.3. Let $(f_t)_{t \geq 0}$ be a solution to equation (6.9), i.e., to

$$\frac{\partial}{\partial t} f_t(z) = \frac{\partial}{\partial z} f_t(z) \cdot M(z,t), \quad f_0(z) = z \in \mathbb{H},$$

where $M(z,t) = \int_{\mathbb{R}} \frac{1}{x-z} \rho_t(dx)$ with ρ_t being a probability measure on $\mathbb{R}$ for a.e. $t \geq 0$, and $t \mapsto M(z,t)$ is measurable for every $z \in \mathbb{H}$.

Let $(K_t)_{t \geq 0}$ be the generated hulls, i.e., $f_t(\mathbb{H}) = \mathbb{H} \setminus K_t$, $K_t \subset \mathbb{H}$. Prove that

$$\sup_{z \in K_t} \operatorname{Im}(z) \leq \sqrt{2t}.$$

6.9. Solutions

S 6.9.1.

(a) Also $(g_t)_{t\geq 0}$ is a normalized radial Loewner chain. We have

$$\frac{\partial}{\partial t} g_t(z) = e^{-i\alpha}\frac{\partial}{\partial t} f_t(e^{i\alpha}z) = -zf_t'(e^{i\alpha}z)p(e^{i\alpha}z,t) = -z\frac{\partial}{\partial z} g_t(z)p(e^{i\alpha}z,t).$$

Thus, the Herglotz vector field of $(g_t)_{t\geq 0}$ is given by $-zp(e^{i\alpha}z,t)$.

(b) Due to Corollary 6.15, we know that there exists exactly one solution of holomorphic mappings $(f_t)_{t\geq 0}$, and that they form a decreasing Loewner chain. Its proof reveals that we can find the solution by solving an ODE (the backward equation (6.4)) instead of a PDE, namely

$$\frac{\partial}{\partial t} g_{s,t}(z) = -g_{s,t}(z)p(g_{s,t}(z),t) = -g_{s,t}(z)(1+g_{s,t}(z)), \quad t\geq s, g_{s,s}(z) = z.$$

This autonomous equation can be solved, e.g., by separation of variables, and the solution is given by

$$g_{s,t}(z) = \frac{z}{e^{t-s}(1+z)-z}.$$

Now fix $t\in[0,\infty)$ and let $T>t$. Then, see the proof of Corollary 6.15, f_t is given by

$$f_t(z) = g_{T-t,T}(z) = \frac{z}{e^t(1+z)-z} = \frac{z}{z(e^t-1)+e^t}.$$

As f_t is a Möbius transformation, the image of $\mathbb{D}$ under f_t is a disk, symmetric with respect to $\mathbb{R}$ with boundary points $f_t(-1)=-1$, $f_t(1) = \frac{1}{2e^t-1}\in(0,1]$.

(c) For $t\in[0,1)$, f_t coincides with the solution from (b). Let $t=1+\varepsilon$, $\varepsilon\in[0,1)$. Then $f_t = f_1\circ f_{1,1+\varepsilon}$ and (see Remark 6.13) $g_t := f_{1,1+t}$ satisfies

$$\frac{\partial}{\partial t} g_t(z) = -z\frac{\partial}{\partial z} g_t(z)(1-z), \quad t\in[0,1), \quad g_0(z) = z.$$

By (a), $g_t(z) = -f_t(-z)$ for $t\in[0,1)$, and g_t is a Möbius transformation mapping $\mathbb{D}$ onto the symmetric disk with boundary points $g_t(1)=1$, $g_t(-1) = \frac{-1}{2e^t-1}\in[-1,0)$. As the composition of Möbius transformations is again a Möbius transformation, we get, by induction, the following description of f_t:

For every $t\geq 0$, $f_t(\mathbb{D})$ is a disk symmetric with respect to $\mathbb{R}$. Within $[0,1)\cup[2,3)\cup[4,5)\ldots$, its left point stays fixed and the right point moves toward 0. Conversely, for $t\in[1,2)\cup[3,4)\cup\ldots$, the disk shrinks while its right point stays fixed and the left point moves toward 0.

S 6.9.2. Similar to the previous exercise, we consider the backward equation (see equation (6.18))

$$\frac{\partial}{\partial t} g_{s,t}(z) = 1 - g_{s,t}(z)^2, \quad t\geq s, \quad g_{s,s}(z) = z.$$

The solution to this autonomous equation is given by

$$g_{s,t}(z) = \frac{z-1+e^{2(t-s)}(1+z)}{1-z+e^{2(t-s)}(1+z)}.$$

Now fix $t \in [0, \infty)$ and let $T > t$. Then f_t is given by

$$f_t(z) = g_{T-t,T}(z) = \frac{z - 1 + e^{2t}(1+z)}{1 - z + e^{2t}(1+z)}.$$

Again, f_t is a Möbius transformation and the image of $\mathbb{D}$ under f_t is a disk, symmetric with respect to $\mathbb{R}$, with boundary points $f_t(-1) = -1$, $f_t(1) = 1$. So $f_t(\mathbb{D}) = \mathbb{D}$ and f_t is an automorphism of $\mathbb{D}$ for every $t \geq 0$.

S 6.9.3. We fix some $T > 0$ and consider the backward flow equation (6.14), i.e.,

$$\frac{d}{dt} w_t = \int_{\mathbb{R}} \frac{1}{u - w_t} \rho_{T-t}(du), \quad t \in [0, T].$$

Split w_t into real and imaginary part, $w_t = x_t + iy_t$. Then

$$\frac{d}{dt} y_t = \int_{\mathbb{R}} \frac{y_t}{|u - x_t - iy_t|^2} \rho_{T-t}(du) \leq \int_{\mathbb{R}} \frac{y_t}{y_t^2} \rho_{T-t}(du) = \frac{1}{y_t}.$$

This implies $y_t \leq \sqrt{2t + y_0^2}$. For $t = T$, we obtain

$$\operatorname{Im}(f_T(x_0 + iy_0)) \leq \sqrt{2T + y_0^2}.$$

So no point of the set K_T can have an imaginary part greater than $\sqrt{2T + y_0^2}$ for any $y_0 > 0$. Hence,

$$\sup_{z \in K_T} \operatorname{Im}(z) \leq \sqrt{2T}.$$

CHAPTER 7

Radial Loewner Chains Revisited

In this chapter, we revisit radial Loewner chains, which can be interpreted as the multiplicative analog of additive Loewner chains. These correspond to the multiplicative monotone convolution for probability measures on the unit circle, which can be viewed as distributions of unitary operators. Furthermore, we will examine kernel convergence of domains, providing a geometric interpretation of locally uniform convergence of univalent functions.

7.1. Multiplicative monotone convolution

Let H be a complex Hilbert space and let $\varphi : B(H) \to \mathbb{C}$ be an expectation. Let $U \in B(H)$ be *unitary*, i.e.,

$$\langle Uv, Uw \rangle = \langle v, w \rangle \text{ for all } v, w \in H.$$

In other words: $UU^* = U^*U = I$, i.e., $U^{-1} = U^*$. Then there exists a probability measure μ on $\partial\mathbb{D}$, the distribution of U, such that $\varphi(U^n) = \int_{\partial\mathbb{D}} x^n \mu(dx)$ for all $n \in \mathbb{N}$. This can be shown similarly to the self-adjoint case or by using the spectral theorem for normal operators (see [**55**, Chapter IX]). For $z \in \mathbb{D}$, we have

$$\begin{aligned}\varphi\left(\frac{Uz}{1-Uz}\right) &= \varphi\left(\sum_{n=1}^{\infty}(Uz)^n\right) = \sum_{n=1}^{\infty}\varphi\left(U^n\right)z^n \\ &= \sum_{n=1}^{\infty}\int_{\partial\mathbb{D}} x^n\mu(dx)z^n = \int_{\partial\mathbb{D}}\sum_{n=1}^{\infty}(xz)^n\mu(dx) = \int_{\partial\mathbb{D}}\frac{xz}{1-xz}\mu(dx).\end{aligned}$$

We define the transforms

$$\begin{aligned}\psi_\mu(z) &= \int_{\partial\mathbb{D}}\frac{xz}{1-xz}\,\mu(dx) = \varphi\left(\frac{Uz}{1-Uz}\right) \quad \text{and} \\ \eta_\mu(z) &= \frac{\psi_\mu(z)}{1+\psi_\mu(z)}, \quad z \in \mathbb{D}.\end{aligned}$$

EXAMPLE 7.1. For $\mu = \delta_p$, $p \in \partial\mathbb{D}$, we get

$$\psi_\mu(z) = \frac{pz}{1-pz} \quad \text{and} \quad \eta_\mu(z) = pz.$$

EXAMPLE 7.2. The uniform distribution μ_H on $\partial\mathbb{D}$ (the normalized Haar measure of $\partial\mathbb{D}$) is special in this setting as

$$\psi_{\mu_H}(z) = \sum_{n=1}^{\infty}\int_{\partial\mathbb{D}} x^n\mu_H(dx)z^n = \sum_{n=1}^{\infty}\frac{1}{2\pi}\int_0^{2\pi} e^{ni\varphi}d\varphi z^n = 0,$$

and thus, also $\eta_{\mu_H} = 0$.

THEOREM 7.3. *Let μ be a probability measure on $\partial\mathbb{D}$. Then η_μ maps $\mathbb{D}$ into $\mathbb{D}$ with $\eta_\mu(0) = 0$, and any holomorphic $f : \mathbb{D} \to \mathbb{D}$ with $f(0) = 0$ has the form $f = \eta_\mu$ for a unique probability measure μ on $\partial\mathbb{D}$.*

PROOF. We have

$$1 + 2\psi_\mu(z) = \int_{\partial\mathbb{D}} \frac{1 + xz}{1 - xz}\,\mu(dx) = \int_{\partial\mathbb{D}} \frac{x^{-1} + z}{x^{-1} - z}\,\mu(dx) =: A_\mu(z),$$

which maps $\mathbb{D}$ into the right half-plane and 0 onto 1. Furthermore, $\eta_\mu(z) = \frac{A_\mu(z)-1}{A_\mu(z)+1}$, which maps $\mathbb{D}$ into $\mathbb{D}$ and fixes 0.

Conversely, if $f : \mathbb{D} \to \mathbb{D}$ is holomorphic with $f(0) = 0$, then $\frac{1+f}{1-f}$ maps $\mathbb{D}$ into the right half-plane and 0 onto 1. The Herglotz representation implies that

$$\frac{1+f}{1-f} = 1 + 2\psi_\mu$$

for a unique probability measure μ on $\partial\mathbb{D}$, and thus, $f = \eta_\mu$. □

If $U, V \in B(H)$ are both unitary, then also UV is unitary. If we assume that (U, V) is monotonically independent (defined as in Theorem 4.21), then

$$\varphi((UV)^n) = \varphi(UVUV \dots UV) = \varphi(U^n)\varphi(V)^n, \quad n \in \mathbb{N}, \tag{7.1}$$

which is not too interesting, as only the first moment of V appears here. However, the assumption of monotone independence of $(U - I, V - I)$ yields an interesting convolution, introduced by Bercovici in [**31**]. Again, monotone independence of $(U - I, V - I)$ is defined just as in Theorem 4.21.

THEOREM 7.4. *Let $U, V \in B(H)$ be unitary with distributions μ and ν, and let α be the distribution of UV. Assume that $(U - I, V - I)$ is monotonically independent. Then*

$$\eta_\alpha = \eta_\mu \circ \eta_\nu.$$

We define the monotone multiplicative convolution as $\mu \rhd \nu := \alpha$.

PROOF. As $(U - I, I)$ is trivially monotonically independent, the monotone independence of $(U - I, V - I)$ implies monotone independence of $(U - I, V)$. We need to show that $\psi_\alpha = \psi_\mu \circ \eta_\nu$. With $X = U - I$, we obtain

$$\begin{aligned}
\frac{UVz}{I - UVz} &= \sum_{n=1}^\infty (UVz)^n = \sum_{n=1}^\infty (X + I)Vz \dots (X + I)Vz \\
&= \sum_{n=1}^\infty \sum_{k=1}^n \sum_{\substack{\nu_1 + \dots + \nu_k = n \\ \nu_1, \dots, \nu_k \geq 1}} (X + I)(Vz)^{\nu_1} X (Vz)^{\nu_2} X \dots (Vz)^{\nu_k}.
\end{aligned}$$

Now we can use monotone independence:

$$\begin{aligned}
\psi_\alpha(z) &= \varphi\left(\frac{UVz}{I-UVz}\right)\\
&= \sum_{n=1}^{\infty}\sum_{k=1}^{n}\sum_{\substack{\nu_1+\cdots+\nu_k=n\\ \nu_1,\dots,\nu_k\geq 1}} \varphi\left((X+I)(Vz)^{\nu_1}X(Vz)^{\nu_2}X\dots(Vz)^{\nu_k}\right)\\
&= \sum_{n=1}^{\infty}\sum_{k=1}^{n}\sum_{\substack{\nu_1+\cdots+\nu_k=n\\ \nu_1,\dots,\nu_k\geq 1}} \varphi\left((X+I)\varphi((Vz)^{\nu_1})X\varphi((Vz)^{\nu_2})X\dots\varphi((Vz)^{\nu_k})\right).
\end{aligned}$$

Next, we have

$$\begin{aligned}
\frac{U\eta_\nu(z)}{I-U\eta_\nu(z)} &= \left(U\frac{\psi_\nu(z)}{I+\psi_\nu(z)}\right)\left(I-U\frac{\psi_\nu(z)}{I+\psi_\nu(z)}\right)^{-1}\\
&= \left(U\psi_\nu(z)\right)\left(I+\psi_\nu(z)I-U\psi_\nu(z)\right)^{-1}\\
&= (X+I)\psi_\nu(z)\left(I-X\psi_\nu(z)\right)^{-1}.
\end{aligned}$$

Hence,

$$\begin{aligned}
\psi_\mu(\eta_\nu(z)) &= \varphi\left(\frac{U\eta_\nu(z)}{I-U\eta_\nu(z)}\right)\\
&= \varphi\left((X+I)\sum_{n=1}^{\infty}(\varphi((zV)^n))\sum_{k=0}^{\infty}(X(\sum_{m=1}^{\infty}\varphi((zV)^m)))^k\right).
\end{aligned}$$

We see that this sum is indeed equal to $\psi_\alpha(z)$. Note that the expansion $(I - X\psi_\nu(z))^{-1} = \sum_{k=0}^{\infty}(X\psi_\nu(z))^k$ only holds for $|z|$ small enough, but the uniqueness of analytic continuation implies that the two functions are indeed identical. □

LEMMA 7.5. *Let μ and $\mu_1, \mu_2, \dots$ be probability measures on $\partial\mathbb{D}$. Then ψ_{μ_n} converges to ψ_μ locally uniformly in $\mathbb{D}$ if and only if η_n converges locally uniformly in $\mathbb{D}$ to η_μ if and only if μ_n converges weakly to μ.*

PROOF. By noting that locally uniform convergence of the η-transforms is equivalent to locally uniform convergence of the ψ-transforms, the statement follows by passing to probability measures on $\hat{\mathbb{R}}$ (as in the proof of Theorem 3.12) and Theorem 3.17. □

THEOREM 7.6. *Let $(\mu_{s,t})_{0\leq s\leq t}$ be a family of probability measures on $\partial\mathbb{D}$ which form a $\rhd$-hemigroup on $\partial\mathbb{D}$, i.e.,*

(1) *$\mu_{s,t} \neq \mu_H$ for all $0 \leq s \leq t$,*
(2) *$\mu_{s,s} = \delta_1$ for all $s \geq 0$,*
(3) *$(s,t) \mapsto \mu_{s,t}$ is continuous with respect to weak convergence,*
(4) *$\mu_{s,u} = \mu_{s,t} \rhd \mu_{t,u}$ for all $0 \leq s \leq t \leq u$.*

Then $(\eta_{\mu_{0,t}})_{t\geq 0}$ is a radial Loewner chain with transition mappings $\eta_{\mu_{s,t}}$. Conversely, if $(f_t)_{t\geq 0}$ is a radial Loewner chain with transition mappings $f_{s,t}$, then $f_{s,t} = \eta_{\mu_{s,t}}$ for a $\rhd$-hemigroup $(\mu_{s,t})_{0\leq s\leq t}$ on $\partial\mathbb{D}$.

PROOF. The first statement follows directly from Theorems 7.3, 7.4 and Lemma 7.5. Note that all η-transforms are non-constant as μ_H is excluded. For the second statement, we note that all $f_{s,t}$ fix 0 and Theorem 7.3 implies that each $f_{s,t}$ is the

η-transform of a unique probability measure $\mu_{s,t}$ on $\partial\mathbb{D}$. By definition, each $f_{s,t}$ is non-constant, and thus, $\mu_{s,t} \neq \mu_H$. □

Theorem 6.17 implies the following unitary analog to the additive case (Corollary 6.18).

COROLLARY 7.7. *Let $(\mu_{s,t})_{0\leq s\leq t}$ be a $\rhd$-hemigroup on $\partial\mathbb{D}$. Then $\eta_{\mu_{s,t}}$ is univalent for all $0 \leq s \leq t$.*

Conversely, each univalent $\eta : \mathbb{D} \to \mathbb{D}$ with $\eta(0) = 0$ can be embedded into the η-transforms of a $\rhd$-hemigroup on $\partial\mathbb{D}$. The proof will be given in Section 7.7.

THEOREM 7.8.

(a) *Let μ be a probability measure on $\partial\mathbb{D}$ such that η_μ is univalent. Then there exists a radial Loewner chain $(f_t)_{t\geq 0}$ such that $f_1 = \eta_\mu$.*

(b) *Let μ be a probability measure on $\partial\mathbb{D}$ such that η_μ is univalent and*

$$\eta_\mu'(0) = \int_{\partial\mathbb{D}} x\, \mu(dx) = e^{-T}, \quad T > 0.$$

Then there exists a normalized radial Loewner chain $(f_t)_{t\geq 0}$ with $f_T = \eta_\mu$.

7.2. Infinitesimal arrays

We now give an alternative proof for the fact that $\rhd$-hemigroups have univalent η-transforms [**70**, Section 7.2], which we initially proved in Theorem 6.14 via Loewner's ordinary differential equation. Similar to Theorem 2.63 (d), we consider infinitesimal arrays of probability measures.

DEFINITION 7.9. A family $(\mu_{n,j})_{n\in\mathbb{N},1\leq j\leq k_n}$ of probability measures on $\partial\mathbb{D}$ is called an *infinitesimal (triangular) array* if $k_n \to \infty$ and for any $\delta \in (0, \pi)$,

$$\lim_{n\to\infty} \max_{1\leq j\leq k_n} \mu_{n,j}(\partial\mathbb{D} \setminus \exp(i[-\delta, \delta])) = 0. \tag{7.2}$$

THEOREM 7.10. *Let μ be a probability measure on $\partial\mathbb{D}$, $\mu \neq \mu_H$. Then the following two statements are equivalent:*

(1) *There exists an infinitesimal array $(\mu_{n,j})_{n\in\mathbb{N},1\leq j\leq k_n}$ on $\partial\mathbb{D}$ such that*

$$\mu_{n,1} \rhd \cdots \rhd \mu_{n,k_n} \to \mu \quad \text{as } n \to \infty.$$

(2) *η_μ is univalent.*

COROLLARY 7.11. *Let $(\mu_{s,t})_{0\leq s\leq t}$ be a $\rhd$-hemigroup on $\partial\mathbb{D}$. Then each $\mu_{s,t}$ has a univalent η-transform.*

PROOF. It suffices to show the statement for $\mu_{0,T}$, $T > 0$. Let

$$\mu_{n,j} = \mu_{(j-1)T/n,jT/n}, \quad n \in \mathbb{N}, j = 1, \dots, n.$$

These measures define an infinitesimal array. We need to verify (7.2). Let $\delta \in (0, \pi)$. As $(s,t) \mapsto \mu_{s,t}$ is continuous, $(s,t) \mapsto \mu_{s,t}(\exp(i[-\delta,\delta]))$ is uniformly continuous on the set of all (s,t) with $0 \leq s \leq t \leq T$. For any $\varepsilon > 0$, we find $N \in \mathbb{N}$ such that $|\delta_1(\exp(i[-\delta,\delta])) - \mu_{n,j}(\exp(i[-\delta,\delta]))| = |1 - \mu_{n,j}(\exp(i[-\delta,\delta]))| = \mu_{n,j}(\partial\mathbb{D} \setminus \exp(i[-\delta,\delta])) < \varepsilon$ for all $n \geq N$ and all $j = 1, \dots, n$. Since

$$\mu_{0,T} = \mu_{0,T/n} \rhd \mu_{T/n,2T/n} \rhd \cdots \rhd \mu_{T-T/n,T} = \mu_{n,1} \rhd \cdots \rhd \mu_{n,n},$$

(1) is shown and as $\mu_{0,T} \neq \mu_H$, $\mu_{0,T}$ has a univalent η-transform. □

LEMMA 7.12. *Let μ be a probability measure on $\partial\mathbb{D}$, $\delta \in (0, \pi/2)$, $n \in \mathbb{N}$, $C_\delta = \partial\mathbb{D} \setminus \exp(i[-\delta, \delta])$, and let $m_n(\mu)$ be the n-th moment of μ. Then we have*

$$|1 - m_1(\mu)|^2 \leq 4\delta^2 + 10(\mu(C_\delta))^2, \tag{7.3}$$

$$|1 - m_n(\mu)| \leq cn|1 - m_1(\mu)|^{1/2}, \quad \text{for some } c > 0. \tag{7.4}$$

PROOF. We regard μ as a probability measure on the interval $(-\pi, \pi]$ and then $1 - m_1(\mu) = \int_{(-\pi,\pi]} 1 - \cos(x) - i\sin(x)\mu(dx)$. So

$$\begin{aligned}
|1 - m_1(\mu)|^2 &= \left(\int_{|x|\leq\delta} (1 - \cos(x))\,\mu(dx) + \int_{(-\pi,\pi]\setminus[-\delta,\delta]} (1 - \cos(x))\,\mu(dx) \right)^2 \\
&\quad + \left(\int_{|x|\leq\delta} \sin(x)\,\mu(dx) + \int_{(-\pi,\pi]\setminus[-\delta,\delta]} \sin(x)\,\mu(dx) \right)^2 \\
&\leq (1 - \cos(\delta) + 2\mu(C_\delta))^2 + (\sin(\delta) + \mu(C_\delta))^2 \\
&\leq 2(1 - \cos(\delta))^2 + 2(2\mu(C_\delta))^2 + 2\sin(\delta)^2 + 2(\mu(C_\delta))^2 \\
&\leq 4\delta^2 + 10(\mu(C_\delta))^2,
\end{aligned}$$

as $\sin(\delta) \leq \delta$ and $1 - \cos(\delta) \leq \delta$. Next, we use that

$$1 - \cos(x) \geq x^2/c_0, \qquad x \in [-\pi, \pi], \tag{7.5}$$

$$|1 - e^{ix}| \leq 2|x|, \qquad x \in \mathbb{R}, \tag{7.6}$$

for some $c_0 > 0$. From (7.5), we get

$$|1 - m_1(\mu)|^2 \geq \left(\int_{(-\pi,\pi]} (1 - \cos(x))\,\mu(dx) \right)^2 \geq c_0^{-2} \left(\int_{(-\pi,\pi]} x^2\,\mu(dx) \right)^2.$$

On the other hand, using (7.6) and the Cauchy–Schwarz inequality shows that

$$\begin{aligned}
|1 - m_n(\mu)| &\leq \int_{(-\pi,\pi]} |1 - e^{inx}|\,\mu(dx) \leq 2n \int_{(-\pi,\pi]} |x|\,\mu(dx) \\
&\leq 2n \left(\int_{(-\pi,\pi]} x^2\,\mu(dx) \right)^{1/2} \leq 2nc_0^{1/2}|1 - m_1(\mu)|^{1/2}.
\end{aligned}$$

□

PROOF OF THEOREM 7.10. Suppose that μ satisfies (1) for an infinitesimal array $(\mu_{n,j})_{n\in\mathbb{N},1\leq j\leq k_n}$. Estimate (7.3) implies

$$\max_{1\leq j\leq k_n} |1 - m_1(\mu_{n,j})| \to 0, \quad n \to \infty. \tag{7.7}$$

Let ν be a probability measure on $\partial\mathbb{D}$. Recall that $\psi_{\delta_1}(z) = z/(1-z)$. For $z \in \mathbb{D}$, (7.4) shows that

$$\begin{aligned}
\left| \psi_\nu(z) - \frac{z}{1-z} \right| &= \left| \sum_{n=1}^{\infty} (1 - m_n(\nu)) z^n \right| \leq c|1 - m_1(\nu)|^{1/2} \sum_{n=1}^{\infty} n|z|^n \\
&= \frac{c|1 - m_1(\nu)|^{1/2}|z|}{(1-|z|)^2} \leq \frac{c|1 - m_1(\nu)|^{1/2}}{(1-|z|)^2}.
\end{aligned} \tag{7.8}$$

Combining (7.7) and (7.8) yields

$$\sup_{1\leq j\leq k_n, z\in r\mathbb{D}} |\psi_{\mu_{n,j}}(z) - \psi_{\delta_1}(z)| \to 0 \quad \text{as } n \to \infty$$

for any $0 < r < 1$, and thus, also $\sup_{1\leq j\leq k_n, z\in r\mathbb{D}} |\eta_{\mu_{n,j}}(z) - z| \to 0$ as $n \to \infty$. Let $0 < r < r' < 1$. Applying Cauchy's integral formula to the derivative yields that

$$\eta'_{\mu_{n,j}}(z) - 1 = \frac{1}{2\pi i} \int_{\partial(r'\mathbb{D})} \frac{\eta_{\mu_{n,j}}(w) - w}{(w-z)^2}\, dw, \qquad z \in r\mathbb{D},$$

and hence, $\sup_{1\leq j\leq k_n, z\in r\mathbb{D}} |\eta'_{\mu_{n,j}}(z) - 1| \to 0$ as $n \to \infty$. In particular,

$$\inf_{1\leq j\leq k_n, z\in r\mathbb{D}} \operatorname{Re}\left(\eta'_{\mu_{n,j}}(z)\right) > 0$$

for all sufficiently large n. In this case, $\eta_{\mu_{n,j}}$ is univalent in $r\mathbb{D}$ (with the same argument as in the proof of Theorem 6.31, with Im replaced by Re). Since $\eta_{\mu_{n,j}}(r\mathbb{D}) \subset r\mathbb{D}$, the composition $\eta_{\mu_{n,1}} \circ \cdots \circ \eta_{\mu_{n,k_n}}$ is also univalent in $r\mathbb{D}$, and hence, its limit η_μ is univalent in $r\mathbb{D}$ since $\mu \neq \mu_H$. Since $0 < r < 1$ was arbitrary, η_μ is univalent in $\mathbb{D}$.

Conversely, if η_μ is univalent, we can embed μ into a $\rhd$-hemigroup, which implies (1) as in the proof of Corollary 7.11. □

7.3. Semigroups with respect to monotone convolution

We now consider $\rhd$-infinitely divisible probability measures on $\partial\mathbb{D}$.

THEOREM 7.13. *Let μ be a probability measure on $\partial\mathbb{D}$, $\mu \neq \mu_H$. Then μ is $\rhd$-infinitely divisible if and only if there exists a continuous $\rhd$-semigroup $(\mu_t)_{t\geq 0}$ of probability measures on $\partial\mathbb{D}$ such that $\mu_1 = \mu$.*

PROOF. We follow [**31**, Theorem 4.4].

Clearly, if μ can be embedded into a $\rhd$-semigroup, then it is $\rhd$-infinitely divisible.

We now assume that μ is $\rhd$-infinitely divisible. For every $n \in \mathbb{N}$, we can write $\mu = \nu_n^{\rhd 2^n}$ for some probability measure ν_n on $\partial\mathbb{D}$. As $\eta_\mu = \eta_{\nu_n}^{\circ 2^n}$, we have $\int_{\partial\mathbb{D}} x\mu(dx) = \eta'_\mu(0) = \eta'_{\nu_n}(0)^{2^n}$. We see that $\eta'_\mu(0) \neq 0$. Otherwise, η'_μ would have a zero of arbitrarily large multiplicity, i.e., η_μ would be constant 0, and thus, $\mu = \mu_H$, which we excluded. Thus, $\eta'_\mu(0) = re^{i\alpha}$, $\alpha \in \mathbb{R}$, $r \in (0, 1]$. Possibly after a rotation, we can assume that $\eta'_{\nu_n}(0) = r^{1/2^n} e^{i\alpha/2^n}$.

Fix $n \in \mathbb{N}$. Then we find a sequence $m_1 < m_2 < \ldots$ such that $\nu_{m_j}^{\rhd 2^{m_j - k}}$, $j \geq n$, converges weakly for each $k = 1, \ldots, n$. We can take the sequence related to $n+1$ as a subsequence of that corresponding to n. By a diagonalization argument, we find a sequence $n_1 < n_2 < \ldots$ such that $\nu_{n_j}^{\rhd 2^{n_j - n}}$, $j \geq n$, converges weakly for each n. Denote the limit by $\mu_{1/2^n}$. Then $\eta'_{\mu_{1/2^n}}(0) = r^{1/2^n} e^{i\alpha/2^n}$, $\mu_{1/2^n}^{\rhd 2^n} = \mu$, and for $m < n$,

$$\mu_{1/2^n}^{\rhd 2^m} = \mu_{1/2^{n-m}}. \tag{7.9}$$

As $n \to \infty$, $\eta'_{\mu_{1/2^n}}(0) \to 1$. This implies that $\mu_{1/2^n} \to \delta_1$ as $n \to \infty$. [Helly's selection theorem gives us a converging subsequence $(\mu_{1/2^{n_k}})_k$ with limit ρ. The integral $\int_{\partial\mathbb{D}} x\rho(dx)$ is equal to 1, which is only possible if $\rho = \delta_1$. Hence, every converging subsequence has the same limit, and we conclude that $\mu_{1/2^n}$ converges to δ_1.]

For $m, n \in \mathbb{N}$, we define $\mu_{m/2^n} = \mu_{1/2^n}^{\rhd m}$, which is well-defined due to (7.9). For a sequence t_k of the form $m/2^n$ with limit 0, we also have $\mu_{t_k} \to \delta_1$. So we define $\mu_0 = \delta_1$.

Let $t \geq 0$, and let s_k and t_k be sequences of the form $m/2^n$, which both converge to t. By passing to subsequences, we may assume that $\mu_{s_k} \to \alpha$ and $\mu_{t_k} \to \beta$. We may assume that $s_k \leq t_k$ for all k. Then $\mu_{t_k} = \mu_{t_k - s_k} \rhd \mu_{s_k}$. Sending k to ∞ yields $\beta = \delta_1 \rhd \alpha = \alpha$. So the limit only depends on t and we denote it by μ_t. The measures $(\mu_t)_{t\geq 0}$ now form a $\rhd$-semigroup with $\mu_1 = \mu$. The continuity of $t \mapsto \mu_t$ follows from the continuity at $t = 0$, which follows from a diagonalization argument and the fact that $\mu_{t_k} \to \delta_1$ for any sequence t_k of the form $m/2^n$ with limit 0. □

This theorem is the multiplicative analog of Theorem 5.22. But here, we can go one step further and state a nice geometric property for η_μ that is equivalent to μ being $\rhd$-infinitely divisible [**70**, Section 7.3].

Let $\alpha \in [-\pi/2, \pi/2]$. A univalent function $f : \mathbb{D} \to \mathbb{C}$ with $f(0) = 0$ is called *α-spirallike* if $e^{-e^{i\alpha}t}w \in f(\mathbb{D})$ for every $t \geq 0$ and $w \in f(\mathbb{D})$. In the special case $\alpha = 0$, f is a *starlike* mapping. If $\alpha = \pm\pi/2$, then f maps $\mathbb{D}$ onto a disk, which implies $f(z) = az$ for some $a \in \mathbb{C} \setminus \{0\}$.

A holomorphic function $f : \mathbb{D} \to \mathbb{C}$ with $f(0) = 0$ is univalent and α-spirallike, $\alpha \in (-\pi/2, \pi/2)$, if and only if

$$\operatorname{Re}\left[e^{-i\alpha}\frac{zf'(z)}{f(z)}\right] > 0 \tag{7.10}$$

on $\mathbb{D}$, see [**151**, Theorem 6.6]. (Note that in [**151**], f is called spirallike of type α if $e^{-e^{-i\alpha}t}w \in f(\mathbb{D})$.)

THEOREM 7.14. *Let μ be a probability measure on $\partial\mathbb{D}$, $\mu \neq \mu_H$. Then μ is $\rhd$-infinitely divisible if and only if there exist $\beta \in [-\pi/2, \pi/2]$, $R \geq 0$, and a β-spirallike mapping $f : \mathbb{D} \to \mathbb{C}$ with $f(0) = 0$, $f'(0) = 1$, such that*

$$\eta_\mu = f^{-1} \circ (e^{-Re^{i\beta}} \cdot f).$$

If this holds true, then $e^{-Re^{i\beta}}$ is the first moment of μ.

PROOF. First, we consider the special case $\mu = \delta_\alpha$. Then $\eta_\mu(z) = \alpha z$. We can take $f(z) = z$.

Now assume that μ is $\rhd$-infinitely divisible and not a point measure (and thus, η_μ is not a rotation). By Theorem 7.13 and Theorem 7.6, η_μ can be embedded into a radial Loewner chain $(\eta_t)_{t\geq 0}$ at $t = 1$ and the transition mappings $\eta_{s,t}$ satisfy $\eta_{s,t} = \eta_{0,t-s} = \eta_{t-s}$. Furthermore we have $\eta_t'(0) = e^{-at}$ for some $a \in \mathbb{C}$. By the Schwarz lemma we must have $\operatorname{Re}(a) > 0$, and thus, $a = Re^{i\beta}$ for some $R > 0$ and $\beta \in (-\pi/2, \pi/2)$.

Let M be the infinitesimal generator of the semigroup, so

$$\frac{\partial}{\partial t}\eta_{s,t}(z) = M(\eta_{s,t}(z)), \quad s \leq t,$$

where $M(z) = -zp(z)$ for a holomorphic $p : \mathbb{D} \to \mathbb{C}$ with $p(0) = a$ and $\mathrm{Re}(p(z)) > 0$ for all $z \in \mathbb{D}$. Then the locally uniform limit

$$\lim_{t\to\infty} e^{at}\eta_{s,t}(z) =: f_s(z)$$

exists and $(f_t)_{t\geq 0}$ is an increasing Loewner chain with $f_t \circ \eta_{s,t} = f_s$, and $f_0(0) = 0$, $f_0'(0) = 1$. This can be shown similarly to the considerations in Section 8.6; see also [**78**, Theorem 2.3] or [**23**, Lemma 1]. As $\eta_{s,t} = \eta_{0,t-s}$, we have

$$f_0 = \lim_{t\to\infty} e^{at-as}\eta_{0,t-s}(z) = e^{-as}\lim_{t\to\infty} e^{at}\eta_{s,t}(z) = e^{-as}f_s.$$

Hence, the Loewner chain has the simple form $(e^{at}f_0)_{t\geq 0}$, which implies that f_0 is a β-spirallike mapping. We conclude that $f_0 = f_1 \circ \eta_{0,1} = (e^a f_0) \circ \eta_{0,1}$, and thus, $\eta_{0,1} = \eta_\mu = f_0^{-1} \circ (e^{-Re^{i\beta}} f_0)$.

Conversely, assume that η_μ has the form $\eta_\mu = f^{-1} \circ (e^{-Re^{i\beta}} \cdot f)$. Then $(f_t)_{t\geq 0} := (e^{at}f)_{t\geq 0}$ with $a = Re^{i\beta}$ is an increasing Loewner chain and $\eta_{0,t} = f_t^{-1} \circ f_0$ defines a semigroup of univalent functions on $\mathbb{D}$ with $\eta_{0,t}(0) = 0$ and $\eta_{0,1} = \eta_\mu$. Theorem 7.13 implies that μ is $\rhd$-infinitely divisible. □

As η_μ is univalent if and only if ψ_μ is univalent, and as $\psi_\mu(0) = 0$, we might also look at the case of starlike ψ-transforms. We call ψ_μ *starlike* if ψ_μ is univalent and $c\psi_\mu(\mathbb{D}) \subset \psi_\mu(\mathbb{D})$ for all $c \in [0, 1]$.

We let $\mathcal{P}_\times(\partial\mathbb{D})$ be the set of all probability measures on $\partial\mathbb{D}$ with non-zero mean. For $\mu \in \mathcal{P}_\times(\partial\mathbb{D})$, the series expansion of $\psi_\mu(z)$ is $m_1(\mu)z + \dots$, where $m_1(\mu)$ is the first moment of μ, and thus, the inverses ψ_μ^{-1} and η_μ^{-1} exist in an open neighborhood of 0.

The following can be seen as the multiplicative analog to Theorem 6.33, the correspondence between starlike F-transforms and self-decomposable probability measures on $\mathbb{R}$.

THEOREM 7.15. *Let $\mu \in \mathcal{P}_\times(\partial\mathbb{D})$. Then the following statements are equivalent:*

(1) *ψ_μ is starlike.*
(2) *The function $z\frac{\psi_\mu'(z)}{\psi_\mu(z)}$ has positive real part.*
(3) *For every $c \in (0, 1]$ there exists a probability measure ν_c on $\partial\mathbb{D}$ such that $\mu \rhd \nu_c = c\mu + (1-c)\mu_H$.*

For $c \in (0, 1]$, let $H_c\mu := c\mu + (1-c)\mu_H$. Note that $\psi_{\mu_H} = 0$, and thus, $\psi_{H_c\mu} = c\psi_\mu$, and that for $\mu \in \mathcal{P}_\times(\partial\mathbb{D})$, ν_c in (3) is uniquely determined by

$$\eta_{\nu_c}(z) = \psi_\mu^{-1}(c\psi_\mu(z)) \tag{7.11}$$

in a neighborhood of 0.

PROOF. As $\psi_\mu'(0) \neq 0$, the equivalence between (1) and (2) follows from Theorem 3.42.

Now let ψ_μ be starlike and $c \in (0, 1]$. From the inclusion $\psi_{H_c\mu}(\mathbb{D}) = c\psi_\mu(\mathbb{D}) \subset \psi_\mu(\mathbb{D})$, we can define the univalent function $\eta_c = \psi_\mu^{-1} \circ \psi_{H_c\mu} : \mathbb{D} \to \mathbb{D}$. Thus $\eta_c = \eta_{\nu_c}$ for some ν_c, which satisfies $\mu \rhd \nu_c = H_c\mu$.

Conversely, assume that for every $c \in (0,1]$ there exists a probability measure ν_c on $\partial\mathbb{D}$ such that $\mu \rhd \nu_c = c\mu + (1-c)\mu_H$. Then $c\psi_\mu(\mathbb{D}) = \psi_{\mu\rhd\nu_c}(\mathbb{D}) \subset \psi_\mu(\mathbb{D})$, and so it suffices to prove the univalence of ψ_μ. Within a neighborhood of 0, we have (7.11), which implies that $\nu_c \rhd \nu_d = \nu_{cd}$. With the reparametrization $\mu_t := \nu_{e^{-t}}$, $t \in [0,\infty)$, we have

$$\mu_s \rhd \mu_t = \mu_{s+t}, \qquad s,t \geq 0, \qquad \mu_0 = \nu_1 = \delta_1. \tag{7.12}$$

Due to (7.11), $c \mapsto \eta_{\nu_c}$ is continuous locally at 0 and Vitali's theorem implies that $c \mapsto \eta_{\nu_c}$ is continuous with respect to locally uniform convergence on $\mathbb{D}$. Thus, $t \mapsto \mu_t$ is continuous and $(\mu_t)_{t\geq 0}$ is a $\rhd$-semigroup, and each η_{μ_t} is univalent.

Fix $\varepsilon \in (0,1)$ such that ψ_μ is univalent in $\varepsilon\mathbb{D}$. Take $0 < r < 1$. Since $\mu_t \to \mu_H$ as $t \to \infty$, ψ_{μ_t} converges locally uniformly to 0, and thus, there exists $t = t(\varepsilon, r) > 0$ such that $\sup_{z\in r\mathbb{D}} |\eta_{\mu_t}(z)| < \varepsilon$. Since η_{μ_t} is univalent on $r\mathbb{D}$, the function $\psi_\mu = e^t\psi_\mu \circ \eta_{\mu_t}$ is also univalent on $r\mathbb{D}$. Since $0 < r < 1$ was arbitrary, we conclude that ψ_μ is univalent on $\mathbb{D}$. □

7.4. Monotone multiplicative processes

Just as in the additive case, monotone hemigroups on $\partial\mathbb{D}$ can be related to certain Markov processes, which then allow us to construct corresponding quantum stochastic processes [**70**, Section 5].

DEFINITION 7.16. A probability kernel k on $\partial\mathbb{D}$ is called $\rhd$-*homogeneous* if it satisfies

$$\delta_x \rhd k(y,\,\cdot\,) = k(x\cdot y,\,\cdot\,)$$

for all $x, y \in \partial\mathbb{D}$. A Markov process $(M_t)_{t\geq 0}$ with values in $\partial\mathbb{D}$ is called a $\rhd$-*homogeneous Markov process* if its transition kernels $(k_{s,t})_{0\leq s\leq t}$ satisfy the following conditions:

(a) The mapping $(s,t) \mapsto k_{s,t}(x,\cdot)$ is continuous with respect to weak convergence for all $x \in \partial\mathbb{D}$.
(b) $k_{s,t}(1,\cdot) \neq \mu_H$ for all $0 \leq s \leq t$.
(c) The kernel $k_{s,t}$ is $\rhd$-*homogeneous* for all $0 \leq s \leq t$.

THEOREM 7.17. *Let $(\mu_{s,t})_{0\leq s\leq t}$ be a $\rhd$-hemigroup on $\partial\mathbb{D}$ and let $\eta_{s,t} = \eta_{\mu_{s,t}}$. Then there exists a $\rhd$-homogeneous Markov process $(M_t)_{t\geq 0}$ with transition kernels $(k_{s,t})_{0\leq s\leq t}$ such that*

$$\int_{\partial\mathbb{D}} \frac{zy}{1-zy} k_{s,t}(x,dy) = \frac{\eta_{s,t}(z)x}{1-\eta_{s,t}(z)x} \; (= \psi_{\delta_x}(\eta_{s,t}(z))) \tag{7.13}$$

for all $0 \leq s \leq t$ and $x \in \partial\mathbb{D}$.

Conversely, let $(M_t)_{t\geq 0}$ be a $\rhd$-homogeneous Markov process on $\partial\mathbb{D}$ with transition kernels $k_{s,t}$. Then $(k_{s,t}(1,\cdot))_{0\leq s\leq t}$ is a $\rhd$-hemigroup on $\partial\mathbb{D}$.

PROOF. The proof is similar to the proof of Theorem 5.52. We only need to use the kernel $xz/(1-xz)$ on $\partial\mathbb{D}$ instead of $1/(z-x)$ on $\mathbb{R}$, and the Herglotz representation formula. We can apply the Stieltjes–Perron inversion formula by passing to probability measures on $\hat{\mathbb{R}}$ (as in the proof of Theorem 3.12). The property $\mu_{s,u} = \mu_{s,t} \rhd \mu_{t,u}$ and (7.13) imply the Chapman–Kolmogorov relation, and vice versa. □

Similar to the additive case, we now define monotone multiplicative processes consisting of unitary operators.

DEFINITION 7.18. Let $(B(H), \varphi)$ be a quantum probability space and $(U_t)_{t\geq 0} \subset B(H)$ a family of unitary operators. We call $(U_t)_{t\geq 0}$ a *monotone multiplicative process* if the following conditions are satisfied:

(a) $U_0 = I$ and the distributions μ_t of U_t are all different from μ_H.
(b) The tuple

$$(U_{t_1} - I, U_{t_1}^{-1}U_{t_2} - I, \ldots, U_{t_{n-1}}^{-1}U_{t_n} - I)$$

is monotonically independent for all $n \geq 2$ and all $t_1, \ldots, t_n \in \mathbb{R}$ with $0 \leq t_1 \leq t_2 \leq \cdots \leq t_n$.
(c) The mapping $(s,t) \mapsto \mu_{s,t}$ is continuous with respect to weak convergence, where $\mu_{s,t}$ denotes the distribution of the multiplicative increment $U_s^{-1}U_t = U_s^*U_t$, $0 \leq s \leq t$.

With Theorem 7.4 we immediately get:

THEOREM 7.19. *Let $(U_t)_{t\geq 0}$ be a monotone multiplicative process. For $0 \leq s \leq t$, let $\mu_{s,t}$ be the distribution of $U_s^{-1}U_t$. Then $(\mu_{s,t})_{0\leq s\leq t}$ is a $\rhd$-hemigroup on $\partial\mathbb{D}$.*

The construction of a monotone multiplicative process from a $\rhd$-hemigroup on $\partial\mathbb{D}$ now works by using its corresponding $\rhd$-homogeneous Markov process.

THEOREM 7.20. *Let $(M_t)_{t\geq 0}$ be a $\rhd$-homogeneous Markov process on $\partial\mathbb{D}$ with $M_0 = 1$ and let $(\mathcal{F}_t)_{t\geq 0}$ be its natural filtration ($\mathcal{F}_t = \sigma((M_s)_{s\leq t})$). Denote by $(\Omega, \mathcal{F}, \mathbb{P})$ the underlying probability space and by P_t the conditional expectation*

$$P_t = \mathbb{E}[\,\cdot\,|\mathcal{F}_t].$$

Define the operators $U_t \in B(L^2(\Omega, \mathcal{F}, \mathbb{P}))$, $t \geq 0$, by

$$U_t = I + (M_t - I)P_t, \tag{7.14}$$

where M_t acts by multiplication on $L^2(\Omega, \mathcal{F}, \mathbb{P})$. Then $(U_t)_{t\geq 0}$ is a monotone multiplicative process on $(B(L^2(\Omega, \mathcal{F}, \mathbb{P})), \langle \mathbf{1}_\Omega, \cdot \mathbf{1}_\Omega\rangle)$, where $\mathbf{1}_\Omega$ is the constant function with value 1 on Ω. Let $\mu_{s,t} := k_{s,t}(1, \cdot)$, where $k_{s,t}$ are the transition kernels of $(M_t)_{t\geq 0}$. Then the distribution of $U_s^*U_t$ is equal to $\mu_{s,t}$.

PROOF. We first check that each U_t is unitary. As M_t and P_t commute, we have

$$\begin{aligned} U_s^*U_t &= (I + (M_s^* - I)P_s)(I + P_t(M_t - I)) \\ &= I + (M_s^* - I)P_s + (M_t - I)P_t + (M_s^* - I)P_s(M_t - I) \\ &= I + (M_t - I)P_t + (M_s^* - I)P_s M_t \end{aligned} \tag{7.15}$$

for all $0 \leq s \leq t$. Taking $s = t$ shows that $U_t^*U_t = I$. Similarly, we can prove that $U_tU_t^* = I$.

We define

$$\psi_{s,t}(z) = \psi_{\mu_{s,t}}(z) = \int_{\partial\mathbb{D}} \frac{zy}{1 - zy} k_{s,t}(1, dy) \quad \text{and} \quad \eta_{s,t} = \eta_{\mu_{s,t}}.$$

Next, we show that the process $(U_t)_{t\geq 0}$ satisfies

$$P_s \frac{zU_s^*U_t}{1-zU_s^*U_t} P_s = \psi_{s,t}(z) P_s \tag{7.16}$$

for all $0 \leq s \leq t$ and $z \in \mathbb{D}$.

The $\triangleright$-homogeneity condition implies $k_{s,t}(x, dy) = \delta_x \triangleright k_{s,t}(1, dy)$, and hence,

$$\int_{\partial\mathbb{D}} \frac{zy}{1-zy} k_{s,t}(x, dy) = \psi_{\delta_x}(\eta_{s,t}(z)) = \frac{\eta_{s,t}(z)x}{1-\eta_{s,t}(z)x}.$$

Therefore, by using the Markov property (2.8), we have

$$\mathbb{E}\left[\frac{zM_t}{1-zM_t} \middle| \mathcal{F}_s \right] = \frac{\eta_{s,t}(z)M_s}{1-\eta_{s,t}(z)M_s} a.s.,$$

which can also be expressed as

$$P_s \frac{zM_t}{1-zM_t} P_s = \frac{\eta_{s,t}(z)M_s}{1-\eta_{s,t}(z)M_s} P_s, \qquad 0 \leq s \leq t, \quad z \in \mathbb{D}. \tag{7.17}$$

Similarly,

$$P_s \frac{1}{1-zM_t} P_s = P_s \left(1 + \frac{zM_t}{1-zM_t}\right) P_s = \frac{1}{1-\eta_{s,t}(z)M_s} P_s. \tag{7.18}$$

We begin with

$$\begin{aligned} P_s \frac{zU_s^*U_t}{1-zU_s^*U_t} P_s &= P_s \left(-1 + \frac{1}{1-zU_s^*U_t}\right) P_s = -P_s + \sum_{n=0}^{\infty} z^n P_s \left(U_s^*U_t\right)^n P_s \\ &= -P_s + \sum_{n=0}^{\infty} z^n P_s \left[I + (M_t - I)P_t + (M_s^* - I)P_s + (M_s^* - I)P_s(M_t - I)P_t)\right]^n P_s \\ &= -P_s + \sum_{n=0}^{\infty} z^n P_s \left[I + (M_t - I)P_t + (M_s^* - I)P_s M_t P_t\right]^n P_s \end{aligned}$$

for $z \in \mathbb{D}$. In the expansion of $P_s \left[I + (M_t - I)P_t + (M_s^* - I)P_s M_t P_t\right]^n P_s$, the projection P_t can be replaced by the identity. This gives us

$$P_s \frac{zU_s^*U_t}{1-zU_s^*U_t} P_s = -P_s + \sum_{n=0}^{\infty} P_s \left[zM_t + z(M_s^* - I)P_s M_t\right]^n P_s. \tag{7.19}$$

By defining $Z_t = (M_t^* - I)P_t$, the above expression can be expanded as follows:

$$\begin{aligned}
&(7.19) = -P_s + \\
&\sum_{n=0}^{\infty}\sum_{k=0}^{n} \sum_{\substack{p_1,\ldots,p_{k+1}\geq 0 \\ p_1+\cdots+p_{k+1}=n-k}} P_s(zM_t)^{p_1}(zZ_sM_t)(zM_t)^{p_2}(zZ_sM_t)\cdots(zZ_sM_t)(zM_t)^{p_{k+1}}P_s \\
&= -P_s + \sum_{k=0}^{\infty} \sum_{p_1,\ldots,p_{k+1}\geq 0} z^{p_1+\cdots+p_{k+1}+k} P_s M_t^{p_1} Z_s M_t^{p_2+1} Z_s M_t^{p_3+1}\cdots Z_s M_t^{p_{k+1}+1} P_s \\
&= -P_s + \sum_{k=0}^{\infty}\sum_{q_1\geq 0}\sum_{q_2,\ldots,q_{k+1}\geq 1} z^{q_1+\cdots+q_{k+1}} P_s M_t^{q_1} Z_s M_t^{q_2} Z_s M_t^{q_3}\cdots Z_s M_t^{q_{k+1}} P_s \\
&= -P_s + \sum_{k=0}^{\infty}\sum_{q_1\geq 0}\sum_{q_2,\ldots,q_{k+1}\geq 1} P_s(zM_t)^{q_1} Z_s(zM_t)^{q_2} Z_s(zM_t)^{q_3}\cdots Z_s(zM_t)^{q_{k+1}} P_s \\
&= -P_s + \sum_{k=0}^{\infty} P_s \frac{1}{1-zM_t} Z_s \frac{zM_t}{1-zM_t} Z_s \cdots Z_s \frac{zM_t}{1-zM_t} P_s.
\end{aligned}$$

By using that $Z_sP_s = P_sZ_s = Z_s$, and by equations (7.17) and (7.18), we get

$$\begin{aligned}
(7.19) &= -P_s + \sum_{k=0}^{\infty} P_s \frac{1}{1-\eta_{s,t}(z)M_s}\left(\frac{\eta_{s,t}(z)M_s}{1-\eta_{s,t}(z)M_s}\right)^k (Z_s)^k P_s \\
&= -P_s + \sum_{k=0}^{\infty} P_s \frac{1}{1-\eta_{s,t}(z)M_s}\left(\frac{\eta_{s,t}(z)M_s}{1-\eta_{s,t}(z)M_s}\right)^k (M_s^* - I)^k P_s \\
&= -P_s + \sum_{k=0}^{\infty} P_s \frac{1}{1-\eta_{s,t}(z)M_s}\left(\frac{\eta_{s,t}(z)M_s(M_s^* - I)}{1-\eta_{s,t}(z)M_s}\right)^k P_s \\
&= -P_s + P_s \frac{1}{1-\eta_{s,t}(z)M_s} \frac{1}{1-\frac{\eta_{s,t}(z)(I-M_s)}{1-\eta_{s,t}(z)M_s}} P_s \\
&= -P_s + P_s \frac{1}{1-\eta_{s,t}(z)} P_s = \frac{\eta_{s,t}(z)}{1-\eta_{s,t}(z)} P_s = \psi_{s,t}(z)P_s,
\end{aligned}$$

and (7.16) has been shown.

Now we apply (7.16) to the constant function $\mathbf{1}_\Omega$ and get

$$\left\langle \mathbf{1}_\Omega, \frac{zU_s^*U_t}{1-zU_s^*U_t}\mathbf{1}_\Omega \right\rangle = \psi_{s,t}(z),$$

which shows that the distribution of $U_s^*U_t$ with respect to the state $\langle \mathbf{1}_\Omega, \cdot \mathbf{1}_\Omega\rangle$ is given by $\mu_{s,t} = k_{s,t}(1,\cdot)$.

As $M_0 = 1$, we have $U_0 = I$. It remains to show that the process has independent multiplicative increments. Formula (7.15) implies that $(U_s^*U_t - I)P_u = P_u(U_s^*U_t - I) = U_s^*U_t - I$ for all $0 \leq s \leq t \leq u$, and hence

$$p(U_s^*U_t - I)P_u = P_u p(U_s^*U_t - I) = p(U_s^*U_t - I) \tag{7.20}$$

for every polynomial $p : \mathbb{R} \to \mathbb{R}$ with $p(0) = 0$.

Next, we can extend (7.16) to

$$P_r \frac{zU_s^*U_t}{1 - zU_s^*U_t} P_r = \psi_{s,t}(z) P_r$$

for any $r \le s$ because $P_r = P_r P_s = P_s P_r$. Thus,

$$\begin{aligned} P_r \sum_{n=1}^{\infty} (U_s^*U_t)^n z^n P_r &= \sum_{n=1}^{\infty} P_r (U_s^*U_t)^n P_r z^n \\ &= \sum_{n=1}^{\infty} \langle \mathbf{1}_\Omega, (U_s^*U_t)^n \mathbf{1}_\Omega \rangle \, z^n P_r = \psi_{s,t}(z) P_r \end{aligned}$$

for all $z \in \mathbb{D}$. Take any $v, w \in L^2(\Omega, \mathcal{F}, \mathbb{P})$. Then

$$\sum_{n=1}^{\infty} \langle w, P_r (U_s^*U_t)^n P_r v \rangle \, z^n = \sum_{n=1}^{\infty} \langle w, \langle \mathbf{1}_\Omega, (U_s^*U_t)^n \mathbf{1}_\Omega \rangle \, P_r v \rangle \, z^n .$$

The coefficients of the power series are identical, so

$$\langle w, P_r (U_s^*U_t)^n P_r v \rangle = \langle w, \langle \mathbf{1}_\Omega, (U_s^*U_t)^n \mathbf{1}_\Omega \rangle \, P_r v \rangle$$

for all $n \ge 1$. We conclude that

$$P_r (U_s^*U_t)^n P_r v = \langle \mathbf{1}_\Omega, (U_s^*U_t)^n \mathbf{1}_\Omega \rangle \, P_r v$$

and $P_r (U_s^*U_t)^n P_r = \langle \mathbf{1}_\Omega, (U_s^*U_t)^n \mathbf{1}_\Omega \rangle \, P_r$ for all $n \in \mathbb{N}_0$.

We can replace $U_s^*U_t$ in this equality also by $U_s^*U_t - I$, and thus, for $0 \le r \le s \le t$ and any polynomial $p : \mathbb{R} \to \mathbb{R}$, we have

$$P_r p(U_s^*U_t - I) P_r = \langle \mathbf{1}_\Omega, p(U_s^*U_t - I) \mathbf{1}_\Omega \rangle \, P_r . \tag{7.21}$$

We can now prove monotone independence of the tuple

$$(U_{t_1} - I, U_{t_1}^{-1} U_{t_2} - I, \ldots, U_{t_{n-1}}^{-1} U_{t_n} - I)$$

by using (7.20) and (7.21) as in the proof of Theorem 5.57. □

Assume that $(\mu_{0,t})_{t \ge 0}$ is a $\rhd$-semigroup such that the generator of $(\eta_{\mu_t})_{t \ge 0}$ is equal to $-z\frac{1+z}{1-z}$ (see Exercise 7.9.3). Analogous to the additive case, we would refer to the corresponding monotone multiplicative process as a *multiplicative monotone Brownian motion* (or *monotone unitary Brownian motion*). A construction of this process using operators on a Fock space can be found in [**84**].

7.5. Classical, free, and Boolean convolution

Anti-monotone independence for unitary operators just yields the reversed convolution $\eta_{\mu \lhd \nu} = \eta_\nu \circ \eta_\mu$.

So let us consider the remaining three cases. We start with Boolean independence. If U, V are two unitary operators with distributions μ, ν such that $U - I$ and $V - I$ are Boolean independent, then the distribution of UV is given by $\mu \boxtimes\!\!\!\!\cup \nu$ defined by

$$\eta_{\mu \boxtimes\!\!\!\cup \nu}(z)/z = \eta_\mu(z)/z \cdot \eta_\nu(z)/z,$$

which is well-defined by Theorem 7.3 and due to the Schwarz lemma. We refer to [**68**, Theorem 2.2]. Assuming that U, V are Boolean independent is again rather too simple, similar to (7.1).

We can easily obtain a Lévy–Khintchine representation for the Boolean case.

THEOREM 7.21. *A family* $(\mu_t)_{t\geq 0}$ *of probability measures on* $\partial\mathbb{D}$ *is a continuous* $\uplus$*-semigroup if and only if there exists a holomorphic function* $p : \mathbb{D} \to \mathbb{C}$ *with* $\mathrm{Re}(p(z)) \geq 0$ *such that*

$$\eta_{\mu_t}(z)/z = \exp(-tp(z)).$$

A probability measure μ *on* $\partial\mathbb{D}$, $\mu \neq \mu_H$, *is* $\uplus$*-infinitely divisible if and only if it can be embedded into a continuous* $\uplus$*-semigroup* $(\mu_t)_{t\geq 0}$.

PROOF. Let $(\mu_t)_{t\geq 0}$ be a continuous $\uplus$-semigroup of probability measures on $\partial\mathbb{D}$ and let $K_t(z) = \eta_{\mu_t}(z)/z$. Fix some $t > 0$. Assume that K_t has a zero z_0 in $\mathbb{D}$. As $K_t(z) = (K_{t/n}(z))^n$, we see that also $K_{t/n}(z_0) = 0$ for every $n \in \mathbb{N}$. Continuity implies that $K_0(z_0) = 0$. However, $K_0(z) = z/z = 1$ for all $z \in \mathbb{D}$, a contradiction. So K_t has no zeros. As $\mathbb{D}$ is simply connected, we can write $K_t(z) = \exp(-p_t(z))$ for some holomorphic functions $p_t(z)$. These functions are unique modulo adding $2ki\pi$, $k \in \mathbb{Z}$. The function $t \mapsto K_t(0)$ must be of the form $K_t(0) = \exp(-at)$, $a \in \mathbb{C}$, see Exercise 7.9.4. We choose each p_t such that $p_t(0) = at$.

As $\eta_{\mu_t} : \mathbb{D} \to \mathbb{D}$ with $\eta_{\mu_t}(0) = 0$, the Schwarz lemma implies that $|\eta_{\mu_t}(z)| \leq |z|$, and thus, $|K_t(z)| \leq 1$ for all $z \in \mathbb{D}$ and all $t \geq 0$. Thus $\mathrm{Re}(p_t(z)) \geq 0$. As $K_t = (K_{t/n})^n$, we have $K_1^m = K_m = K_{m/n}^n$ for all $m, n \in \mathbb{N}$, i.e., $\exp(-mp_1) = \exp(-np_{m/n})$, which implies $m \cdot p_1 = np_{m/n} + 2k(m,n)i\pi$, with $k(m,n) \in \mathbb{Z}$. With $p_t(0) = at$ we obtain $p_{m/n} = \frac{m}{n}p_1$. The continuity of the semigroup implies that $p_t = tp_1$ for all $t \geq 0$.

Conversely, if $\eta_{\mu_t}(z)/z = \exp(-tp(z))$, then it is easy to see that $(\mu_t)_{t\geq 0}$ is a continuous $\uplus$-semigroup.

If μ can be embedded into a continuous $\uplus$-semigroup $(\mu_t)_{t\geq 0}$, it is clearly $\uplus$-infinitely divisible. Conversely, assume that μ is $\uplus$-infinitely divisible, i.e., for any $n \in \mathbb{N}$ we can write $\mu = \mu_n \uplus \cdots \uplus \mu_n$ for some μ_n. As $\mu \neq \mu_H$, $\eta_\mu(z)/z$ is not constant 0. Furthermore, if $\eta_\mu(z)/z$ had an isolated zero, its multiplicity would be arbitrarily large. So $\eta_\mu(z)/z$ has no zeros and we can write $\eta_\mu(z)/z = \exp(-p(z))$, for some holomorphic $p : \mathbb{D} \to \mathbb{C}$ having non-negative real part. Thus, μ can be embedded into the $\uplus$-semigroup $(\mu_t)_{t\geq 0}$ defined by $\eta_{\mu_t}(z)/z = \exp(-tp(z))$. $\square$

COROLLARY 7.22. *A probability measure* μ *on* $\partial\mathbb{D}$, $\mu \neq \mu_H$, *is* $\uplus$*-infinitely divisible if and only if* $\eta_\mu(z)/z$ *has no zeros in* $\mathbb{D}$.

In the remaining cases of classical and free independence, we simply assume that U, V are independent and obtain the free multiplicative convolution $\mu \boxtimes \nu$ and the classical multiplicative convolution $\mu \circledast \nu$ from the distribution of UV. We denote the moments of a probability measure μ on $\partial\mathbb{D}$ by

$$m_n(\mu) = \int_{\partial\mathbb{D}} x^n \mu(dx).$$

The classical multiplicative convolution $\circledast$ can be characterized by

$$m_n(\mu \circledast \nu) = m_n(\mu)m_n(\nu), \qquad n \in \mathbb{N}.$$

REMARK 7.23. Recall that $\psi_\mu(z) = \sum_{n=1}^{\infty} m_n(\mu)z^n$. So

$$\psi_{\mu \circledast \nu}(z) = \sum_{n=1}^{\infty} m_n(\mu)m_n(\nu)z^n.$$

In complex analysis, this construction is well known. Given two power series $\sum_{n=0}^{\infty} a_n z^n$ and $\sum_{n=0}^{\infty} b_n z^n$, the power series $\sum_{n=0}^{\infty} a_n b_n z^n$ is also called their Hadamard product (or Hadamard convolution). We give an example of a result from geometric function theory: if $f, g \in S$ are both convex functions, then so is their Hadamard product, see [**166, 167**].

Recall that $\mathcal{P}_\times(\partial\mathbb{D})$ denotes the set of all probability measures on $\partial\mathbb{D}$ with non-zero mean. For $\mu \in \mathcal{P}_\times(\partial\mathbb{D})$, we define Σ_μ by

$$\Sigma_\mu(z) = \frac{1}{z}\eta_\mu^{-1}(z)$$

in the neighborhood of 0 where η_μ^{-1} is defined. For $\mu, \nu \in \mathcal{P}_\times(\partial\mathbb{D})$, Voiculescu characterized the free multiplicative convolution $\boxtimes$ by

$$\Sigma_{\mu\boxtimes\nu}(z) = \Sigma_\mu(z)\Sigma_\nu(z)$$

in a neighborhood of 0, see [**190**].

Also in the cases of classical and free multiplicative convolution, infinitely divisible probability measures on $\partial\mathbb{D}$ lead to Lévy–Khintchine representations similar to the additive cases, see [**87**]. The case of free convolution will be needed in Chapter 9.

THEOREM 7.24 (Theorem 6.7 in [**33**]). *Let $\mu \in \mathcal{P}_\times(\partial\mathbb{D})$. Then the following three statements are equivalent:*

(1) *μ is freely infinitely divisible.*
(2) *There exists a $\boxtimes$-semigroup $(\mu_t)_{t\geq 0}$ (i.e., $\mu_0 = \delta_1$, $\mu_{t+s} = \mu_t \boxtimes \mu_s$ for all $s, t \geq 0$, and $t \mapsto \mu_t$ is continuous with respect to weak convergence) such that $\mu_1 = \mu$.*
(3) *There exists a holomorphic function $u : \mathbb{D} \to \mathbb{C}$ with $\operatorname{Re}(u) \geq 0$ such that $\Sigma_\mu = e^u$.*

Conversely, for any holomorphic $u : \mathbb{D} \to \mathbb{C}$ with $\operatorname{Re}(u) \geq 0$, the function e^u is the Σ-transform of some freely infinitely divisible μ.

Furthermore, a family $(\mu_t)_{t\geq 0}$ of probability measures from $\mathcal{P}_\times(\partial\mathbb{D})$ is a $\boxtimes$-semigroup if and only if $\Sigma_{\mu_t} = e^{tu}$ for all $t \geq 0$ and a holomorphic function $u : \mathbb{D} \to \mathbb{C}$ with $\operatorname{Re}(u) \geq 0$.

REMARK 7.25. It follows that if $\mu \in \mathcal{P}_\times(\partial\mathbb{D})$ is freely infinitely divisible, then $(g \circ \eta_\mu)(z) = z$, $g(z) = z\Sigma_\mu(z)$, for all $z \in \mathbb{D}$ due to the identity theorem. Thus, η_μ is univalent.

7.6. Kernel convergence

Let $(D_n)_n$ be a sequence of simply connected domains with $0 \in D_n \subsetneq \mathbb{C}$, and let $(f_n)_n$ be the sequence of the corresponding conformal mappings $f_n : \mathbb{D} \to D_n$, $f_n(0) = 0$, $f_n'(0) > 0$. Can we determine the convergence of $(f_n)_n$ by looking at the domains D_n? The answer is given by a result of Carathéodory.

First, we define the notion of kernel convergence. Let $(D_n)_{n\in\mathbb{N}}$ be a sequence of domains with $0 \in D_n$ for every $n \in \mathbb{N}$. Assume that 0 is an inner point of $\cap_n D_n$. The kernel $\ker(D_n)_n$ of the sequence $(D_n)_n$ is defined as the largest domain D that contains 0 and has the property that every compact subset of D is contained in all but finitely many domains D_n. If 0 is no inner point of $\cap_n D_n$, then we simply define $\ker(D_n)_n = \{0\}$.

DEFINITION 7.26 (Kernel convergence). The sequence $(D_n)_n$ converges to its kernel $D = \ker(D_n)_n$, written as

$$D_n \to D,$$

if the kernel of every subsequence of $(D_n)_n$ is equal to D.

EXAMPLE 7.27. Let $(D_n)_n$ be a monotonically increasing sequence, i.e., $D_n \subset D_{n+1}$. Then D_n converges with respect to kernel convergence and

$$D_n \to \bigcup_{n\geq 1} D_n.$$

EXAMPLE 7.28. Let $D_n = \{z \in \mathbb{C} \,|\, \mathrm{Re}(z) > -1\}$ if n is even and $D_n = \{z \in \mathbb{C} \,|\, \mathrm{Re}(z) < 1\}$ if n is odd. Then $\ker(D_n)_n = \{z \in \mathbb{C} \,|\, -1 < \mathrm{Re}(z) < 1\}$, but the sequence $(D_n)_n$ does not converge.

Carathéodory proved the following relationship [**51**]. We follow [**62**, Theorem 3.1].

THEOREM 7.29 (Carathéodory kernel theorem). *Let $(D_n)_n$ be a sequence of simply connected domains with $0 \in D_n \subsetneq \mathbb{C}$ for all $n \in \mathbb{N}$. Let f_n be the conformal mapping $f_n : \mathbb{D} \to D_n$, $f_n(0) = 0$, $f_n'(0) > 0$, and let $D = \ker(D_n)_n$. Then we have:*

f_n converges locally uniformly in $\mathbb{D}$ to f if and only if $D_n \to D \neq \mathbb{C}$.

In the case of convergence, we have:

1. *If $D = \{0\}$, then $f(z) \equiv 0$.*
2. *If $D \neq \{0\}$, then D is simply connected and f is the conformal mapping $f : \mathbb{D} \to D$ with $f(0) = 0$, $f'(0) > 0$.*

PROOF. First, we assume that f_n converges locally uniformly to f. Then f is either univalent in $\mathbb{D}$ or constant. In the second case, f has to be constant 0, since $f(0) = \lim_{n\to\infty} f_n(0) = 0$.

Case 1: $f(z) \equiv 0$. Assume that $D \neq \{0\}$. Then there exists a disk $r\overline{\mathbb{D}} \subset D$. For almost every $n \in \mathbb{N}$, we have that $g_n := f_n^{-1} : r\mathbb{D} \to \mathbb{D}$, i.e., $g_n(r \cdot z) : \mathbb{D} \to \mathbb{D}$. As $g_n(0) = 0$, the Schwarz lemma implies

$$|g_n'(0)| \leq 1/r, \quad \text{so} \quad |f_n'(0)| \geq r > 0.$$

On the other hand, $f_n'(0)$ converges to 0, a contradiction. Hence, $D = \{0\}$. The same argument holds for every subsequence of $(D_n)_n$, and we conclude that $D_n \to \{0\}$.

Case 2: f is univalent. Then $f(0) = 0$ and $f'(0) > 0$. Let $\Delta = f(\mathbb{D})$. We show that $D = \Delta$.

We first show that $\Delta \subset D$. Let $E \subset \Delta$ be a compact subset and let $\Gamma \subset \Delta \setminus E$ be a smooth Jordan curve that surrounds E. Let $\delta > 0$ be the distance of E to Γ and $\gamma = f^{-1}(\Gamma)$. Let $w_0 \in E$. Then $|f(z) - w_0| \geq \delta$ for all $z \in \gamma$. Because of the uniform convergence of f_n on γ, we obtain $|f_n(z) - f(z)| < \delta$ for all $z \in \gamma$ and $n \geq N$ for some $N \in \mathbb{N}$. We get

$$|f_n(z) - f(z)| < |f(z) - w_0|$$

for all $z \in \gamma$. Rouché's theorem implies that $f(z) - w_0$ has the same number of zeros in the interior of γ as $f_n(z) - w_0$, namely one. Hence, $w_0 \in D_n$ for all $n \geq N$. Note that N does not depend on w_0, so $E \subset D_n$ for all $n \geq N$. This shows that $\Delta \subset D$.

Next, we consider the inverse mapping $\varphi_n = f_n^{-1}$, which is defined on E for all $n \geq N$. By taking a sequence of increasing compact sets E_n, and because $|\varphi_n^{-1}(z)| < 1$, Montel's theorem implies that we can choose a subsequence φ_{n_k} converging locally uniformly in Δ to a univalent function $\varphi : \Delta \to \mathbb{D}$ with $\varphi(0) = 0$, $\varphi'(0) > 0$.

Let $z_0 \in \mathbb{D}$ and $w_0 = f(z_0) \in \Delta$. Let $C \subset \mathbb{D}$ be a circle around z_0 with radius $\varepsilon > 0$. Let δ be the distance from $f(C)$ to w_0. Then $|f_{n_k}(z) - f(z)| < \delta$ on C for all $k \geq K$ large enough. Thus, for $k \geq K$, $f_{n_k}(z_k) = w_0$ for some z_k inside C by Rouché's theorem. So $|z_0 - z_k| < \varepsilon$ and $z_k = \varphi_{n_k}(w_0)$. Thus, $|\varphi(w_0) - z_0| \leq |\varphi(w_0) - \varphi_{n_k}(w_0)| + |z_k - z_0| < 2\varepsilon$ for k large enough. So $\varphi(w_0) = z_0$, and thus, $\varphi = f^{-1}$. The same argument holds for every subsequence of φ_n. Hence, φ_n converges locally uniformly in Δ to f^{-1}.

The same reasoning shows that there exists a subsequence of φ_n that converges locally uniformly in D to a univalent function $\psi : D \to \mathbb{D}$. ψ extends f^{-1} to a univalent function defined on D. As $f^{-1}(\Delta) = \mathbb{D}$, we must have $D = \Delta$. We can apply this argument for any subsequence $(D_{n_k})_k$ to see that the kernel of $(D_{n_k})_k$ is equal to Δ. Hence, $D_n \to D$.

Conversely, assume that $D_n \to D \neq \mathbb{C}$. We have that $f_n/f_n'(0) \in S$. Koebe's 1/4-theorem (Theorem 3.39) implies that

$$\{w \in \mathbb{C} \,|\, |w| < 1/4 \cdot f_n'(0)\} \subset D_n.$$

If the sequence $(f_n'(0))_n$ was unbounded, then there would be a subsequence D_{n_k} with $\ker(D_{n_k})_k = \mathbb{C}$, a contradiction. Hence, $f_n'(0)$ is bounded above by $M > 0$ and the growth theorem (Theorem A.3) implies that

$$|f_n(z)| \leq f_n'(0)\frac{|z|}{(1-|z|)^2} \leq M\frac{|z|}{(1-|z|)^2}$$

for all $z \in \mathbb{D}$. Thus, the sequence f_n is locally bounded.

Case 1: $D = \{0\}$. Assume that $f_n'(0)$ does not converge to 0. Then there is a subsequence f_{n_k} with $f_{n_k}'(0) \geq \varepsilon > 0$. So $\{w \in \mathbb{C} \,|\, |w| < \varepsilon/4\} \subset D_{n_k}$ and $\ker(D_{n_k})_k \neq \{0\}$, a contradiction. Consequently,

$$|f_n(z)| \leq f_n'(0)\frac{|z|}{(1-|z|)^2} \to 0$$

as $n \to \infty$, i.e., f_n converges locally uniformly in $\mathbb{D}$ to 0.

Case 2: $D \neq \{0\}$. We have to show that f_n converges locally uniformly in $\mathbb{D}$. $(f_n)_n$ is a normal family. Let f and g be the limit functions of converging subsequences f_{n_k} and f_{m_k}, respectively. We have already seen that $f(\mathbb{D})$ and $g(\mathbb{D})$ are the kernels of $(D_{n_k})_k$ and $(D_{m_k})_k$, respectively, i.e., we have $D = f(\mathbb{D}) = g(\mathbb{D})$. As $f(0) = g(0) = 0$, $f'(0), g'(0) > 0$, we have $f = g$. Hence, f_n converges locally uniformly in $\mathbb{D}$. □

COROLLARY 7.30. *Let $(\mu_n)_{n\in\mathbb{N}}$ be a sequence of probability measures on $\partial\mathbb{D}$ such that each η_{μ_n} is univalent. If μ_n converges weakly to $\mu \neq \mu_H$, then $\eta_{\mu_n}(\mathbb{D})$ converges with respect to kernel convergence to $\eta_\mu(\mathbb{D})$.*

PROOF. As $\mu \neq \mu_H$, η_μ is univalent and $\eta'_\mu(0) \neq 0$. We apply Theorem 7.29 to $\eta_{\mu_n}/\eta'_{\mu_n}(0)$ to get kernel convergence of the domains $\eta_{\mu_n}(\mathbb{D})/\eta'_{\mu_n}(0)$ to $\eta_\mu(\mathbb{D})/\eta'_\mu(0)$. As $\eta'_{\mu_n}(0) \to \eta'_\mu(0)$, we conclude kernel convergence of $\eta_{\mu_n}(\mathbb{D})$ to $\eta_\mu(\mathbb{D})$. □

REMARK 7.31. The converse statement is not true: For $\mu_n = \delta_{e^{in}}$, we have $\eta_{\mu_n}(z) = e^{in}z$ and kernel convergence of the image domains $\eta_{\mu_n}(\mathbb{D}) = \mathbb{D}$.

We call a function $f \in S$ a *slit mapping* if $f(\mathbb{D}) = \mathbb{C} \setminus \Gamma$, where Γ is the image of a continuous and injective function $\gamma : [0,\infty) \to \mathbb{C}$.

THEOREM 7.32. *Let $f \in S$. Then there exists a sequence $(f_n)_n \subset S$ of slit mappings such that f_n converges locally uniformly in $\mathbb{D}$ to f.*

PROOF. The sequence $g_n(z) = r_n^{-1} f(r_n z)$, where $0 < r_n \nearrow 1$, converges locally uniformly in $\mathbb{D}$ to f. Thus, it suffices to consider the case where $f(\mathbb{D})$ is bounded by an analytic Jordan curve. Let $w_0 \in f(\partial\mathbb{D})$. We can connect w_0 with ∞ within $\mathbb{C}\setminus\overline{f(\mathbb{D})}$ by a curve Γ, i.e., there exists a continuous injective function $\gamma : [0,\infty) \to \mathbb{C}$ with $\gamma(0) = w_0$, $|\gamma(t)| \to \infty$ as $t \to \infty$, and $\gamma(0,\infty) \subset \mathbb{C} \setminus \overline{f(\mathbb{D})}$. Let $(w_n)_n \subset f(\partial\mathbb{D}) \setminus \{w_0\}$ be a sequence converging on $f(\partial\mathbb{D})$ in clockwise order to w_0. We now construct the curve J_n by connecting Γ with the arc of $f(\partial\mathbb{D})$ that connects w_0 to w_n in clockwise orientation, see Figure 7.1. Then

$$\mathbb{C} \setminus J_n \to f(\mathbb{D}).$$

Let h_n be the conformal mapping $h_n : \mathbb{D} \to \mathbb{C} \setminus J_n$ with $h_n(0) = 0$, $h'_n(0) > 0$. Theorem 7.29 implies that h_n converges locally uniformly in $\mathbb{D}$ to f. As $h'_n(0) \to f'(0) = 1$, $f_n := h_n/h'_n(0)$ is a sequence of slit mappings in S that converges locally uniformly in $\mathbb{D}$ to f. □

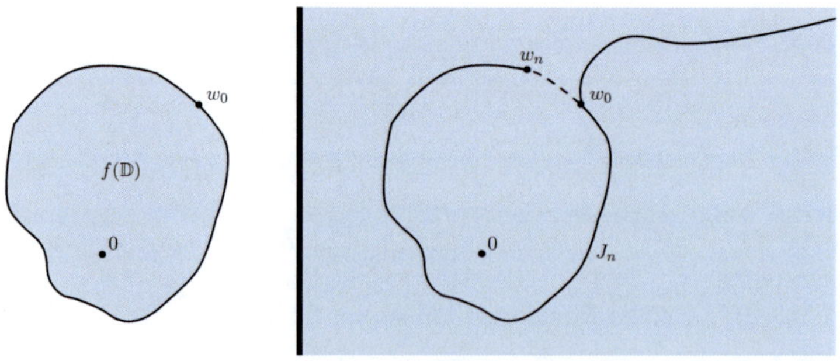

FIGURE 7.1. Proof of Theorem 7.32.

7.7. Embeddings into radial Loewner chains

In this section, we prove Theorem 7.8.

A domain $D \subset \mathbb{C}$ is called *doubly connected* if the complement $\hat{\mathbb{C}} \setminus D$ has two connected components K_1, K_2. If both K_1 and K_2 consist of only one point, then D is conformally equivalent to $\mathbb{C} \setminus \{0\}$. If K_1 consists of one point and $|K_2| \geq 2$, then $D \cup K_1$ is conformally equivalent to $\mathbb{D}$, and thus, D is conformally equivalent to $\mathbb{D} \setminus \{0\}$. In the more complicated case $|K_1| \geq 2, |K_2| \geq 2$, D is also called a *conformal ring domain.*

THEOREM 7.33. *Let D be a conformal ring domain. Then there exists a unique $0 < R < 1$ such that D can be mapped conformally onto $\{z \in \mathbb{C} \,|\, R < |z| < 1\}$.*

PROOF. Step 1 (existence): First, we map $D \cup K_1$ by f_1 conformally onto $\mathbb{D}$. Next we map the simply connected domain $f_1(D) \cup (\mathbb{C} \setminus \mathbb{D}) \cup \{\infty\}$ by a conformal map f_2 onto $\mathbb{D}$. So $(f_2 \circ f_1)(D) \subset \mathbb{D}$ is a ring domain whose boundary consists of $\partial\mathbb{D}$ and an analytic Jordan curve $\gamma \subset \mathbb{D}$. To simplify notation, we assume that D already has this form and that $0 \notin D$ (otherwise, we apply an automorphism of $\mathbb{D}$).

Now we map D by $\log(z)$ into the left half-plane, see Figure 7.2. As $\log(z)$ is a multi-valued function, we obtain, for every $z \in D$, a set of points

$$\log(z) = w + 2k\pi i, k \in \mathbb{Z},$$

and $\exp w = z$. In this sense, $\log(D)$ is a $2\pi i$–periodic domain whose boundary consists of the imaginary axis and the curve $\log(\gamma) \subset \{z \in \mathbb{C} \,|\, \mathrm{Re}(z) < 0\}$.

This domain is simply connected and $\neq \mathbb{C}$. Let f_3 be a conformal mapping from $\log(D)$ onto $S' = \{z \in \mathbb{C} \,|\, -1 < \mathrm{Re}(z) < 0\}$. We can assume that $f_3(0) = 0, f_3(\pm i\infty) = \pm i\infty$. Then $f_3(2\pi i) = w_0$ with $\mathrm{Im}(w_0) > 0$. So the mapping $\varphi = (2\pi)/|w_0| \cdot f_3$ maps $\log(D)$ conformally onto $S'' = \{z \in \mathbb{C} \,|\, -2\pi/|w_0| < \mathrm{Re}(z) < 0\}$. Next, we show that

$$(*) \qquad \varphi(z + 2k\pi i) = \varphi(z) + 2k\pi i, k \in \mathbb{Z}.$$

It suffices to consider the case $k = 1$. For $k = 1$, both functions map $\log(D)$ conformally onto S'', fix $\pm i\infty$, and map 0 onto $2\pi i$. Hence, both functions are identical (see Exercise 3.9.2). Finally, we take f as

$$f(z) = \exp(\varphi(\log(z))).$$

Because of $(*)$, f is a (single-valued) holomorphic function which maps D onto $\{z \in \mathbb{C} \,|\, R < |z| < 1\}$ with $R = \exp(-2\pi/|w_0|)$. The mapping f is injective, as $f(z_1) = f(z_2)$ implies $\varphi(\log(z_1)) = \varphi(\log(z_2)) + 2k\pi i$ for a $k \in \mathbb{Z}$, and thus, $\varphi(\log(z_1)) = \varphi(\log(z_2) + 2k\pi i)$, i.e., $\log(z_1) = \log(z_2) + 2k\pi i$. Hence, $\log(z_1)$ and $\log(z_2)$ are the same sets, which implies $z_1 = z_2$.

Step 2 (uniqueness): For $R \in (0,1)$, let A_R be the annulus $A_R = \{z \in \mathbb{C} \,|\, R < |z| < 1\}$. We need to show that A_R cannot be mapped conformally onto A_r, $r \in (0,1)$, whenever $r \neq R$. Assume that $f : A_R \to A_r$ is conformal. By using the Schwarz reflection principle, we can extend f to a conformal map from $\mathbb{C} \setminus \{0\}$ onto itself. Hence, $f(z) = az$ or $f(z) = a/z$ for some $a \in \mathbb{C} \setminus \{0\}$. As $f(A_R) = A_r$ and $r \neq R$, we obtain contradictions in both cases by computing $|a|$. □

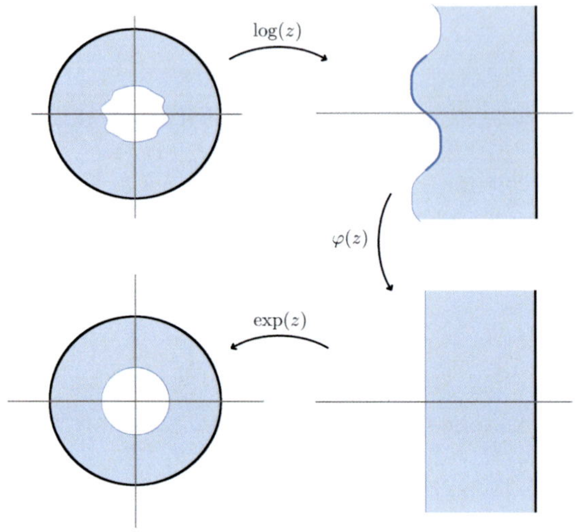

FIGURE 7.2. Mappings from the proof of Theorem 7.33.

The above result can be generalized to multiply connected domains. We refer to [**143**, Chapter VII].

REMARK 7.34. Let $D \subset A_r$, $0 < r < 1$, be a conformal ring domain whose boundary consists of $\partial\mathbb{D}$ and an analytic Jordan curve γ that is symmetric with respect to conjugation. In the proof of Theorem 7.33, we can jump to the step of applying log. The functions $\log, f_3, \exp$ will also be symmetric with respect to conjugation, and we conclude that there exists a conformal mapping $f : D \to A_R$, $0 < R < 1$, such that $f(\overline{z}) = \overline{f(z)}$ for all $z \in D$.

Next, we need the following lemma [**151**, Lemma 6.2].

LEMMA 7.35. *Every sequence $(f_{n,t})_{t\geq 0}$ of normalized radial Loewner chains contains a subsequence that converges for each $t \geq 0$ locally uniformly in $\mathbb{D}$ to a normalized radial Loewner chain $(f_t)_{t\geq 0}$.*

PROOF. Due to estimate (6.1), the functions $(t, z) \mapsto f_{n,t}(z)$ are uniformly equicontinuous on the set of all (t, z) with $t \in [0, m]$ and $|z| \leq 1-1/m$, for each $m \in \mathbb{N}$. The Arzelà–Ascoli theorem implies that there exists a subsequence $(f_{n_k,t})_{t\geq 0}$ that converges uniformly on this set. A diagonalization argument shows that there exists a subsequence that converges locally uniformly on the set of all (t, z) with $t \in [0, \infty), z \in \mathbb{D}$. The limit functions $(f_t)_{t\geq 0}$ are also normalized by $f_t(0) = 0$, $f_t'(0) = e^{-t}$. Consequently, they are all non-constant and thus univalent (as the limit of a sequence of univalent functions). Furthermore, $t \mapsto f_t$ is continuous as locally uniform convergence preserves continuity. For $0 \leq s \leq t$, we have $f_{n_k,t}(\mathbb{D}) \subset f_{n_k,s}(\mathbb{D})$, and thus, also $f_t(\mathbb{D}) \subset f_s(\mathbb{D})$. We conclude that $(f_t)_{t\geq 0}$ is a normalized radial Loewner chain. □

PROOF OF THEOREM 7.8. We need to show the following two statements:

(a) Let μ be a probability measure on $\partial\mathbb{D}$ such that η_μ is univalent. Then there exists a radial Loewner chain $(f_t)_{t\geq 0}$ such that $f_1 = \eta_\mu$.

(b) Let μ be a probability measure on $\partial\mathbb{D}$ such that η_μ is univalent and

$$\eta'_\mu(0) = \int_{\partial\mathbb{D}} x\,\mu(dx) = e^{-T}, \quad T > 0.$$

Then there exists a normalized radial Loewner chain $(f_t)_{t\geq 0}$ with $f_T = \eta_\mu$.

We begin with statement (b). (This is essentially [**151**, Section 6.1, Problem 3].)

Fix $n \in \mathbb{N}$, $n \geq 2$. Consider

$$F_n : \mathbb{D} \to \mathbb{D}, \quad F_n(z) = \eta_\mu(z(1 - 1/n)).$$

Then $F_n(\mathbb{D})$ is bounded by an analytic curve, which is contained in $\mathbb{D}$. Therefore, $\mathbb{D}\setminus\overline{F_n(\mathbb{D})}$ is a conformal ring domain, and Theorem 7.33 yields a conformal mapping G_n from $\mathbb{D} \setminus \overline{F_n(\mathbb{D})}$ onto a normalized ring domain $\{z \in \mathbb{C} \,|\, R_n < |z| < 1\}$, where we can assume that $\partial\mathbb{D}$ is mapped onto $\partial\mathbb{D}$ and $\partial F_n(\mathbb{D})$ onto the circle defined by $|z| = R_n$. (The proof of Theorem 7.33 together with the continuous extension theorem for conformal mappings, see Theorem B.1 from Appendix B, show that G_n can be extended continuously to the boundary.)

For $0 \leq t \leq -\log R_n$, let $D_{t,n}$ be the simply connected domain bounded by the curve $G_n^{-1}(e^{-t}\partial\mathbb{D})$ and let $f_{t,n} : \mathbb{D} \to D_{t,n}$ be the conformal mapping with $f_{t,n}(0) = 0 < f'_{t,n}(0)$. Due to the Schwarz lemma, $t \mapsto f'_{t,n}(0)$ is strictly decreasing and this function is continuous by kernel convergence. So we can reparametrize $f_{t,n}$ such that $f'_{t,n}(0) = e^{-t}$. Then $f_{t,n}$ is defined for all $t \in [0, T_n]$ with $f_{T_n,n} = F_n$. For $t > T_n$, let $f_{t,n}(z) = F_n(e^{T_n - t}z)$.

Each $f_{t,n}$ is univalent and $t \mapsto f_{t,n}$ is continuous due to kernel convergence. Hence, $(f_{t,n})_{t\geq 0}$ is a normalized radial Loewner chain. By Lemma 7.35, there exists a subsequence $(f_{t,n_k})_k$ which converges for each $t \geq 0$ locally uniformly in $\mathbb{D}$ to a normalized radial Loewner chain $(f_t)_{t\geq 0}$. Note that $T_n > T := -\log \eta'_\mu(0)$ for all n and $T_{n_k} \to T$. Hence, for any $\varepsilon > 0$, we have

$$\begin{aligned} f_{T+\varepsilon}(z) &= \lim_{k\to\infty} f_{T+\varepsilon,n_k}(z) = \lim_{k\to\infty} F_{n_k}(e^{T_{n_k} - T - \varepsilon} z) \\ &= \lim_{k\to\infty} \eta_\mu(e^{T_{n_k} - T - \varepsilon} z(1 - 1/n_k)) = \eta_\mu(e^{-\varepsilon} z). \end{aligned}$$

From the continuity of $t \mapsto f_t$, we conclude that $f_T = \eta_\mu$.

Now we can prove (a). If μ is a point measure, then $\eta_\mu(z) = e^{i\alpha}z$, $\alpha \in \mathbb{R}$, and we can simply take the Loewner chain $f_t(z) = e^{it\alpha}z$.

Now assume that η_μ is not a rotation. Then $e^{i\alpha}\eta'_\mu(0) = e^{-T}$ for some $T > 0$ and $\alpha \in \mathbb{R}$. Due to (b), $\eta_\mu(e^{i\alpha}z) = g_T$ for a normalized radial Loewner chain $(g_t)_{t\geq 0}$. Define f_t by $f_t(z) = g_{t\cdot T}(e^{-i\alpha t}z)$. It is easy to see that $(f_t)_{t\geq 0}$ is a radial Loewner chain and $f_1(z) = g_T(e^{-i\alpha}z) = \eta_\mu(z)$. ☐

COROLLARY 7.36. *Let μ be a probability measure on $\partial\mathbb{D}$ such that η_μ is univalent with $\eta'_\mu(0) = e^{-T}$, $T > 0$, and $\overline{\eta_\mu(z)} = \eta_\mu(\overline{z})$ for all $z \in \mathbb{D}$. Then there exists a normalized radial Loewner chain $(f_t)_{t\geq 0}$ such that $f_T = \eta_\mu$ and $\overline{f_t(z)} = f_t(\overline{z})$ for all $z \in \mathbb{D}$ and $t \geq 0$.*

PROOF. In the proof of Theorem 7.8 (b), we see that F_n is symmetric with respect to complex conjugation. Due to Remark 7.34, G_n can be chosen to be symmetric as well. Then also $\overline{f_{t,n}(z)} = f_{t,n}(\overline{z})$ for all $z \in \mathbb{D}$, $t \geq 0$ and $n \in \mathbb{N}$. Hence, the limit Loewner chain $(f_t)_{t\geq 0}$ with $f_T = \eta_\mu$ also satisfies $\overline{f_t(z)} = f_t(\overline{z})$. □

7.8. Embeddings into additive Loewner chains

PROOF OF THEOREM 6.38. Let μ be a probability measure on $\mathbb{R}$ with finite first and second moment such that F_μ is univalent. We need to show the following two statements:

(a) There exists an additive Loewner chain $(f_t)_{t\geq 0}$ such that $f_1 = F_\mu$.
(b) If
$$\int_{\mathbb{R}} x\,\mu(dx) = 0, \qquad \int_{\mathbb{R}} x^2\,\mu(dx) =: T > 0,$$
then there exists an additive Loewner chain $(f_t)_{t\geq 0}$ having the normalization from Theorem 6.35 for $t \in [0, T]$ such that $f_T = F_\mu$.

We start with (b). First, assume that $\mu \in \mathcal{P}(\mathbb{R})$ has compact support, mean 0, and variance $T > 0$. The function F_μ maps $\mathbb{H}$ onto $\mathbb{H} \setminus B$ for a closed bounded set $B \subset \overline{\mathbb{H}}$. We find a closed bounded interval $I \subset \mathbb{R}$ such that $B \cap \mathbb{R} \subset I$. Let $\overline{B}$ be the complex conjugation of B. By extending F_μ to the lower half-plane (via $\overline{F_\mu(\overline{z})}$) we find a closed bounded interval $J \subset \mathbb{R}$ such that F_μ extends to a univalent function on $\hat{\mathbb{C}} \setminus J$ with image domain $\hat{\mathbb{C}} \setminus (B \cup \overline{B} \cup I)$. Next, we find conformal mappings
$$H_1 : \mathbb{D} \to \hat{\mathbb{C}} \setminus J, \quad H_2 : \mathbb{D} \to \hat{\mathbb{C}} \setminus I, \quad \text{with } H_1(0) = H_2(0) = \infty.$$
We can let H_1 and H_2 be of the form $a/k(z) + b$, where k is the Koebe function (which maps $\mathbb{D}$ onto $\mathbb{C} \setminus (-\infty, -1/4])$ and $a, b \in \mathbb{R}$, $a > 0$. Then $H_j(\overline{z}) = \overline{H_j(z)}$ for all $z \in \mathbb{D}$, $j = 1, 2$.

Thus, $H := H_2^{-1} \circ F_\mu \circ H_1$ fixes 0 and maps $\mathbb{D}$ conformally onto $\mathbb{D} \setminus K$ for some set $K \subset \mathbb{D}$, and $H'(0) > 0$. Furthermore, $\overline{H(z)} = H(\overline{z})$ for all $z \in \mathbb{D}$.

Write $H'(0) = e^{-S}$. Due to Corollary 7.36, we find a normalized radial Loewner chain $(g_t)_{t\geq 0}$ with $g_S = H$ and $\overline{g_t(z)} = g_t(\overline{z})$ for all $z \in \mathbb{D}$, $t \geq 0$.

For $t \geq 0$, define G_t by $g_t = H_2^{-1} \circ G_t \circ H_1$. Then each G_t maps $\mathbb{H}$ into $\mathbb{H}$. (This is true for $t = S$ and continuity with respect to t excludes that G_t would map $\mathbb{H}$ into the lower half-plane.) Furthermore, $G_t(\infty) = \infty$ for all $t \geq 0$. Thus we can find a continuous function $t \mapsto a_t > 0$ such that $E_t(z) := G_t(a_t z)$ is an F-transform for every $t \geq 0$. Clearly, $a_S = 1$. Consequently, F_μ is embedded within $(E_t)_{t\geq 0}$ at $t = S$. Furthermore, each E_t is the F-transform of a probability measure with compact support. Let $m(t)$ be its first moment. Then $m(t)$ is continuous and $(F_t)_{t\geq 0} := (E_t(z + m(t)))_{t\geq 0}$ is an additive Loewner chain. We can reparametrize it such that $F_t(z) = z - \frac{t}{z} + \ldots$ at ∞ for all $t \geq 0$, and then $F_T = F_\mu$.

Now, let $\mu \in \mathcal{P}^0(\mathbb{R})$ with variance $T > 0$. We then find a sequence μ_n of probability measures with mean 0, variance T, and compact support such that $\mu_n \to \mu$. For each $n \in \mathbb{N}$, let $(f_{t,n})_{t\geq 0}$ be an additive Loewner chain from above with $f_{T,n} = F_{\mu_n}$.

Let $(K_n)_{n\in\mathbb{N}}$ be an increasing sequence of compact subsets of $\mathbb{H}$ whose union is $\mathbb{H}$. By using (6.12) and (6.11), we conclude that the functions $(t,z)\mapsto f_{n,t}(z)$ are uniformly equicontinuous and uniformly bounded on the set of all (t,z) with $t\in[0,m]$ and $z\in K_m$, for each $m\in\mathbb{N}$. As in the proof of Lemma 7.35, we find a subsequence $(f_{n_k,t})_{t\geq 0}$ of Loewner chains that converges for each $t\geq 0$ locally uniformly. The limit functions are non-constant due to Lemma 6.34, which also implies that the limit is again an additive Loewner chain $(F_{\mu_t})_{t\geq 0}$, where each μ_t has mean 0 and variance $\leq t$. Furthermore, $F_{\mu_T} = \lim_{n\to\infty} F_{\mu_n} = F_\mu$. As $\mathrm{Var}(\mu)=T$, $\mathrm{Var}(\mu_t)=t$ for all $t\leq T$.

In order to show (a), let $\mu\in\mathcal{P}(\mathbb{R})$ with mean $c\in\mathbb{R}$ and finite variance $T\geq 0$. If $T=0$, then $\mu=\delta_c$ and μ can be embedded into the additive Loewner chain $(z-ct)_{t\geq 0}$. Now let $T>0$. Then we find an additive Loewner chain $(f_t)_{t\geq 0}$ from (b) such that $f_T = F_\mu(z+c)$. Hence, F_μ is embedded within the additive Loewner chain $(f_{t\cdot T}(z-ct))_{t\geq 0}$ at $t=1$. □

THEOREM 7.37. *Let $\mu\in\mathcal{P}^0(\mathbb{R})$ be symmetric, i.e., $\mu(B)=\mu(-B)$ for all Borel subsets $B\subset\mathbb{R}$, and assume that F_μ is univalent. Let $T=\mathrm{Var}(\mu)>0$. Then there exists an additive Loewner chain $(F_{\mu_t})_{t\in[0,T]}$ such that $\mu_T=\mu$ and each $\mu_t\in\mathcal{P}^0(\mathbb{R})$ is symmetric as well with $\mathrm{Var}(\mu_t)=t$.*

PROOF. Let $B\subset\mathbb{H}$ be a bounded set such that $\mathrm{Re}(z)>0$ for all $z\in B$ and $\mathbb{H}\setminus B$ is simply connected. By embedding $\mathbb{H}\setminus B$ into a Loewner chain as in the proof of Theorem 6.38, we obtain a family $(B_t)_{t\in[0,S]}$ such that $B_s\subset B_t$ when $s\leq t$, $\mathbb{H}\setminus B_t$ is simply connected for all $t\in[0,S]$, $B_0=\emptyset$, $B_S=B$, and $t\mapsto\mathbb{H}\setminus B_t$ is continuous with respect to kernel convergence (with reference point ∞).

Let C_t be the reflection of B_t with respect to the imaginary axis. For each $t\in[0,S]$, we find $\mu_t\in\mathcal{P}_c(\mathbb{R})$ with mean 0 such that $F_{\mu_t}(\mathbb{H})=\mathbb{H}\setminus(B_t\cup C_t)$ and F_{μ_t} is univalent. Then each μ_t is symmetric. Moreover, $t\mapsto\mathbb{H}\setminus(B_t\cup C_t)$ is continuous, and thus, $t\mapsto F_{\mu_t}$ is continuous. Therefore, $(F_{\mu_t})_{t\in[0,S]}$ is an additive Loewner chain. By reparametrization, we obtain the Loewner chain $(F_{\mu_t})_{t\in[0,T]}$ such that $\mathrm{Var}(\mu_t)=t$. As in the proof of Theorem 6.38, we can now approximate the general case by symmetric, compactly supported probability measures. □

Now let μ be a general probability measure on $\mathbb{R}$ such that F_μ is univalent. In order to embed F_μ into an additive Loewner chain, we need more advanced results on the boundary behavior of conformal mappings.

Let us transfer the problem to the unit disk. We find Möbius transformations α, β such that $G:=\alpha\circ F_\mu\circ\beta:\mathbb{D}\to\mathbb{D}$ is univalent with $G(0)=0$. So we find a radial Loewner chain $(G_t)_{t\geq 0}$ such that $G_1=G$. Now consider the mappings $(F_t)_{t\geq 0}$ on $\mathbb{H}$ defined by $G_t=\alpha\circ F_t\circ\beta$. As $F_0=\alpha^{-1}\circ\beta^{-1}$, F_0 might not be the identity. We find a continuous family $(\gamma_t)_{t\geq 0}$ of automorphisms of $\mathbb{H}$ such that $\gamma_0=F_0^{-1}$ and γ_t is the identity for $t\geq 1$. Define $\hat{f}_t$ by $\hat{f}_t=F_t\circ\gamma_t$. Then $(\hat{f}_t)_{t\geq 0}$ is a Loewner chain on $\mathbb{H}$ with $\hat{f}_1=F_1=F_\mu$. However, not all $\hat{f}_t$ need to be F-transforms.

THEOREM 7.38. *Let μ be a probability measure on $\mathbb{R}$ such that F_μ is univalent. Then there exists an additive Loewner chain $(f_t)_{t\geq 0}$ such that $f_1=F_\mu$.*

PROOF. We construct $(f_t)_{t\geq 0}$ by modifying $(\hat{f}_t)_{t\geq 0}$ from above. For $t \geq 1$, we simply define f_t by $f_t(z) = F_\mu(z + i(t-1))$.

Now let $t < 1$. In order to transform $\hat{f}_t$ into an F-transform, we need further results on the boundary behavior of conformal mappings. The curve $F_\mu^{-1}(iy)$, $y > 0$ large enough, converges to ∞ non-tangentially as $y \to \infty$, see Lemma 5.10. Also $\hat{f}_t^{-1}(iy)$ converges to some $a_t \in \hat{\mathbb{R}}$ by [**152**, Prop. 2.14]. Let us pass to the unit disk by using the Cayley transform, where $1 \in \partial\mathbb{D}$ corresponds to ∞, and denote by $g_t : \mathbb{D} \to \mathbb{D}$ the mapping that corresponds to $\hat{f}_t$. Then $g_t^{-1}(r) \to p_t \in \partial\mathbb{D}$ as $r \in (0,1)$ converges to 1. The Lehto–Virtanen theorem [**152**, Sections 4.1 and 4.2] implies that $g_t(z) \to 1$ as $z \to p_t$ non-tangentially. Then $t \mapsto p_t$ is continuous and we find rotations α_t, continuous in t, such that $h_t := g_t \circ \alpha_t$ satisfies $h_t(z) \to 1$ as $z \to 1$ non-tangentially. Note that α_0 is the identity. The Julia–Wolff lemma [**152**, Prop. 4.13] implies that the angular derivative of $h_t(z)$, as $z \to 1$ non-tangentially, exists and belongs to $(0, \infty]$. As $F_\mu(\mathbb{H}) \subset \hat{f}_t(\mathbb{H})$, it cannot be ∞ due to [**152**, Theorem 4.14]. Now we go back to the upper half-plane. Let $\tilde{f}_t : \mathbb{H} \to \mathbb{H}$ be the mapping that corresponds to h_t. Then $\tilde{f}_t(z)/z$ converges to some $c_t > 0$ as $z \to \infty$ non-tangentially. We define f_t by $f_t(z) = \tilde{f}_t(z/c_t)$. Then $(f_t)_{t\geq 0}$ is an additive Loewner chain with $f_1 = F_\mu$.

We note that the statement can also be obtained from a more general result in [**48**]. Theorem 1.2 in [**48**], with $\Lambda(t) \equiv 0$, implies that we can write $F_\mu = f_{0,1}$, where $(f_{s,t})_{0\leq s\leq t\leq 1}$ is an evolution family on $\mathbb{H}$ in the sense of Remark 6.6, and

(i) $f_{s,t}$ has a boundary regular fixed point at ∞ for all $0 \leq s \leq t$,
(ii) $\lim_{y\to\infty} f_{s,t}(iy)/iy = 1$ for all $0 \leq s \leq t$.

We conclude that every $f_{s,t}$ is an F-transform. The family $(f_t)_{t\geq 0}$ with $f_t = f_{1-t,1}$ for $t \in [0,1]$, $f_t = f_{0,1}$ for $t > 1$, is an additive Loewner chain with $f_1 = f_{0,1} = F_\mu$. □

7.9. Exercises

E 7.9.1. Let $\mu = \frac{1}{2}\delta_1 + \frac{1}{2}\delta_i$. Compute η_μ and determine the distributions α and β defined by

$$\eta_\alpha(z) = \eta_\mu(-z), \quad \eta_\beta(z) = -\eta_\mu(z).$$

E 7.9.2. Let $f : \mathbb{D} \to \mathbb{C}$ be holomorphic with $\mathrm{Re}(f(z)) > 0$ for all $z \in \mathbb{D}$ and $f(0) = 1$. The Herglotz representation yields a unique probability measure ρ on $\partial\mathbb{D}$ such that

$$f(z) = \int_{\partial\mathbb{D}} \frac{w+z}{w-z}\rho(dw).$$

Show that

$$\rho\left(\{e^{i\alpha}\}\right) = \frac{1}{2}\lim_{r\nearrow 1}(1-r)f(re^{i\alpha}), \quad \alpha \in \mathbb{R}.$$

E 7.9.3. Let

$$G(z) = \frac{(1+z)^2}{z} = \frac{-1}{k(-z)},$$

where k is the Koebe function.

(a) For $t \geq 0$ and $z \in \mathbb{D}$, let $f_t(z) = G^{-1}(e^t G(z))$. Verify that $(f_t)_{t\geq 0}$ is a normalized radial Loewner chain and compute its Herglotz vector field as well as the image domain $f_t(\mathbb{D})$.

(b) Define μ_t via $\eta_{\mu_t} = f_t$. Compute the support of μ_t.

E 7.9.4. Let $F : [0, \infty) \to \mathbb{C}$ be continuous with $F(0) = 1$ and

$$F(s+t) = F(s)F(t) \quad \text{for all } s, t \geq 0.$$

Show that $F(t) = e^{at}$ for all $t \geq 0$ and some $a \in \mathbb{C}$. (See [**65**, Cauchy's Functional Equation] for more on this problem.)

7.10. Solutions

S 7.10.1. We have

$$\psi_\mu(z) = \frac{1}{2}\frac{z}{1-z} + \frac{1}{2}\frac{iz}{1-iz} \quad \text{and} \quad \eta_\mu(z) = \frac{(1+i)z^2 - z}{z - 1 + i}.$$

α is simply given by $\frac{1}{2}\delta_{-1} + \frac{1}{2}\delta_{-i}$. Solving the equation $-\eta_\mu(z) = 1$ leads to $z = \pm\frac{1-i}{\sqrt{2}}$. Recalling that $\int_{\partial\mathbb{D}} \frac{x^{-1}+z}{x^{-1}-z}\beta(dx) = 1 + 2\psi_\beta(z) = \frac{1+\eta_\beta(z)}{1-\eta_\beta(z)}$, we need to conjugate the poles (due to the x^{-1} in the integral) to obtain the points $\frac{1+i}{\sqrt{2}}$ and $\frac{-1-i}{\sqrt{2}}$. So define $\beta = a\delta_{\frac{1+i}{\sqrt{2}}} + (1-a)\delta_{\frac{-1-i}{\sqrt{2}}}$, then

$$\psi_\beta(z) = \frac{-\frac{1}{\sqrt{2}}(z((-1+i) + (2-2i)a) - z^2}{z^2 + i}.$$

From $\eta_\beta(z) = -\eta_\mu(z)$, we obtain

$$\psi_\beta(z) = \frac{\frac{1}{2}(1-i)z - z^2}{z^2 + i},$$

which leads to $a = \frac{1}{2} - \frac{1}{2\sqrt{2}}$.

S 7.10.2. Let $C : \mathbb{H} \to \mathbb{D}$ be the Cayley map. The function $g(z) = if(C(z))$ maps $\mathbb{H}$ into $\mathbb{H}$ with $g(i) = i$. We have

$$g(z) = a + \int_{\hat{\mathbb{R}}} \frac{1 + xz}{x - z}\hat{\gamma}(dx),$$

where $\hat{\gamma}$ is the push-forward of ρ to $\hat{\mathbb{R}}$ via C^{-1}, see the proof of Theorem 3.12. Furthermore, $g(i) = a + i(\hat{\gamma}(\mathbb{R})) = i$, i.e., $a = 0$ and

$$\hat{\gamma}(\{\infty\}) = \lim_{y\to\infty} \frac{g(iy)}{iy}.$$

Let $\alpha \in \mathbb{R}$. First, assume that $e^{i\alpha} = 1$. Then

$$\begin{aligned}\rho(\{e^{i\alpha}\}) = \hat{\gamma}(\{\infty\}) &= \lim_{y\to\infty} \frac{g(iy)}{iy} = \lim_{y\to\infty} \frac{if(C(iy))}{iy} \\ &= \lim_{r\nearrow 1} \frac{if(r)}{i(1+r)/(1-r)} = \lim_{r\nearrow 1} \frac{1}{2}(1-r)f(r).\end{aligned}$$

If $e^{i\alpha} \neq 1$, then we consider $f(e^{i\alpha}z)$ and obtain $\rho\left(\{e^{i\alpha}\}\right) = \lim_{r\nearrow 1} \frac{1}{2}(1-r)f(re^{i\alpha})$.

S 7.10.3.

(a) As k maps $\mathbb{D}$ conformally onto $\mathbb{C}\setminus(-\infty,-1/4]$, G is a conformal mapping from $\mathbb{D}$ onto $\hat{\mathbb{C}}\setminus[0,4]$. We have $G(z)=z+2+1/z$, and G maps 0 onto ∞. It follows that f_t fixes 0 with $f_t(z)=e^{-t}z+\dots$.

The domains $f_t(\mathbb{D})$ are clearly shrinking. Consequently, $(f_t)_{t\geq 0}$ is a normalized radial Loewner chain.

We have $f_t(\mathbb{D})=\mathbb{D}\setminus K_t$, where $K_t\subset\mathbb{D}$ is mapped by G onto the line segment $(4,4e^t]$. As G maps $\mathbb{R}\cup\{\infty\}$ onto itself, K_t must have the form $K_t=[x_t,1)\subset(0,1)$. The equation $G(x_t)=4e^t$ leads to

$$x_t=2e^t-1-2\sqrt{e^{2t}-e^t}\quad\text{or}\quad x_t=2e^t-1+2\sqrt{e^{2t}-e^t}.$$

The second solution is always >1 and does not belong to $\mathbb{D}$. Consequently, $K_t=[2e^t-1-2\sqrt{e^{2t}-e^t},1)$. Finally,

$$\frac{\partial}{\partial t}f_t(z)=(G^{-1})'(e^tG(z))\cdot e^tG(z),\qquad \frac{\partial}{\partial z}f_t(z)=(G^{-1})'(e^tG(z))\cdot e^tG'(z).$$

Hence,

$$\frac{\partial}{\partial t}f_t(z)=\frac{\partial}{\partial z}f_t(z)M(z)$$

for all $t\geq 0$ with

$$M(z)=G(z)/G'(z)=-z\frac{1+z}{1-z}.$$

(b) We recall that

$$\int_{\partial\mathbb{D}}\frac{x^{-1}+z}{x^{-1}-z}\mu_t(dx)=1+2\psi_{\mu_t}(z)=\frac{1+\eta_{\mu_t}(z)}{1-\eta_{\mu_t}(z)}.$$

As the function f_t extends continuously to $\overline{\mathbb{D}}$, we need to consider the points $x\in\partial\mathbb{D}$ with $f_t(x)=1$ and those with $|f_t(x)|<1$.

There is a closed arc $C_t\subset\mathbb{D}$ which is mapped by f_t onto $[2e^t-1-2\sqrt{e^{2t}-e^t},1]$, and $f_t(\partial\mathbb{D}\setminus C_t)=\partial\mathbb{D}\setminus\{1\}$. Therefore, the support of μ_t is C_t. (Note the x^{-1} in the above formula. However, C_t is symmetric with respect to $\mathbb{R}$.)

The arc C_t is mapped onto $[4,4e^t]$ by $G\circ f_t=e^tG(z)$. Therefore, C_t is mapped by G onto $[4e^{-t},4]$. The equation $G(z)=4e^{-t}$ leads to

$$z=2e^{-t}-1\pm 2i\sqrt{e^{-t}-e^{-2t}}.$$

As G maps 1 onto 4, C_t is the closed arc with endpoints $2e^{-t}-1-2i\sqrt{e^{-t}-e^{-2t}}$ and $2e^{-t}-1+2i\sqrt{e^{-t}-e^{-2t}}$ that contains 1.

S 7.10.4. Let $I(t)=\int_0^t F(s)ds$. As $I(0)=0$ and $I'(0)=F(0)=1$, $I(\varepsilon)\neq 0$ for some $\varepsilon>0$ small enough and then

$$\begin{aligned}F(t)&=I(\varepsilon)^{-1}F(t)I(\varepsilon)=I(\varepsilon)^{-1}\int_0^\varepsilon F(t+s)ds=I(\varepsilon)^{-1}\int_t^{t+\varepsilon}F(s)ds\\&=I(\varepsilon)^{-1}(I(t+\varepsilon)-I(t)).\end{aligned}$$

Hence, $F(t)$ is differentiable and

$$\frac{d}{dt}F(t) = \lim_{h \searrow 0} \frac{F(t+h) - F(t)}{h} = F(t) \lim_{h \searrow 0} \frac{F(h) - F(0)}{h} = F(t)F'(0).$$

This differential equation and the initial condition $F(0) = 1$ imply $F(t) = e^{at}$ for $a = F'(0) \in \mathbb{C}$.

CHAPTER 8

Slit Mappings

If a univalent function $f : \mathbb{H} \to \mathbb{H}$ has the property that $\mathbb{H} \setminus f(\mathbb{H})$ is a simple curve, then f is called a slit mapping. These mappings are of particular interest for various reasons. Loewner already considered slit mappings on the unit disk, because they can approximate all other univalent functions, see Section 8.6. We will first consider the upper half-plane, and derive a special differential equation for additive Loewner chains consisting of slit mappings.

8.1. Loewner equation for slit mappings

For $T > 0$, let $\gamma : [0, T] \to \mathbb{H} \cup \mathbb{R}$ be continuous and injective with $\gamma(0) \in \mathbb{R}$ and $\gamma(0, T] \subset \mathbb{H}$. Such a curve is also called a *slit.*

By Theorem 6.19, for any $t \in [0, T]$, there exists a unique $\mu_t \in \mathcal{P}_c(\mathbb{R})$ with mean 0 such that F_{μ_t} is univalent and $F_{\mu_t}(\mathbb{H}) = \mathbb{H} \setminus \gamma(0, t]$. Then the function $t \mapsto F_{\mu_t}(\mathbb{H})$ is continuous with respect to kernel convergence, and thus, $t \mapsto \mu_t$ is continuous. (We can pass to the unit disk to apply Theorem 7.29 as in the proof of Theorem 6.38.) In particular, $t \mapsto \mathrm{Var}(\mu_t)$ is continuous and strictly increasing. Therefore, we can reparametrize γ such that $\mathrm{Var}(\mu_t) = t$, and then, F_{μ_t} has the normalization

$$F_{\mu_t}(z) = z - \frac{t}{z} + \dots \tag{8.1}$$

at ∞. This parametrization is also called the *hydrodynamic parametrization* of the slit. We redefine T such that γ is defined on $[0, T]$.

The boundary behavior of F_{μ_t} (see Figure 8.1) is quite simple. By Theorem B.1, F_{μ_t} can be extended continuously to $\mathbb{H} \cup \mathbb{R}$, and by Theorem B.3, there exists a closed bounded interval $I_t \subset \mathbb{R}$ with $F_{\mu_t}(I_t) = \gamma[0, t]$ and $F_{\mu_t}(\mathbb{R} \setminus I_t) = \mathbb{R} \setminus \{\gamma(0)\}$. For any $x \in \mathbb{R} \setminus I_t$, F_{μ_t} can be extended analytically to a neighborhood of x (by $z \mapsto \overline{F_{\mu_t}(\overline{z})}$ for $z \in \mathbb{H}^-$). Write

$$\mathbb{R} \setminus I_t = (-\infty, c_1) \cup (c_2, +\infty).$$

As $F_{\mu_t}(x) \to +\infty$ for $x \to +\infty$, F_{μ_t} maps $(-\infty, c_1)$ onto $(-\infty, \gamma(0))$ and $(c_2, +\infty)$ onto $(\gamma(0), +\infty)$.

Furthermore, there exists a unique $U(t) \in I_t$ such that $F_{\mu_t}(U(t)) = \gamma(t)$. Thus, $F_{\mu_t}^{-1}$ can be extended continuously to $(\mathbb{H} \setminus \gamma(0, t]) \cup \{\gamma(t)\}$, and

$$U(t) = F_{\mu_t}^{-1}(\gamma(t)).$$

Each $p \in \gamma[0, t)$ has two preimages $a, b \in I_t$, $a < U(t) < b$.
The size of the interval I_t can be related to the size of $\gamma[0, t]$.

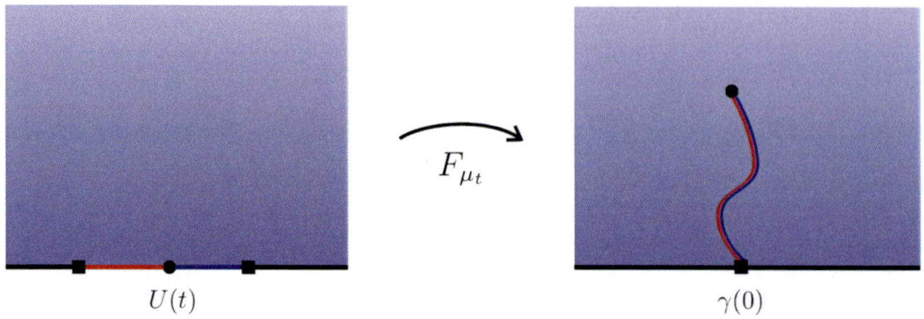

FIGURE 8.1. The slit mapping F_{μ_t}.

LEMMA 8.1. *Let $t \in (0,T]$. The following relations hold:*

(a) *Let $R_t = \sup\{|z-\gamma(0)| \mid z \in \gamma[0,t]\}$. Then*
$$I_t \subset \{x \in \mathbb{R} \mid |x-\gamma(0)| \le 2R_t\}.$$

(b) *Let $S_t = \sup\{|x-U(t)| \mid x \in I_t\}$. Then*
$$\gamma[0,t] \subset \{z \in \mathbb{C} \mid |z-U(t)| \le 2S_t\}.$$

PROOF. Consider the function $G_t := F_{\mu_t}^{-1}(R_t z + \gamma(0))/R_t$. It can be extended to a univalent function in $\mathbb{C} \setminus \overline{\mathbb{D}}$ with expansion
$$G_t(z) = z + b_0 + b_1/z + \dots$$
at ∞. Therefore, G_t belongs to the class Σ (defined in Section 3.7). As $F_{\mu_t}(z) = z - \frac{t}{z} + \dots$, we have $b_0 = \gamma(0)/R_t$, and Corollary 3.36 implies that
$$E_1 := \mathbb{C} \setminus G_t(\mathbb{C} \setminus \overline{\mathbb{D}}) \subset \{z \in \mathbb{C} \mid |z-\gamma(0)/R_t| \le 2\}.$$
As $I_t \subset R_t \cdot E_1$, we have $I_t \subset \{x \in \mathbb{R} \mid |x-\gamma(0)| \le 2R_t\}$, and (a) is proven.

Next, consider $H_t := F_{\mu_t}(S_t z + U(t))/S_t$. Just like G_t, H_t can be extended to a univalent function in $\mathbb{C} \setminus \overline{\mathbb{D}}$ with expansion
$$H_t(z) = z + U(t)/S_t + \dots,$$
and Corollary 3.36 implies that
$$E_2 := \mathbb{C} \setminus H_t(\mathbb{C} \setminus \overline{\mathbb{D}}) \subset \{z \in \mathbb{C} \mid |z-U(t)/S_t| \le 2\}.$$
As $\gamma[0,t] \subset S_t \cdot E_2$, we have $\gamma[0,t] \subset \{z \in \mathbb{C} \mid |z-U(t)| \le 2S_t\}$, which proves (b). □

The family $(F_{\mu_t})_{t\in[0,T]}$ is (a part of) an additive Loewner chain. We already know that it satisfies Loewner's partial differential equation almost everywhere, but now we can derive a much more precise result. The idea of describing a simple curve by a differential equation for conformal mappings is due to Loewner. He considered slit mappings defined on $\mathbb{D}$ in his original paper, see Section 8.6. The following theorem is due to Kufarev et al. [**112**]. We follow the presentation in [**60**].

THEOREM 8.2. *The function $t \mapsto U(t)$, $t \in [0,T]$, is continuous with $U(0) = \gamma(0)$, and F_{μ_t} satisfies*

$$\frac{\partial}{\partial t} F_{\mu_t}(z) = \frac{\partial}{\partial z} F_{\mu_t}(z) \cdot \frac{1}{U(t)-z} \quad \textit{for all } t \in [0,T] \textit{ and } z \in \mathbb{H}. \tag{8.2}$$

Let $g_t = F_{\mu_t}^{-1}$. *Then*

$$\frac{\partial}{\partial t} g_t(z) = \frac{1}{g_t(z) - U(t)} \quad \text{for all } t \in [0,T] \text{ and } z \in F_{\mu_T}(\mathbb{H}). \tag{8.3}$$

The function $U : [0,T] \to \mathbb{R}$ *is also called the* driving function *of the slit* γ.

PROOF. We define the transition mapping $F_{s,t}$ by $F_{s,t} = F_{\mu_s}^{-1} \circ F_{\mu_t}$, $s \leq t$. For $s < t$, this function maps $\mathbb{H}$ conformally onto $\mathbb{H} \setminus C_{s,t}$, where $C_{s,t} \subset \overline{\mathbb{H}}$ is a simple curve with starting point $U(s)$. $F_{s,t}$ has a continuous extension to $\overline{\mathbb{H}}$, and we let $I_{s,t} \subset \mathbb{R}$ be the closed bounded interval that is mapped onto $C_{s,t}$ (and $F_{s,t}(\mathbb{R} \setminus I_{s,t}) = \mathbb{R} \setminus \{U(s)\}$). Then $U(t)$ is an inner point of $I_{s,t}$, which is mapped by $F_{s,t}$ to the tip of $C_{s,t}$. The mappings $(F_{s,u})_{s \leq u \leq t}$ are simply the unique conformal mappings with normalization (8.1) for the slit $C_{s,t}$.

By Lemma 8.1 (a), $I_{s,t} \subset \{x \in \mathbb{R} \,|\, |x - U(s)| \leq 2\sup\{|z - U(s)| \,|\, z \in C_{s,t}\}$. Hence, if we fix s and let $t \to s$, then the left and right end point of $I_{s,t}$ both converge to $U(s)$, and thus, also $U(t) \to U(s)$.

Conversely, we now fix t. Then $\gamma(s) \to \gamma(t)$ as $s \to t$. The interval $I_{s,t}$ is mapped by F_{μ_t} onto $\gamma[s,t]$. Therefore, $I_{s,t}$ shrinks to $U(t)$ as $s \to t$. Lemma 8.1 (b) implies that $C_{s,t} \subset \{z \in \mathbb{C} \,|\, |z - U(t)| \leq 2\sup\{|x - U(t)| \,|\, x \in I_{s,t}\}$. Since $U(s) \in C_{s,t}$, $U(s) \to U(t)$ as $s \to t$. It follows that U is continuous.

Next, we derive (8.2). Due to Theorem 5.7, $F_{s,t}$ can be written as

$$F_{s,t}(z) = z - \int_{\mathbb{R}} \frac{1}{z - x} \tau_{s,t}(dx)$$

for a compactly supported, finite, non-negative measure $\tau_{s,t}$ on $\mathbb{R}$ and $\tau_{s,t}(\mathbb{R}) = t - s$. Since $F_{s,t}$ extends continuously to $\overline{\mathbb{H}}$, the Stieltjes–Perron inversion formula implies that

$$F_{s,t}(z) = z - \frac{1}{\pi} \int_{I_{s,t}} \frac{\operatorname{Im}(F_{s,t}(x))}{z - x} dx.$$

Furthermore, as the measure $\tau_{s,t}$, given by $\frac{1}{\pi}\operatorname{Im}(F_{s,t}(x))dx$, has the mass $t - s$, we have

$$t - s = \frac{1}{\pi} \int_{I_{s,t}} \operatorname{Im}(F_{s,t}(x))dx.$$

We have

$$\frac{F_{s,t}(z) - z}{t - s} = \frac{\frac{1}{\pi}\int_{I_{s,t}} \frac{\operatorname{Im}(F_{s,t}(x))}{x - z} dx}{\frac{1}{\pi}\int_{I_{s,t}} \operatorname{Im}(F_{s,t}(x))dx}.$$

As $t \searrow s$, $I_{s,t}$ shrinks to $\{U(s)\}$ by Lemma 8.1. By applying the mean value theorem for integrals, applied to the real and imaginary part separately, we obtain

$$\lim_{t \searrow s} \frac{F_{s,t}(z) - z}{t - s} = \frac{1}{U(s) - z}.$$

The case $s \nearrow t$ can be treated similarly. From

$$\frac{F_{\mu_t}(z) - F_{\mu_s}(z)}{t - s} = \frac{F_{\mu_s}(F_{s,t}(z)) - F_{\mu_s}(z)}{t - s} = \frac{F_{\mu_s}(F_{s,t}(z)) - F_{\mu_s}(z)}{F_{s,t}(z) - z} \frac{F_{s,t}(z) - z}{t - s},$$

we obtain (8.2), and differentiating $(F_{\mu_t} \circ F_{\mu_t}^{-1})(z) = z$ yields (8.3) (as in the proof of Corollary 6.15). □

EXAMPLE 8.3. The arcsine distributions $(\mu_t)_{t\geq 0} = (Arc(0,t))_{t\geq 0}$ lead to the Loewner chain $(F_{\mu_t})_{t\geq 0} = (\sqrt{z^2-2t})_{t\geq 0}$, see Example 6.8, and F_t satisfies (8.2) with $U(t)\equiv 0$.

The constant driving function $U(t)\equiv u$ leads to

$$F_{\mu_t}(z) = \sqrt{(z-u)^2-2t}+u,$$

i.e., $\mu_t = \delta_{-u}\rhd Arc(0,t)\rhd\delta_u$. In contrast to $u=0$, this measure has an atom when $u\neq 0$, see Example 5.17.

Lemma 6.37 immediately yields the following property of the slit equation.

LEMMA 8.4. *Let $c,d>0$ and let $f_t = F_{\mu_t}$ be the solution to the slit equation (8.2). Consider the scaled measures $\nu_t(B) = \mu_{d\cdot t}(c\cdot B)$. Let $h_t = F_{\nu_t}$. Then h_t solves*

$$\frac{\partial}{\partial t}h_t(z) = \frac{\partial}{\partial z}h_t(z)\cdot\frac{d/c^2}{U(d\cdot t)/c - z}.$$

For $c=\sqrt{d}$, we get the same normalization as in (8.2). As $F_{\nu_t}(\mathbb{H}) = F_{\mu_{c^2\cdot t}}(\mathbb{H})/c$, we see that if γ is generated by $U:[0,T]\to\mathbb{R}$, then the scaled curve $\gamma[0,T]/c$ is generated by $U(c^2\cdot t)/c$, $t\in[0,T/c^2]$.

EXAMPLE 8.5. If γ is self-similar in the sense that $\gamma[0,T]/c\subset\gamma[0,T]$ for some $T>0$ and $c>1$, then $U(t) = U(c^2\cdot t)/c$ for all $t\in[0,T/c^2]$.

EXAMPLE 8.6. Assume that γ is a straight line segment starting at 0. Let $U:[0,T]\to\mathbb{R}$ be the corresponding driving function. Then $\gamma[0,T]/c\subset\gamma[0,T]$ for all $c>1$. So $U(T/c^2) = U(T)/c$ for all $c>1$, and with $t = T/c^2$, we have

$$U(t) = \frac{U(T)}{\sqrt{T}}\sqrt{t}\quad\text{for all } t\in[0,T].$$

One can calculate the conformal mappings in this case explicitly and derive a relation between the angle of γ and the value C of the corresponding driving function $C\sqrt{t}$, see Exercise 8.7.2.

The encoding

$$\text{simple curve } \gamma \quad\Longrightarrow\quad \text{continuous driving function } U, \tag{8.4}$$

given by Theorem 8.2, is remarkable, as the two-dimensional curve γ is represented by a one-dimensional function. This can also be used to transform complicated random curves in $\mathbb{H}$ into Brownian motions on $\mathbb{R}$, which we briefly describe in Section 8.4.

Next, we will show that the map (8.4) is not surjective.

8.2. Solutions which are not slit mappings

Since (8.2) is a special case of (6.9), it follows that, conversely, every continuous function $U:[0,\infty)\to\mathbb{R}$ generates an additive Loewner chain $(f_t)_{t\geq 0}$ via equation (8.2). Let $K_t = \overline{\mathbb{H}\setminus f_t(\mathbb{H})}$. We will see that the growing sets K_t do not necessarily describe a slit.

Assume there exists some $T>0$ such that $U:[0,T)\to\mathbb{R}$ describes a slit γ, i.e., $K_t = \gamma[0,t]$ for all $t\in[0,T)$. As $t\to T$, we could have $\gamma(t)\to c$ for some $c\in\mathbb{R}$ or $c\in\gamma[0,T)$. Since $\mathbb{H}\setminus K_T$ is simply connected, K_T will consist of $\gamma[0,T]$ plus the bounded connected components of $(\mathbb{H} \cup \mathbb{R}) \setminus \gamma[0,T]$. As $t \to T$, $\gamma(t)$ could also have several accumulation points, for example, if it describes an infinite spiral, see Figure 8.2.

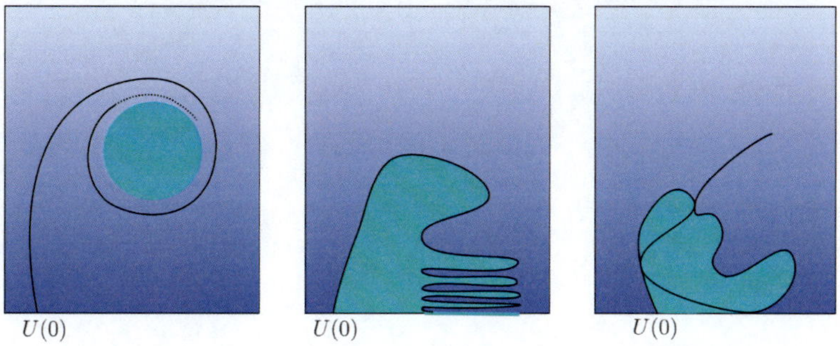

FIGURE 8.2. Three cases, where K_t is a slit only for t small enough.

We now show that even though U is continuous on $[0,T]$, K_T may not be a slit. A first example was given by Kufarev in [**111**]. In order to find such an example, we consider the inverse equation (8.3):

$$\frac{d}{dt}w(t) = \frac{1}{w(t) - U(t)}, \quad w(0) = z_0. \tag{8.5}$$

For $z_0 \in f_T(\mathbb{H})$, the solution exists for all $t \in [0,T]$, and we have $w(t; z_0) = f_t^{-1}(z_0)$.

However, for every $z_0 \in \mathbb{H}$, this equation can be solved at least for t small enough. In fact, we can solve it also for every $z_0 \in \mathbb{R} \setminus \{U(0)\}$. Assume that U generates a simple curve. Then each f_t^{-1} can be extended analytically to a neighborhood of every $x \in \mathbb{R} \setminus \{U(0)\}$. Thus, the solution $w(t;x)$ exists up to $t = T$ for all $x \in \mathbb{R} \setminus \{U(0)\}$ in this case.

THEOREM 8.7. *Let $U(t) = p\sqrt{1-t}$ with $|p| \geq 2\sqrt{2}$. Then K_1 is not a slit.*

PROOF. We may assume that $p > 0$, and thus, p can be expressed as

$$p = c + 2/c$$

for some $c > 0$. For $t \in [0,1)$, the function $w(t) = c\sqrt{1-t}$ satisfies

$$\frac{d}{dt}w(t) = \frac{-c}{2\sqrt{1-t}} = \frac{1}{-\frac{2}{c}\sqrt{1-t}} = \frac{1}{w(t) - p\sqrt{1-t}}.$$

Therefore, $w(t)$ solves (8.5) for $U(t) = p\sqrt{1-t}$ and all $t \in [0,1)$. The initial value of this function is given by $w(0) = c < p = U(0)$. Both $w(t)$ and $U(t)$ are strictly decreasing, and $w(t) < U(t)$ for all $t < 1$, but $w(1) = 0 = U(1)$. Thus, w does not solve (8.5) at $t = 1$. Since $w(t) = f_t^{-1}(w(0)) = f_t^{-1}(c)$, we have $c \in K_1$, and because $c < U(0)$, K_1 cannot be a slit. $\square$

One can show that K_t is a slit for all $t \in [0,1)$, which hits $\mathbb{R}$ as $t \to 1$ (see Figure 8.3), and that for $|p| < 2\sqrt{2}$, $U(t) = p\sqrt{1-t}$ generates a slit also at $t = 1$, which can be related to a logarithmic spiral, see [**120**, Section 3] for more details.

The previous example can be used to obtain a whole class of functions that do not generate slits.

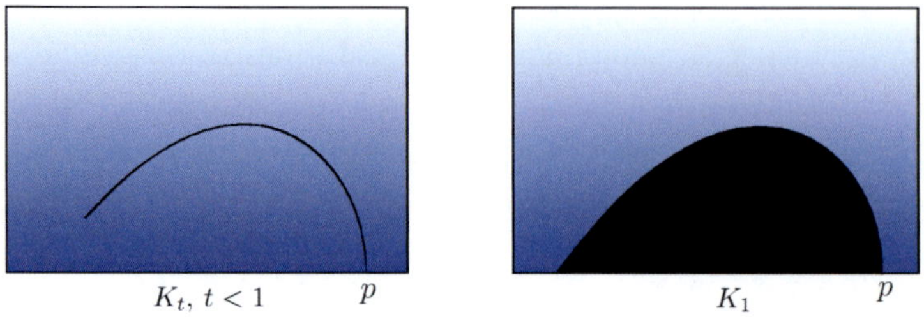

FIGURE 8.3. Growth process for the driving function $p\sqrt{1-t}$, $p \geq 2\sqrt{2}$.

THEOREM 8.8. *For $T > 0$, let $U : [0,T] \to \mathbb{R}$ be continuous with $U(T) = 0$ and*

$$U(t) \geq 2\sqrt{2}\sqrt{T-t} \quad \textit{for all } t \in [T-\varepsilon, T],$$

for some $\varepsilon > 0$. Then K_T is not a slit.

PROOF. We first consider the case where $U(t) \geq 2\sqrt{2}\sqrt{T-t}$ for all $t \in [0,T]$. Assume that K_T is a slit. Then the solution $w(t)$ to

$$\frac{d}{dt}w(t) = \frac{1}{w(t)-U(t)}, \quad w(0) = \sqrt{2}\sqrt{T} < 2\sqrt{2}\sqrt{T} \leq U(0),$$

exists for all $t \in [0,T]$. In particular, $w(t) < U(t)$ for all $t \in [0,T]$.

Consequently, $\frac{d}{dt}w(t) < 0$, and $w(t)$ is strictly decreasing. We have $0 < U(t) - w(t) \geq 2\sqrt{2}\sqrt{T-t} - w(t)$ for all $t \in [0,T]$. Therefore,

$$0 > \frac{d}{dt}w(t) \geq \frac{1}{w(t) - 2\sqrt{2}\sqrt{T-t}}, \quad w(0) = \sqrt{2}\sqrt{T}.$$

This differential inequality yields $w(t) \geq \sqrt{2}\sqrt{T-t}$ (let $c = \sqrt{2}$ and $p = 2\sqrt{2}$ in the proof of the previous theorem). Hence, $w(T) \geq 0 = U(T)$, which is a contradiction as $w(t) < U(t)$ for all $t \in [0,T]$.

Next, we consider the case where $U(t) \geq 2\sqrt{2}\sqrt{T-t}$ for all $t \in [T-\varepsilon, T]$ for some $\varepsilon > 0$. Assume that K_T is a slit. Then also the transition mappings $(f_{T-\varepsilon,t})_{T-\varepsilon \leq t \leq T}$ are slit mappings. As $f_{0,t} = f_{0,s} \circ f_{s,t}$, we have $f_{0,t}^{-1} = f_{s,t}^{-1} \circ f_{0,s}^{-1}$, and the solution to

$$\frac{d}{dt}w(t) = \frac{1}{w(t)-U(t)}, \quad t \geq T-\varepsilon, \quad w(T-\varepsilon) = z_0,$$

generates the inverse mappings $f_{T-\varepsilon,t}^{-1}(z_0)$, whose corresponding slit starts at $U(T-\varepsilon)$. We now consider $z_0 = \sqrt{2}\sqrt{\varepsilon} < 2\sqrt{2}\sqrt{\varepsilon} \leq U(T-\varepsilon)$, and obtain a contradiction just as in the previous case. □

As shown in [**122**], there is a converse statement to the previous theorem under the assumption that U is locally (left) $\frac{1}{2}$-Hölder continuous.

THEOREM 8.9. *Let $T > 0$ and $U : [0,T] \to \mathbb{R}$ be continuous, and assume that there exist $\varepsilon, c > 0$ such that*

$$|U(T) - U(t)| \leq c\sqrt{T-t} \quad \textit{for all } t \in [T-\varepsilon, T].$$

Assume that the solution $w(t)$ *to* (8.5) *with* $w(0) = x_0 \in \mathbb{R} \setminus \{U(0)\}$ *exists for all* $t \in [0, T)$, *and that* $\lim_{t \nearrow T} w(t) = U(T)$. *Then*

$$c \geq 2\sqrt{2}.$$

We begin with the following lemma.

LEMMA 8.10. *Let* $h_1 : \mathbb{R} \to \mathbb{R}$, $h_1(x) = x$, *and define* h_n *for* $n \geq 2$ *inductively via*

$$h_{n+1} : \mathbb{R} \setminus \{x \in \mathbb{R} \,|\, h_n(x) = 0\} \to \mathbb{R}, \quad h_{n+1}(x) = x - \frac{2}{h_n(x)}.$$

Let x_n *be the largest zero of* h_n. *Then* $(x_n)_{n\in\mathbb{N}}$ *is a strictly increasing sequence converging to* $2\sqrt{2}$. *If* $h_n(c) \geq 0$ *for some* $c \geq 0$ *and all* $n \in \mathbb{N}$, *then* $c \geq 2\sqrt{2}$.

PROOF. By induction, we show that h_{n+1} maps $(x_n, +\infty)$ strictly increasingly onto $\mathbb{R}$ for all $n \in \mathbb{N}$. This can be easily verified for $h_2(x) = x - \frac{2}{x}$. So assume that the statement is true for some h_{n+1}, $n \in \mathbb{N}$. Then x_{n+1} is the unique zero of h_{n+1} in $(x_n, +\infty)$. We have $\lim_{x \searrow x_{n+1}} h_{n+2}(x) = -\infty$ and $\lim_{x \nearrow +\infty} h_{n+2}(x) = +\infty$. Furthermore, for $y > x > x_{n+1}$, $h_{n+2}(y) = y - \frac{2}{h_{n+1}(y)} > x - \frac{2}{h_{n+1}(x)} = h_{n+2}(x)$.

Consequently, $(x_n)_n$ is a strictly increasing sequence. We prove that it is bounded above by $2\sqrt{2}$ by showing inductively that $h_n(x) \geq \sqrt{2}$ for all $x \geq 2\sqrt{2}$. Clearly, $h_1(x) = x \geq 2\sqrt{2} \geq \sqrt{2}$, and for $n \geq 1$, $x \geq 2\sqrt{2}$, we have

$$h_{n+1}(x) = x - \frac{2}{h_n(x)} \geq 2\sqrt{2} - \frac{2}{h_n(x)} \geq 2\sqrt{2} - \frac{2}{\sqrt{2}} = \sqrt{2}.$$

Hence,

$$\lim_{n\to\infty} x_n =: x_0 \leq 2\sqrt{2}. \tag{8.6}$$

Obviously, we have $h_n(x_0) > 0$ for all $n \in \mathbb{N}$. Now suppose $h_n(x_0) \leq 1/\sqrt{2}$ for some n, then $h_{n+1}(x_0) = x_0 - \frac{2}{h_n(x_0)} \leq x_0 - 2\sqrt{2} \leq 0$, a contradiction. Hence, $h_n(x_0) > 1/\sqrt{2}$ for all $n \in \mathbb{N}$. Furthermore,

$$\begin{aligned} h_n(x_0) - h_{n+1}(x_0) &= h_n(x_0) - x_0 + \frac{2}{h_n(x_0)} \geq h_n(x_0) - 2\sqrt{2} + \frac{2}{h_n(x_0)} \\ &= \frac{h_n(x_0)^2 - 2\sqrt{2}h_n(x_0) + 2}{h_n(x_0)} = \frac{(h_n(x_0) - \sqrt{2})^2}{h_n(x_0)} \geq 0. \end{aligned}$$

It follows that the sequence $h_n(x_0)$ is decreasing, and thus, it converges with limit $\tilde{h}$ satisfying

$$\tilde{h} = x_0 - \frac{2}{\tilde{h}}.$$

So $\tilde{h} = \frac{x_0 \pm \sqrt{x_0^2 - 8}}{2}$, and thus, $x_0^2 \geq 8$. As x_0 must be positive, we conclude $x_0 \geq 2\sqrt{2}$. Together with (8.6), this implies $x_0 = 2\sqrt{2}$.

Finally, assume that $h_n(c) \geq 0$ for all $n \in \mathbb{N}$. Then $h_1(c) = c \geq 0 = x_1$. As h_{n+1} maps $(x_n, +\infty)$ strictly increasingly onto $\mathbb{R}$, we conclude that $c \geq x_n$ for all $n \in \mathbb{N}$, and thus, $c \geq x_0 = 2\sqrt{2}$. □

PROOF OF THEOREM 8.9. We use the functions h_n from the previous lemma.

First, we may assume that $T = 1$ and that $x_0 < U(0)$. Then $w(t)$ is strictly decreasing, and for $t \in [1-\varepsilon, 1]$ we have

$$U(t) - w(t) \leq U(t) - w(1) = U(t) - U(1) \leq c\sqrt{1-t} = h_1(c)\sqrt{1-t}.$$

We show by induction that $U(t) - w(t) \leq h_n(c)\sqrt{1-t}$ for all $n \in \mathbb{N}$ and $t \in [1-\varepsilon, 1]$. So assume that the inequality holds for some $n \in \mathbb{N}$, and let $t \in [1-\varepsilon, 1]$. Then

$$\frac{d}{dt}w(t) = \frac{1}{w(t) - U(t)} \leq \frac{-1}{h_n(c)\sqrt{1-t}}.$$

Integrating yields

$$w(1) - w(t) \leq -\frac{2}{h_n(c)}\sqrt{1-t}.$$

Therefore,

$$U(t) - w(t) = U(t) - U(1) + w(1) - w(t) \leq \left(c - \frac{2}{h_n(c)}\right)\sqrt{1-t} = h_{n+1}(c)\sqrt{1-t}.$$

As $U(t) - w(t)$ is positive for $t \in [1-\varepsilon, 1)$, we have $h_n(c) \geq 0$ for all $n \in \mathbb{N}$, and Lemma 8.10 implies that $c \geq 2\sqrt{2}$. □

It follows that a local $\frac{1}{2}$-Hölder continuity of U with $c < 2\sqrt{2}$ prevents $w(t)$ from hitting the driving function $U(t)$, and then this necessary condition for generating a slit is satisfied.

Curiously, this can also be achieved if U is "irregular enough".

EXAMPLE 8.11. Let $T > 0$ and $U : [0, T] \to \mathbb{R}$ be continuous such that there exists a strictly increasing sequence $(t_n)_{n\in\mathbb{N}}$ converging to T with $U(t_n) = U(T)$ for all $n \in \mathbb{N}$. Let $t_0 \in [0, T)$, and assume that the solution $w(t)$ to (8.5) with $w(t_0) = x_0 \in \mathbb{R} \setminus \{U(t_0)\}$ exists for all $t \in [t_0, T)$. Then $\lim_{t\nearrow T} w(t) \neq U(T)$:

Assume that $x_0 < U(t_0)$. Then $w(t) < U(t)$ for all $t \in [t_0, T)$, and $w(t)$ is strictly decreasing. Let N be large enough such that $t_N > t_0$. Then $\lim_{t\nearrow T} w(t) < w(t_N) < U(t_N) = U(T)$. The case $x_0 > U(t_0)$ can be treated in the same way.

Let us remain in the regular case as in Theorem 8.9.

THEOREM 8.12. *Let $T > 0$ and $U : [0, T] \to \mathbb{R}$ be continuous, and assume that for every $t \in (0, T]$ there exist $\delta > 0$ and $c < 2\sqrt{2}$ such that*

$$|U(t) - U(t-h)| \leq c\sqrt{h} \quad \textit{for all } h \in [0, \delta].$$

Suppose that $w(t)$ is a solution of (8.5) with initial value $w(t_0) = x_0 \in \mathbb{R} \setminus \{U(t_0)\}$, $t_0 \in [0, T)$. Then there exists $\varepsilon > 0$, independent of x_0, such that

$$|w(T) - U(T)| > \varepsilon.$$

Note that the solution w exists locally and due to Theorem 8.9, it exists for all $t \in [t_0, T]$, with $|w(T) - U(T)| > 0$. Theorem 8.12 states that this value is bounded from below by a positive constant independent of x_0.

PROOF. We prove the statement by contradiction. Hence, we assume that for every $\varepsilon > 0$ there is an $x_0^\varepsilon \in \mathbb{R} \setminus \{U(t_0)\}$ such that $|w(T) - U(T)| \leq \varepsilon$. Without loss of generality we may assume that $x_0^\varepsilon < U(t_0)$. Then $U(T) - w(T) \leq \varepsilon$, and the solution $w(t)$ is strictly decreasing.

Let $\delta > 0$ and $c < 2\sqrt{2}$ such that $|U(T) - U(T-h)| \leq c\sqrt{h}$ for all $h \in [0, \delta]$. Again, we denote by h_n the sequence from Lemma 8.10. As $c < 2\sqrt{2}$, there exists an $N \in \mathbb{N}$ such that

$$h_N(c) < 0. \tag{8.7}$$

We take the smallest such N and assume that $h_n(c) > 0$ for all $n < N$. (If there is an n with $h_n(c) = 0$, one can pass on to a slightly larger c.)

Next, define e_n by

$$\begin{aligned} e_1 &= \varepsilon, \\ e_{n+1} &= \varepsilon + \frac{2e_n}{(h_n(c))^2} \log\left(1 + \frac{h_n(c)\sqrt{T}}{e_n}\right) \quad \text{for} \quad 1 \leq n < N. \end{aligned}$$

Inductively, we can show that for every $n \leq N$, we have $e_n > 0$ and

$$\lim_{\varepsilon \to 0} e_n = 0. \tag{8.8}$$

We prove by induction that

$$w(T) - w(t) \leq e_n - \varepsilon + (h_n(c) - c)\sqrt{T-t} \quad \text{for all} \quad t \in [T-\delta, T] \quad \text{and} \quad n \leq N.$$

The case $n = 1$ states $w(T) - w(t) \leq 0$, which is true because $w(t)$ is decreasing. Next, assume that the statement holds for an $n < N$, and let $t \in [T - \delta, T]$. Then

$$-w(t) \leq -w(T) - \varepsilon + e_n + (h_n(c) - c)\sqrt{T-t}.$$

Consequently,

$$\begin{aligned} U(t) - w(t) &\leq U(t) - U(T) + U(T) - w(T) - \varepsilon + e_n - (c - h_n(c))\sqrt{T-t} \\ &\leq e_n + h_n(c)\sqrt{T-t}, \end{aligned}$$

which implies

$$\dot{w}(t) = \frac{1}{w(t) - U(t)} \leq -\frac{1}{e_n + h_n(c)\sqrt{T-t}}.$$

Integrating gives

$$\begin{aligned} w(T) - w(t) &\leq -\frac{2}{h_n(c)}\sqrt{T-t} + \frac{2e_n}{(h_n(c))^2}\log(1 + \frac{h_n(c)}{e_n}\sqrt{T-t}) \\ &= (c - \frac{2}{h_n(c)} - c)\sqrt{T-t} + \frac{2e_n}{(h_n(c))^2}\log(1 + \frac{h_n(c)}{e_n}\sqrt{T-t}) \\ &= (h_{n+1}(c) - c)\sqrt{T-t} + \varepsilon + \frac{2e_n}{(h_n(c))^2}\log(1 + \frac{h_n(c)}{e_n}\sqrt{T-t}) - \varepsilon. \end{aligned}$$

Since $h_n(c) > 0$ and $e_n > 0$, we conclude that

$$\begin{aligned} w(T) - w(t) &\leq (h_{n+1}(c) - c)\sqrt{T-t} + \varepsilon + \frac{2e_n}{(h_n(c))^2}\log(1 + \frac{h_n(c)\sqrt{T}}{e_n}) - \varepsilon \\ &= (h_{n+1}(c) - c)\sqrt{T-t} + e_{n+1} - \varepsilon. \end{aligned}$$

For $n = N$, we get

$$w(T) - w(t) \le e_N - \varepsilon + (h_N(c) - c)\sqrt{T-t}.$$

On the other hand,

$$w(T) - w(t) = w(T) - U(T) + U(T) - w(t) \ge -\varepsilon + U(T) - U(t) \ge -\varepsilon - c\sqrt{T-t}.$$

Thus,

$$h_N(c)\sqrt{T-t} + e_N \ge 0.$$

We can now send $\varepsilon \to 0$, and obtain $h_N(c) \ge 0$ by (8.8), which is a contradiction to (8.7). □

In the above setting, let $M^-(t_0, T) > 0$ be the infimum of $U(T) - w(T)$ taken over all solutions with $w(t_0) = x_0 < U(t_0)$, and let $M^+(t_0, T)$ be the infimum of $w(T) - U(T)$ taken over all solutions with $w(t_0) = x_0 > U(t_0)$.

The functions $t_0 \mapsto M^-(t_0, T), M^+(t_0, T)$ are both continuous and strictly decreasing:

We can apply Theorem 8.12 also for every $t \in (0, T]$ instead for T, and obtain $M^-(t_0, t)$, $M^+(t_0, t)$, $t_0 \in [0, t)$. Let $0 \le s_0 < t_0$. Consider a solution w with initial value $w(s_0) < U(s_0)$. Then $U(t_0) - w(t_0) > M^-(s_0, t_0)$, and thus, $U(T) - w(T) \ge M^-(t_0, T) + c$ for some $c > 0$, so $M^-(s_0, T) > M^-(t_0, T)$.

Consider the backward flow equation

$$\frac{d}{dt}x(t) = \frac{1}{U(T-t) - x(t)}, \qquad x(0) = x_0 \in \mathbb{R} \setminus \{U(T)\}.$$

Let $x_0 = U(T) - M^-(t_0, T) - 1/n$. Then the solution must exist until $t_1 = T - t_0$. As n increases, also $x(t_1)$ increases, and the limit $\lim_{n\to\infty} x(t_1) \le U(t_0)$ exists. Due to the definition of $M^-(t_0, T)$, the limit cannot be $< U(t_0)$.

Thus, the solutions with initial value $x_0 = U(T) - M^-(t_0, T)$ and $x_0 = U(T) + M^+(t_0, T)$ will both exist in $[0, T - t_0)$, and then they hit $U(T - (T - t_0)) = U(t_0)$. Let

$$I_T = [U(T) - M^-(0, T), U(T) + M^+(0, T)].$$

We can define a decreasing homeomorphism $h : I_T \to I_T$ by

$$h(U(T)) = U(T) \quad \text{and} \quad h(U(T) \pm M^{\pm}(t_0, T)) = U(T) \mp M^{\mp}(t_0, T), t_0 \in [0, T).$$

Roughly speaking, this means that f_T maps I_T in a 2–to–1 way onto the boundary of K_T, that is, $x \in I_T$ and $h(x)$ are mapped onto the same set of points. Note that we do not know whether f_T extends continuously to I_T. Thus, we still do not know whether K_T is a slit under the conditions of Theorem 8.12, but it shares the symmetry property of slits of being "welded".

REMARK 8.13. If K_T is a slit, then f_T extends continuously to I_T and both $x \in I_T$ and $h(x)$ are mapped onto the same point. The homeomorphism h is then called the *welding homeomorphism* of the slit γ. The theory of conformal welding implies: γ is a "quasislit", i.e., $\mathbb{H} \setminus \gamma$ is the image of $\mathbb{H} \setminus [0, i]$ under a quasiconformal homeomorphism of $\hat{\mathbb{C}}$ fixing $\mathbb{H}$ and ∞, if and only if h is quasisymmetric, see [**122**, Lemma 6] and [**128**, Lemma 2.2].

It is shown in [**122, 128**] that if

$$|U(t) - U(s)| \leq c\sqrt{|t-s|}$$

for all $s, t \in [0, T]$ and a constant $c < 2\sqrt{2}$, then U generates a quasislit. (The constant $2\sqrt{2}$ corresponds to 4 in [**122**], which considers a different time normalization.) Conversely, if U generates a quasislit, then $|U(t) - U(s)| \leq c\sqrt{|t-s|}$ for some constant $c > 0$, see [**128**]. We also refer to [**120**] and to [**121**], which considers space-filling curves. Applying [**128**, Theorem 1.1] and [**122**, Theorem 2] locally yields the following statement.

THEOREM 8.14. *Assume that for every $t \in [0, T]$ there exists an $\varepsilon > 0$ such that*

$$\sup_{\substack{r,s \in [0,T] \\ r \neq s \\ |r-t|,|s-t|<\varepsilon}} \frac{|U(r) - U(s)|}{\sqrt{|r-s|}} < 2\sqrt{2}.$$

Then $U : [0, T] \to \mathbb{R}$ generates a simple curve.

For example, if U is continuously differentiable, then $\varepsilon > 0$ can be chosen to make the supremum in Theorem 8.14 arbitrarily small.

Now let $U : [0, T] \to \mathbb{R}$ be an arbitrary continuous function. In order to approximate the generated sets K_t numerically, one can approximate U by a piecewise constant function:

$$U_n(t) = u_k \quad \text{for all } t \in \left(\frac{kT}{n}, \frac{(k+1)T}{n}\right], k = 0, \dots, n-1.$$

U_n is not continuous, but we can still solve the corresponding Loewner equation, which is then of the form (6.9). Denote the transition mappings of the solution by $f_{n,s,t}$. Then

$$f_{n,0,T} = f_{n,0,T-T/n} \circ f_{n,T-T/n,T} = \cdots = f_{n,0,T/n} \circ f_{n,T/n,2T/n} \circ \cdots \circ f_{n,T-T/n,T}.$$

Each conformal map in the last expression is of the form $\sqrt{(z-u_k)^2 - 2\Delta t} + u_k$, with $\Delta t = T/n$, see Example 8.3. We will see in Lemma 8.25 that if U_n converges to U in L^1-norm, then $f_{n,0,T}$ converges to f_T with respect to locally uniform convergence.

We note that an explicit algorithm for approximating the set K_T directly can be found in [**106, 107**]. Figure 8.4 shows examples of driving functions and approximations of the corresponding generated set K_T.

Let $\mu \in \mathcal{P}^0(\mathbb{R})$ with $\text{Var}(\mu) = T$, such that F_μ is univalent. We know that we can embed F_μ into a normalized Loewner chain at $t = T$. It is clear that, in general, this embedding cannot be given by (8.2) for a continuous U, for example, if $F_\mu(\mathbb{H}) = \mathbb{H} \setminus ([0, i] \cup [1, 1+i])$. This mapping can be generated, for example, by first growing $[0, i]$ for $t \in [0, S]$, with some $S < T$, and then $[1, 1+i]$ for $t \in (S, T]$, i.e., with a function $U : [0, T] \to \mathbb{R}$ that is discontinuous at S. The following question was asked in [**150**] (for the case of the slit equation on the unit disk).

QUESTION 8.15. Let $\mu \in \mathcal{P}^0(\mathbb{R})$ such that F_μ is univalent. Is it true that F_μ can be generated by (8.2) with a measurable driving function U?

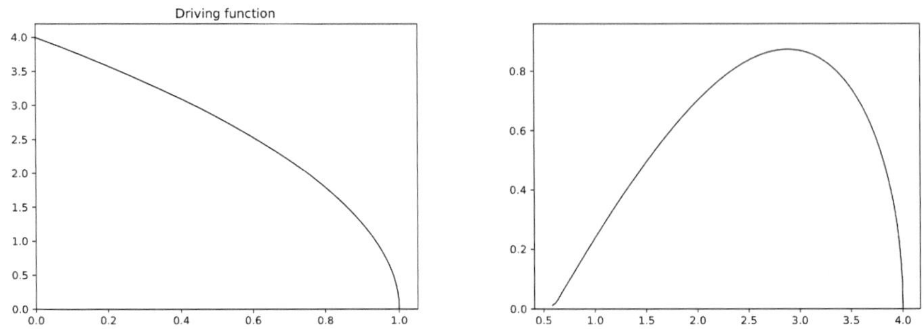

$U(t) = 4\sqrt{1-t}$, $t \in [0,1]$.

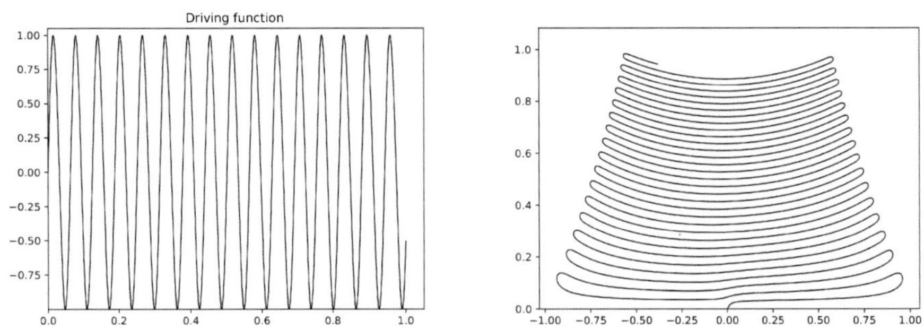

$U(t) = \sin(100 \cdot t)$, $t \in [0,1]$.

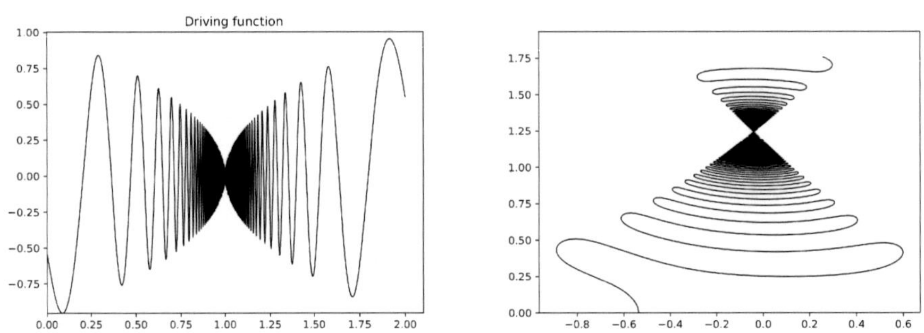

$U(t) = \sqrt{|1-t|}\sin\left(\frac{10}{1-t}\right)$, $t \in [0,2]$.

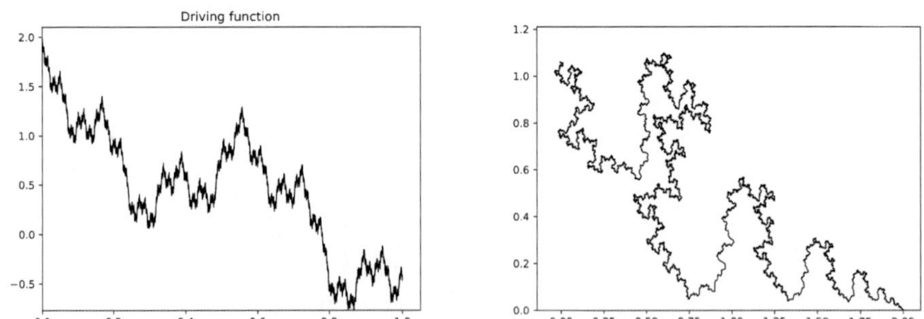

$U(t) = \sum_{k=0}^{20} \cos((3.3)^k \pi t)/2^k$, $t \in [0,1]$.

FIGURE 8.4. Numerical approximations of K_T for some driving functions.

8.3. Further properties of slit mappings

If the F-transform of a probability measure μ is a univalent slit mapping, then μ has some special properties [**95**].

THEOREM 8.16. *Let μ be a probability measure on $\mathbb{R}$ such that F_μ is univalent and maps $\mathbb{H}$ conformally onto $\mathbb{H}\setminus\gamma$, where γ is a slit starting at $C\in\mathbb{R}\setminus\{0\}$. Then μ has the following properties:*

(a) $\operatorname{supp}\mu = \{x_0\}\cup[a,b]$, *where μ has a continuous density $d(x)$ on the compact interval $[a,b]$ and an atom at some $x_0\in\mathbb{R}\setminus[a,b]$. Furthermore, $d(a)=d(b)=0$ and $d(x)>0$ in (a,b).*

(b) *The Hilbert transform $\mathcal{H}_\mu$ is defined and continuous on $\mathbb{R}\setminus\{x_0\}$ with $\mathcal{H}_\mu(a)=\mathcal{H}_\mu(b)=\frac{1}{\pi C}$.*

(c) *There exists a decreasing homeomorphism $h:[a,b]\to[a,b]$ with*

$$d(h(x))=d(x) \qquad \text{and} \qquad \mathcal{H}_\mu(h(x))=\mathcal{H}_\mu(x)$$

for all $x\in[a,b]$.

PROOF. As the domain $\mathbb{H}\setminus\gamma$ has a locally connected boundary, the mapping F_μ can be extended continuously to $\overline{\mathbb{H}}$, see Theorem B.1.

There exists an interval $[a,b]$ such that $F_\mu([a,b])=\gamma$, and there exists a unique $u\in(a,b)$ such that $F_\mu(u)$ is the tip of the slit. All points $[a,u]$ correspond to the left side, all points $[u,b]$ to the right side of γ. (This orientation follows from the behavior of $F_\mu(x)$ as $x\to\pm\infty$.) Hence, there exists a unique homeomorphism $h:[a,b]\to[a,b]$ with $h(u)=u$ and $h[a,u]=[u,b]$, such that $F_\mu(h(x))=F_\mu(x)$ for all $x\in[a,b]$.

Furthermore, F_μ has exactly one zero $x_0\in\mathbb{R}\setminus[a,b]$ on $\mathbb{R}$, as the slit does not start at 0. As $C=F_\mu(a)=F_\mu(b)$, we have $x_0<a$ if and only if $C>0$.

It follows from the Stieltjes–Perron inversion formula that $\operatorname{supp}\mu=\{x_0\}\cup[a,b]$, and that μ is absolutely continuous on $[a,b]$. Its density $d(x)$ satisfies

$$d(x)=\lim_{\varepsilon\to 0}-\frac{1}{\pi}\operatorname{Im}(1/F_\mu(x+i\varepsilon))=-\frac{1}{\pi}\operatorname{Im}(1/F_\mu(x)).$$

Hence, $d(h(x))=d(x)$ for all $x\in[a,b]$, $d(x)>0$ on (a,b), and $d(a)=d(b)=0$.

Let $\lambda=\mu(\{x_0\})$. We have

$$\frac{1}{\pi}\operatorname{Re}(1/F_\mu(x))=\hat{\mathcal{H}}_\mu(x)=\hat{\mathcal{H}}_d(x)+\frac{\lambda}{\pi(x-x_0)}=\mathcal{H}_d(x)+\frac{\lambda}{\pi(x-x_0)}=\mathcal{H}_\mu(x)$$

for every $x\in\mathbb{R}\setminus\{x_0\}$, due to Exercise 3.9.4. Here, $\mathcal{H}_d$ and $\hat{\mathcal{H}}_d$ are defined by replacing $\mu(dt)$ by $d(t)dt$ in the integration, and formally, we apply Exercise 3.9.4 to the probability measure defined by the density $d(t)/(1-\lambda)$.

Thus, $\mathcal{H}_\mu(x)$ is continuous on $\mathbb{R}\setminus\{x_0\}$, $\mathcal{H}_\mu(a)=\mathcal{H}_\mu(b)=\frac{1}{\pi C}$, and $\mathcal{H}_\mu(h(x))=\mathcal{H}_\mu(x)$ on $[a,b]$. □

The case of a slit starting at 0 is quite similar.

THEOREM 8.17. *Let μ be a probability measure on $\mathbb{R}$ such that F_μ is univalent and maps $\mathbb{H}$ conformally onto $\mathbb{H}\setminus\gamma$, where γ is a slit starting at $C=0$. Then μ has the following properties:*

(a) $\operatorname{supp}\mu=[a,b]$, *where μ has a continuous density $d(x)>0$ on (a,b).*

(b) $\mathcal{H}_\mu$ *is defined and continuous on* $\mathbb{R} \setminus \{a, b\}$, *with*

$$\lim_{x \searrow a} |\mathcal{H}_\mu(x)| = \lim_{x \nearrow b} |\mathcal{H}_\mu(x)| = \infty \qquad \text{or} \qquad \lim_{x \searrow a} d(x) = \lim_{x \nearrow b} d(x) = \infty.$$

(c) *There exists a decreasing homeomorphism* $h : [a, b] \to [a, b]$ *with*

$$d(h(x)) = d(x) \qquad \text{and} \qquad \mathcal{H}_\mu(h(x)) = \mathcal{H}_\mu(x)$$

for all $x \in (a, b)$.

PROOF. The statements (a) and (c) are shown just as in the previous theorem. We can restrict μ to compact subintervals of (a, b) to use Exercise 3.9.4, and then show by approximation that $\frac{1}{\pi} \operatorname{Re}(1/F_\mu(x)) = \mathcal{H}_\mu(x)$ for $x \in \mathbb{R} \setminus \{a, b\}$. As $x \searrow a$ or $x \nearrow b$, we have $F_\mu(x) \to 0$, so $|G_\mu(x)| \to \infty$. Hence, $|\operatorname{Re}(G_\mu(x))| \to \infty$ or $|\operatorname{Im}(G_\mu(x))| \to \infty$. □

EXAMPLE 8.18. Let $\mu = W(0, 1)$ be the semicircle distribution. Then μ is symmetric, i.e., $\mu(B) = \mu(-B)$. Exercise 3.9.3 shows that G_μ extends continuously to $\mathbb{H} \cup \mathbb{R}$ with $\operatorname{Re} G_\mu(x) = \frac{x}{2}$ for $x \in [-2, 2]$,

$$\operatorname{Re} G_\mu(x) = \tfrac{x - \sqrt{x^2 - 4}}{2} \text{ for } x > 2, \text{ and } \operatorname{Re} G_\mu(x) = \tfrac{x + \sqrt{x^2 - 4}}{2} \text{ for } x < -2.$$

Exercise 3.9.4 implies $\mathcal{H}_\mu(x) = \frac{1}{\pi} \operatorname{Re} G_\mu(x)$ for all $x \in \mathbb{R}$, and we see that $\mathcal{H}_\mu(-x) = -\mathcal{H}_\mu(x)$ for all $x \in \mathbb{R}$. In contrast to the case of slit mappings, $\mathcal{H}_\mu$ is antisymmetric instead of being "welded".

Next, we can prove that there are more $\rhd$-hemigroup distributions than $\rhd$-infinitely divisible distributions, unlike the cases of classical, Boolean and free independence.

If μ is $\rhd$-infinitely divisible, then, by Theorem 5.22, F_μ can be embedded into an additive Loewner chain $(F_{\nu_t})_{t \geq 0}$ that is a semigroup. If μ has compact support, then also ν_t has compact support for all $t \geq 0$. If, in addition, μ has mean 0, then also ν_t has mean 0 for all $t \geq 0$.

THEOREM 8.19. *There exists a* $\rhd$*-hemigroup distribution* μ*, which is not* $\rhd$*-infinitely divisible.*

PROOF. Choose μ with mean 0 such that F_μ is univalent and $F_\mu(\mathbb{H}) = \mathbb{H} \setminus \gamma(0, 1]$ for a simple curve $\gamma : [0, 1] \to \overline{\mathbb{H}}$, with $\gamma(0) \in \mathbb{R}$ and $\gamma(0, 1] \subset \mathbb{H}$. This is possible for any such curve due to Theorem 6.19. Then μ is a $\rhd$-hemigroup distribution. Assume that γ is not a vertical line segment.

Let $(F_{\mu_t})_{t \geq 0}$ be the corresponding additive Loewner chain constructed in Section 8.1, with $F_{\mu_t}(z) = z - \frac{t}{z} + \dots$ at ∞. Then F_{μ_t} satisfies (8.2). Any additive Loewner chain generating F_μ and having this normalization is identical to $(F_{\mu_t})_{t \geq 0}$.

Now assume that μ is monotonically infinitely divisible and let F_μ be embedded into an additive Loewner chain $(F_{\nu_t})_{t \geq 0}$, which is a semigroup. Then each ν_t has mean 0 and we have $\int_{\mathbb{R}} x^2 \nu_t(dx) = ct$ for some $c > 0$. So $F_{\nu_t} = z - \frac{ct}{z} + \dots$ A time change yields $c = 1$ and then $F_{\nu_t} = F_{\mu_t}$ for all $t \geq 0$.

The Loewner chain $(F_{\nu_t})_{t \geq 0}$ satisfies (8.2), and as $(F_{\nu_t})_{t \geq 0}$ is a semigroup, the Herglotz vector field $\frac{1}{U(t) - z}$ does not depend on t, i.e., $U(t) \equiv u \in \mathbb{R}$. In other words, $\gamma[0, 1]$ must be a vertical line segment connecting u to some $u + iT$,

$T > 0$, which is a contradiction to our assumption. Hence, μ is not $\rhd$-infinitely divisible. $\square$

THEOREM 8.20. *There exist $\rhd$-infinitely divisible probability measures μ, ν such that $\mu \rhd \nu$ is not $\rhd$-infinitely divisible.*

PROOF. Let $\mu = Arc(0,1)$. Then F_μ maps $\mathbb{H}$ onto the complement of the vertical line segment $[0, \sqrt{2}i]$. The interval $[-\sqrt{2}, \sqrt{2}]$ is mapped onto $[0, \sqrt{2}i]$. Let $\nu = \delta_{-1} \rhd \mu \rhd \delta_{+1}$. Then F_ν maps $\mathbb{H}$ onto the complement of the vertical line segment from 1 to $1 + \sqrt{2}i$, and $F_\mu \circ F_\nu$ maps $\mathbb{H}$ onto the complement of $K = [0, \sqrt{2}i] \cup \gamma$, where γ is a simple curve in $\mathbb{H}$ that starts at $F_\mu(1) \in (0, \sqrt{2}i)$, and $\gamma \cap [0, \sqrt{2}i] = \{F_\mu(1)\}$.

As μ is generated by 8.2 with $U(t) \equiv 0$, and ν by $U(t) \equiv 1$, both μ and ν are $\rhd$-infinitely divisible. (Alternatively, since $Arc(0,1) = Arc(0,1/n)^{\rhd n}$, we have $\nu = \delta_{-1} \rhd Arc(0,1/n)^{\rhd n} \rhd \delta_{+1} = (\delta_{-1} \rhd Arc(0,1/n) \rhd \delta_{+1})^{\rhd n}$.)

Now assume that $\mu \rhd \nu$ is $\rhd$-infinitely divisible as well. Then there exists an additive Loewner chain $(F_{\mu_t})_{t\geq 0}$ that is a semigroup and satisfies $F_{\mu_1} = F_\mu \circ F_\nu$. For $t = 1/n$ small enough, $\mathbb{H} \setminus F_{\mu_t}(\mathbb{H})$ must be a subset of $[0, \sqrt{2}i]$. But then $F_{\mu_1} = F_{\mu_{1/n}}^{\circ n}$ also maps $\mathbb{H}$ onto $\mathbb{H} \setminus [0, Ti]$ for some $T > 0$, a contradiction. $\square$

8.4. SLE

In this section, we briefly describe a stochastic version of Loewner's slit equation, the Schramm-Loewner evolution (SLE), which has important applications in stochastic geometry and statistical physics. It was discovered by O. Schramm and his idea represented nothing less than a breakthrough in the research of conformal invariance.

Consider a random walk on the lattice $\{a + bi \mid a, b \in \mathbb{Z}\}$ that starts at $z = 0$, where at each step it moves to one of the 4 neighboring points, each with probability 1/4. If we stop such a random walk after N steps, we can define the corresponding *loop-erased random walk* (LERW) as the simple curve obtained by removing all loops in chronological order, see Figures 8.5 and 8.6.

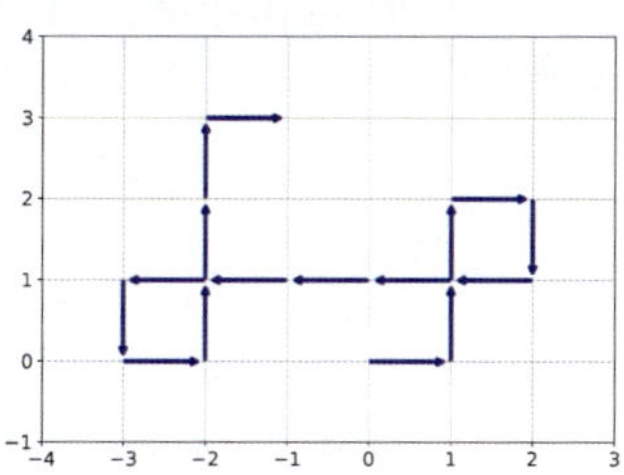

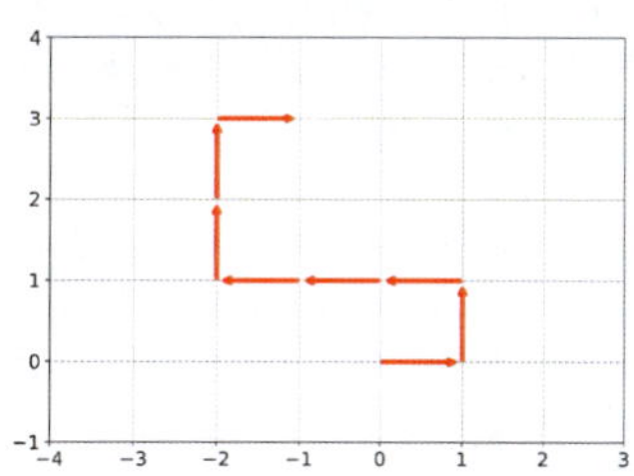

FIGURE 8.5. Random walk (left) and the corresponding loop-erased random walk (right).

For every $N \in \mathbb{N}$, let $S_N : [0,1] \to \mathbb{C}$ be the linear interpolation of a random walk with N steps. Then $S_N/\sqrt{N}$ converges to a two-dimensional Brownian motion as $N \to \infty$ (Donsker's theorem). Provided the limit exists, how can we describe the distribution of the limit of the loop-erased random walk? In contrast to the random walk $(M_n)_n$, its loop-erased subcurve $(M_{n_k})_k$ is far away from being Markovian. If we know the first N points, then $\{M_{n_k} \mid k > N\}$ must be disjoint from $\{M_{n_1}, \ldots, M_{n_N}\}$.

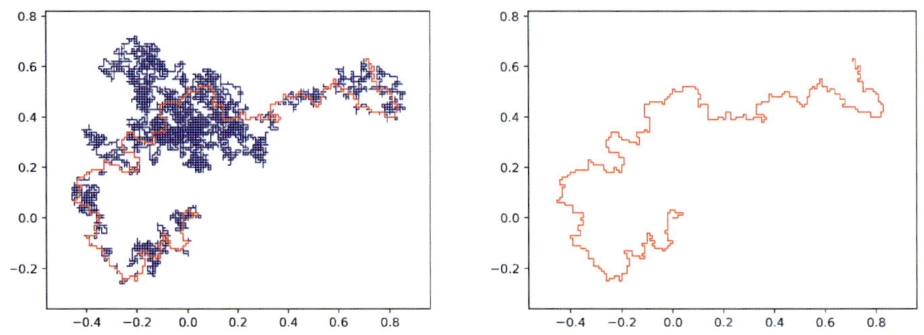

FIGURE 8.6. A simulation of a scaled random walk with $N = 10000$ steps and the corresponding loop-erased random walk.

O. Schramm had a brilliant idea to solve such problems, published in [**172**]. A two-dimensional random simple curve can be transformed by Loewner's differential equation into its real-valued driving function. If such a curve is conditioned to stay in the upper half-plane, we can use the slit-equation (8.2) to transform the random slit into a real-valued random function U. It turns out that the loop-erased random walk, as well as random curves from various other models, simply correspond to U being a scaled Brownian motion.

DEFINITION 8.21. For $\kappa > 0$, *SLE(κ)* (Schramm-Loewner evolution) is defined as the evolution described by the Loewner equation

$$\frac{\partial}{\partial t} f_t(z) = \frac{\partial}{\partial z} f_t(z) \cdot \frac{2}{\sqrt{\kappa} B_t - z}, \quad f_0(z) = z \in \mathbb{H},$$

where B_t is a Brownian motion.

REMARK 8.22. SLE, as described here, is also called *chordal SLE*. The factor 2 leads to the normalization $f_t(z) = z - \frac{2t}{z} + \ldots$ at ∞. This normalization is more useful when comparing chordal SLE to other versions of SLE, such as radial SLE in the unit disk, see [**173**].

The random conformal mappings f_t map $\mathbb{H}$ onto $\mathbb{H} \setminus K_t$, where $(K_t)_{t\geq 0}$ is a family of growing bounded random sets. The random growth process $(K_t)_{t\geq 0}$ has been shown to be the scaling limit of random curves from different models, depending on the value of κ:

- $SLE(2)$: loop-erased random walk,
- $SLE(3)$: critical Ising model,
- $SLE(4)$: harmonic explorer,
- $SLE(4)$: contour lines of the discrete Gaussian free field,
- $SLE(6)$: critical percolation,
- $SLE(8)$: uniform spanning tree.

These results are remarkable. The highly non-Markovian curves in the complex plane can be encoded by Loewner's differential equation via a Brownian motion.

The random sets K_t do not necessarily describe a random simple curve. If $\kappa \in [0,4]$, then for any $t > 0$, K_t is almost surely the image of a simple curve. If $\kappa \in (4,8)$, then K_t is almost surely the image of a curve that touches itself, plus the compact components of its complement. For $\kappa \geq 8$, K_t is almost surely the image of a space-filling curve; see Figures 8.7 and 8.8. We refer the reader to [**115**].

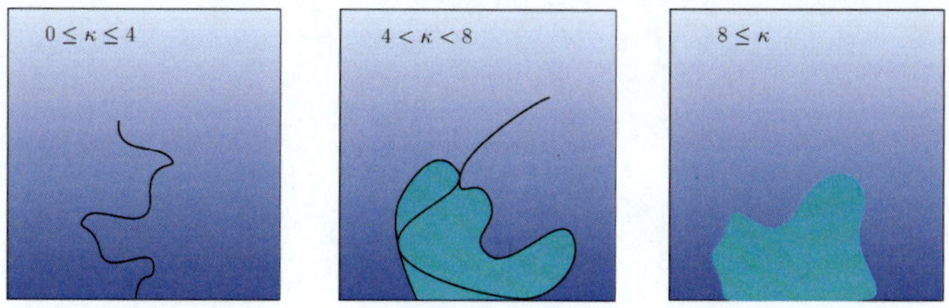

FIGURE 8.7. A sketch for the three cases of SLE curves.

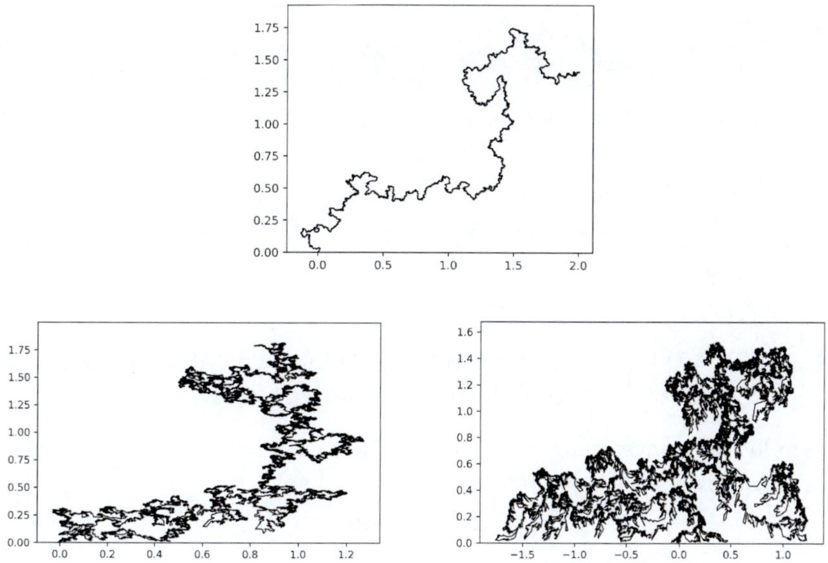

FIGURE 8.8. Simulations of SLE for the cases $\kappa = 2$ (top), $\kappa = 6$ (bottom, left), and $\kappa = 9$ (bottom, right).

8.5. Approximation by the slit equation

In this section, we regard the following type of Herglotz vector fields.

DEFINITION 8.23. Let $(\nu_t)_{t\geq 0}$ be a family of probability measures on $\mathbb{R}$ such that $t \mapsto H(z,t) = \int_{\mathbb{R}} \frac{\nu_t(du)}{u-z}$ is measurable for every $z \in \mathbb{H}$, and assume that there exists $M > 0$ such that

$$\operatorname{supp} \nu_t \subset [-M, M]$$

for all $t \geq 0$. We denote the set of all such functions by $\mathcal{H}_M$.

For $H \in \mathcal{H}_M$, we consider the Loewner equation (6.9), i.e.,

$$\frac{\partial}{\partial t} f_t(z) = \frac{\partial}{\partial z} f_t(z) \cdot H(z,t) \quad \text{for a.e. } t \geq 0,\ f_0(z) = z \in \mathbb{H}. \tag{8.9}$$

From Theorem 6.35, we know that the solution is an additive Loewner chain $(f_t)_{t\geq 0} = (F_{\mu_t})_{t\geq 0}$, where μ_t has mean 0 and variance t.

THEOREM 8.24. *Let $(f_t)_{t\geq 0}$ be the solution of equation (8.9). Then each f_t maps $\mathbb{H}$ conformally onto $\mathbb{H} \setminus K_t$ for a bounded set $K_t \subset \mathbb{H}$, and there exists a bound $C(t,M) > 0$ such that $\sup_{z\in K_t} |z| < C(t,M)$.*

Furthermore, there exists a bound $D(t,M) > 0$ such that

$$\operatorname{supp} \mu_t \subset [-D(t,M), D(t,M)].$$

PROOF. For $T > 0$, we consider the inverse Loewner equation (6.15):

$$\frac{d}{dt} w(t) = -H(w(t),t), \quad w(0) = M + 1 \in \mathbb{R}.$$

As $\operatorname{supp} \nu_t \subset [-M, M]$, the solution is strictly increasing and exists for all $t \geq 0$. Since

$$\frac{d}{dt} w(t) \leq \frac{1}{w(t) - M},$$

we obtain an upper bound $A(T,M)$ for $w(T)$ that only depends on M and T. We have $w(T) = f_T^{-1}(M+1) \leq A(T,M)$. So f_T extends conformally onto $\mathbb{C} \setminus [-A(T,M), A(T,M)]$.

The function $f_t(A(t,M)z)/A(t,M)$ extends to a univalent function in $\mathbb{C} \setminus \overline{\mathbb{D}}$ with expansion $f_t(A(t,M)z)/A(t,M) = z - t/(A(t,M)^2 z) + \ldots$ at ∞. Corollary 3.36 implies $K_t \subset \{z \in \mathbb{C} \,|\, |z| \leq 2A(t,M)\}$.

The existence of the uniform bound $D(t,M)$ follows from the bound $A(t,M)$ and the Stieltjes–Perron inversion formula. □

The following convergence result is standard in Loewner theory, see e.g., [**79**, Lemma 4.12] for a slightly different setting.

LEMMA 8.25. *Fix $T > 0$. For every $n \in \mathbb{N}$, let $H_n(z,t) \in \mathcal{H}_M$. Assume that there exists $H(z,t) \in \mathcal{H}_M$ such that*

$$\int_0^t H_n(z,s)ds \to \int_0^t H(z,s)ds$$

for every $t \in [0,T]$ locally uniformly in $\mathbb{H}$ as $n \to \infty$. Let $f_{n,t}$ and f_t be the solutions to (8.9) for the Herglotz vector fields $H_n(z,t)$ and $H(z,t)$ respectively. Then $f_{n,t} \to f_t$ for every $t \in [0,T]$ locally uniformly in $\mathbb{H}$.

PROOF. Let $\nu_{n,t}$ and ν_t be the measures associated with $H_n(z,t)$ and $H(z,t)$ respectively.

The set $\{\int_{\mathbb{R}} \frac{\nu(du)}{u-z} \,|\, \nu \in \mathcal{P}(\mathbb{R}) \text{ with } \operatorname{supp} \nu \subset [-M,M]\}$ is a normal family. Thus, if $G \in \mathcal{H}_M$ and $K \subset \mathbb{H}$ is a compact set, then there exists $L(K) > 0$ such that

$$|G(z,t) - G(w,t)| \leq L(K)|z - w| \quad \text{for all } z, w \in K \text{ and all } t \in [0,T].$$

We consider $g_{n,t} := f_{n,t}^{-1}$, $g_t := f_t^{-1}$. These functions satisfy

$$\frac{\partial g_{n,t}(z)}{\partial t} = \int_{\mathbb{R}} \frac{\nu_{n,t}(du)}{g_{n,t}(z) - u}, \quad \frac{\partial g_t(z)}{\partial t} = \int_{\mathbb{R}} \frac{\nu_t(du)}{g_t(z) - u} \quad \text{for a.e. } t \geq 0,$$

and we have

$$g_{n,t}(z) = z - \int_0^t H_n(g_{n,s}(z), s) ds, \quad g_t(z) = z - \int_0^t H(g_s(z), s) ds.$$

Let $K \subset \mathbb{H}$ be a compact set on which all $g_{n,t}$ and g_t are defined. Due to Theorem 8.24, there exists a second compact set $K' \subset \mathbb{H}$, $K \subset K'$, such that $g_{n,t}(z), g_t(z) \in K'$ for all $z \in K$, $n \in \mathbb{N}$, and $t \in [0, T]$.

We know that $\int_0^t H_n(s, z) ds$ converges uniformly on K' to $\int_0^t H(z, s) ds$ for all $t \in [0, T]$. Now fix $t \in [0, T]$. For $z \in K$, we have

$$|g_{n,t}(z) - g_t(z)| \leq \left| \int_0^t H_n(g_{n,s}(z), s) - H_n(g_s(z), s) ds \right|$$
$$+ \left| \int_0^t H_n(g_s(z), s) - H(g_s(z), s) ds \right| \leq L(K') \int_0^t |g_{n,s}(z) - g_s(z)| ds + \varepsilon_n,$$

for a sequence ε_n converging to 0. Gronwall's lemma implies that $g_{n,t} \to g_t$ uniformly on K. Hence, also $f_{n,t} \to f_t$ locally uniformly in $\mathbb{H}$. □

LEMMA 8.26. *Let $H(z, t) = \int_{\mathbb{R}} \frac{\nu_t(du)}{u - z} \in \mathcal{H}_M$, and let $(f_t)_{t \geq 0}$ be the corresponding solution to (8.9). Fix $T > 0$. Then there exists a sequence $U_n : [0, T] \to [-M, M]$ of continuous driving functions such that the corresponding solutions $(f_{n,t})_{t \geq 0}$ to (8.2) converge locally uniformly to f_t, for every $t \in [0, T]$, as $n \to \infty$.*

PROOF. Step 1: Assume that $H(z, t) = \frac{1}{U(t) - z}$ for a piecewise continuous driving function $U : [0, T] \to [-M, M]$. We can then approximate $H(z, t)$ by a sequence $H_n(z, t) = \frac{1}{U_n(t) - z}$ with continuous driving functions $U_n : [0, T] \to [-M, M]$ in the sense of Lemma 8.25.

Step 2: It suffices to approximate a piecewise constant $t \mapsto H(\cdot, t)$, and thus, a constant function $H(z, t) = \int_{\mathbb{R}} \frac{\nu(du)}{u - z}$ by piecewise continuous driving functions. Since we can approximate ν by a convex combination of point measures, we may assume that H has the form $H(z, t) = \sum_{k=1}^N \frac{\lambda_k}{u_k - z}$, where $u_1, \ldots, u_N \in [-M, M]$, and $\lambda_1, \ldots, \lambda_N \in [0, 1]$ with $\sum_{k=1}^N \lambda_k = 1$.

This Herglotz vector field can be approximated by a piecewise continuous driving function as follows. We choose $m \in \mathbb{N}$ and divide the interval $[0, T]$ into m intervals $I_1 := [0, \frac{T}{m}]$, $I_2 := (\frac{T}{m}, \frac{2T}{m}]$, $\ldots$, $I_m := (T - \frac{T}{m}, T]$. We define the driving function U_m on I_1 as follows:

$$U_m(t) = u_1 \quad \text{on} \quad [0, \lambda_1 \cdot T/m],$$
$$U_m(t) = u_2 \quad \text{on} \quad (\lambda_1 \cdot T/m, (\lambda_1 + \lambda_2) \cdot T/m], \ldots,$$
$$U_m(t) = u_N \quad \text{on} \quad ((\lambda_1 + \cdots + \lambda_{N-1}) \cdot T/m, T/m].$$

We now repeat this construction for $I_2, \ldots, I_m$. Define $H_m(z, t) = \frac{1}{U_m(t) - z}$. Then $H_m(z, t)$ approximates $H(z, t)$ in the sense of Lemma 8.25. □

8.6. The origin of Loewner chains

Loewner introduced Loewner chains in [**123**] in order to prove the Bieberbach conjecture for $n = 3$. Due to Theorem 7.32, it is sufficient to prove it only for slit mappings in S.

So let $f \in S$ be a slit mapping with $f(\mathbb{D}) = \mathbb{C} \setminus \Gamma$, where Γ is the image of a continuous and injective function $\gamma : [0, \infty) \to \mathbb{C}$. For every $t > 0$, the domain $\mathbb{C} \setminus \gamma[t, \infty)$ is simply connected, and we find a conformal mapping

$$f_t : \mathbb{D} \to \mathbb{C} \setminus \gamma[t, \infty) \quad \text{with} \quad f_t(0) = 0,\ f_t'(0) > 0.$$

LEMMA 8.27. *The function $\alpha(t) := f_t'(0)$ has the following properties:*

a) *$\alpha(t)$ is strictly increasing.*
b) *$\alpha(t) \to \infty$ as $t \to \infty$.*
c) *$t \mapsto \alpha(t)$ is continuous.*

PROOF.

a) The function $\varphi_{s,t} := f_t^{-1} \circ f_s$, $0 \le s < t$, is a conformal mapping from $\mathbb{D}$ onto a proper subdomain of $\mathbb{D}$, and $\varphi_{s,t}(0) = 0$. The Schwarz lemma implies that $\varphi_{s,t}'(0) = f_s'(0)/f_t'(0) = \alpha(s)/\alpha(t) < 1$, i.e., $\alpha(s) < \alpha(t)$.
b) Let $M > 0$. Then $M \cdot \mathbb{D} \subset f_t(\mathbb{D})$ for all $t \ge T$ and some $T > 0$. Hence, for $t \ge T$, $f_t^{-1}(Mz)$ maps $\mathbb{D}$ into $\mathbb{D}$, and the Schwarz lemma implies $M/\alpha(t) < 1$, i.e., $\alpha(t) > M$.
c) We have $\mathbb{C} \setminus \gamma[s, \infty) \to \mathbb{C} \setminus \gamma[t, \infty)$ with respect to kernel convergence as $s \to t$. The statement now follows from Theorem 7.29.

□

Consequently, Γ can be parametrized such that $f_t'(0) = e^t$. For $t \in [0, \infty)$, we define the mapping φ_t by $\varphi_t : \mathbb{D} \to \mathbb{D}$, $\varphi_t(z) = f_t^{-1}(f(z))$. It follows that φ_t has a power series expansion of the form

$$\varphi_t(z) = e^{-t} \left(z + \sum_{n=2}^{\infty} a_n(t) z^n \right). \tag{8.10}$$

LEMMA 8.28. *With the notation from above, we have*

$$\lim_{t \to \infty} e^t \varphi_t(z) = f(z)$$

locally uniformly in $\mathbb{D}$.

PROOF. As $e^{-t} f_t \in S$, the growth theorem, Theorem A.3, implies that

$$\frac{e^t |z|}{(1 + |z|)^2} \le |f_t(z)| \le \frac{e^t |z|}{(1 - |z|)^2}$$

for all $z \in \mathbb{D}$. For an arbitrary $w \in \mathbb{C}$, $w \neq 0$, and a sufficiently large t, we can consider $z = f_t^{-1}(w)$, and we obtain

$$(1 - |f_t^{-1}(w)|)^2 \le e^t \frac{|f_t^{-1}(w)|}{|w|} \le (1 + |f_t^{-1}(w)|)^2.$$

In particular,

$$|f_t^{-1}(w)| \le 4|w|e^{-t}.$$

This implies that $f_t^{-1}(w) \to 0$ locally uniformly, and that $e^t |f_t^{-1}(w)/w| \to 1$ locally uniformly as $t \to \infty$.

We conclude that $(e^t f_t^{-1}(w)/w)_{t\geq 0}$ is a normal family (defined on an arbitrarily large disk for sufficiently large t). Let $(e^{t_n} f_{t_n}^{-1}(w)/w)_n$ be a convergent subsequence, where $t_n \to \infty$, with limit G. Then $|G(w)| \equiv 1$, so G is a constant function. Because of $G(0) = \lim_{n\to\infty} e^{t_n}(f_{t_n}^{-1})'(0) = 1$, we have $G(w) \equiv 1$. This limit is independent of the choice of the subsequence, and consequently, $e^t f_t^{-1}(w)/w$ converges locally uniformly to the constant function 1 as $t \to \infty$. This implies

$$e^t \varphi_t(z) = e^t f_t^{-1}(f(z)) = e^t f_t^{-1}(f(z))/f(z) \cdot f(z) \to f(z)$$

locally uniformly in $\mathbb{D}$. □

$(f_t)_{t\geq 0}$ is a normalized increasing Loewner chain and f_t satisfies (6.3) with a minus sign, i.e.,

$$\frac{\partial}{\partial t} f_t(z) = z \frac{\partial}{\partial z} f_t(z) p(z,t) \quad \text{for almost all } t \geq 0 \text{ and all } z \in \mathbb{D}.$$

For this special case of removing a slit, Loewner proved a more precise differential equation.

THEOREM 8.29. *For every $z \in \mathbb{D}$, $t \mapsto \varphi_t(z)$ satisfies the differential equation*

$$\text{(8.11)} \qquad \frac{\partial \varphi_t(z)}{\partial t} = -\varphi_t(z) \frac{\kappa(t) + \varphi_t(z)}{\kappa(t) - \varphi_t(z)} \qquad \text{for all } t \geq 0, \quad \varphi_0(z) = z,$$

where $\kappa : [0,\infty) \to \partial\mathbb{D}$ is a continuous function, the driving function of Γ.

PROOF. Consider the mapping $\varphi_{s,t}(z) = f_t^{-1}(f_s(z))$, $s \leq t$. It maps $\mathbb{D}$ conformally onto $\mathbb{D} \setminus J_{s,t}$, where $J_{s,t} \subset \mathbb{D}$ is a curve. We have $\varphi_{s,t}(0) = 0$ and $\varphi_{s,t}'(0) = e^{s-t}$. Due to Theorem B.1, $\varphi_{s,t}$ has a continuous extension to $\overline{\mathbb{D}}$, and we let $C_{s,t} \subset \partial\mathbb{D}$ be the circular arc that is mapped onto $J_{s,t}$.

For each $\tau \geq 0$, we let $\kappa(\tau) \in \partial\mathbb{D}$ be the point that is mapped onto $\gamma(\tau)$ by f_τ, which is the endpoint of the slit corresponding to f_τ. Then $\kappa(s)$ is an interior point of $C_{s,t}$. Similar to the case of the upper half-plane, one can now show that $C_{s,t}$ shrinks to $\{\kappa(s)\}$ as $t \searrow s$ for fixed s, and to $\{\kappa(t)\}$ as $s \nearrow t$ for fixed t, and that $t \mapsto \kappa(t)$ is continuous; see [**62**, Theorem 3.3].

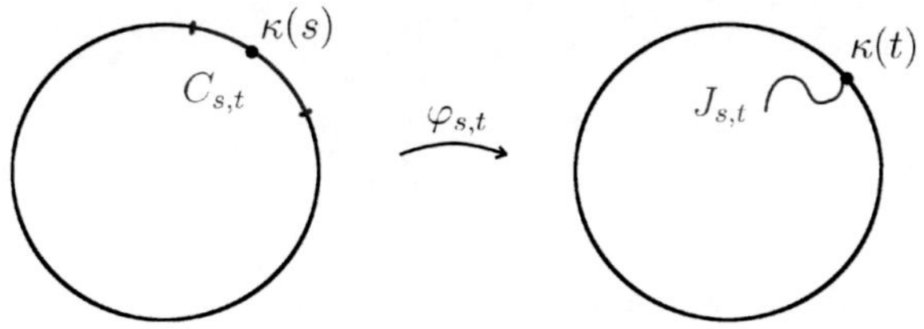

The Loewner equation can now be derived as follows. We consider the function

$$L_{s,t}(z) = \log\left(\frac{\varphi_{s,t}(z)}{z}\right),$$

where we choose the branch of the logarithm such that $L_{s,t}(0) = s - t$. We can extend $L_{s,t}$ continuously to $\overline{\mathbb{D}}$, with $\operatorname{Re} L_{s,t}(z) = 0$ for all $z \in \partial\mathbb{D}$ that do not belong to the interior of the circular arc $C_{s,t}$, and $\operatorname{Re} L_{s,t}(z) < 0$ for all z in the interior of $C_{s,t}$.

By Theorem 3.10, the function

$$z \mapsto \frac{1}{2\pi}\int_0^{2\pi} \frac{e^{i\tau}+z}{e^{i\tau}-z}\operatorname{Re} L_{s,t}(e^{i\tau})\,d\tau$$

is holomorphic on $\mathbb{D}$, its real part extends continuously to $\overline{\mathbb{D}}$, and coincides with $\operatorname{Re} L_{s,t}$ on $\partial\mathbb{D}$. The maximum principle implies that the difference between this function and $L_{s,t}$ is equal to 0, i.e.,

$$L_{s,t}(z) = \frac{1}{2\pi}\int_0^{2\pi} \frac{e^{i\tau}+z}{e^{i\tau}-z}\operatorname{Re} L_{s,t}(e^{i\tau})\,d\tau = \frac{1}{2\pi}\int_{e^{i\tau}\in C_{s,t}} \frac{e^{i\tau}+z}{e^{i\tau}-z}\operatorname{Re} L_{s,t}(e^{i\tau})\,d\tau.$$

In particular,

$$s-t = L_{s,t}(0) = \frac{1}{2\pi}\int_{e^{i\tau}\in C_{s,t}} \operatorname{Re} L_{s,t}(e^{i\tau})\,d\tau. \tag{8.12}$$

Because of $\varphi_{s,t}\circ\varphi_s = \varphi_t$, we have

$$\log\left(\frac{\varphi_t(z)}{\varphi_s(z)}\right) = \frac{1}{2\pi}\int_{e^{i\tau}\in C_{s,t}} \frac{e^{i\tau}+\varphi_s(z)}{e^{i\tau}-\varphi_s(z)}\operatorname{Re} L_{s,t}(e^{i\tau})\,d\tau.$$

The mean value theorem for integrals, applied to the real and imaginary part separately, implies that we can write

$$\log\left(\frac{\varphi_t(z)}{\varphi_s(z)}\right) = \frac{1}{2\pi}\left(\operatorname{Re}\frac{e^{i\tau_1}+\varphi_s(z)}{e^{i\tau_1}-\varphi_s(z)} + i\operatorname{Im}\frac{e^{i\tau_2}+\varphi_s(z)}{e^{i\tau_2}-\varphi_s(z)}\right)\int_{e^{i\tau}\in C_{s,t}} \operatorname{Re} L_{s,t}(e^{i\tau})\,d\tau,$$

where $e^{i\tau_1}, e^{i\tau_2} \in C_{s,t}$ depend on z.

For $z=0$, $\varphi_t(0)$ is obviously differentiable with $\frac{\partial}{\partial t}\varphi_t(0) = 0$. For $z\neq 0$, we can locally write $\log\left(\frac{\varphi_t(z)}{\varphi_s(z)}\right) = \log(\varphi_t(z)) - \log(\varphi_s(z))$. Now we divide by $t-s$ and take the limit $t\searrow s$. Together with (8.12), we obtain the right derivative

$$\lim_{t\searrow s}\frac{\log(\varphi_t(z))-\log(\varphi_s(z))}{t-s} = -\frac{\kappa(s)+\varphi_s(z)}{\kappa(s)-\varphi_s(z)},$$

as $C_{s,t}$ shrinks to $\{\kappa(s)\}$. The case $s\nearrow t$ can be treated similarly. From $\varphi_t(z) = e^{\log(\varphi_t(z))}$, we obtain

$$\frac{\partial\varphi_t(z)}{\partial t} = -\varphi_t(z)\frac{\kappa(t)+\varphi_t(z)}{\kappa(t)-\varphi_t(z)}.$$

□

REMARK 8.30. It follows that

$$\frac{\partial}{\partial t}f_t(z) = z\frac{\partial}{\partial z}f_t(z)\frac{\kappa(t)+z}{\kappa(t)-z} \qquad \text{for all } t\geq 0 \text{ and } z\in\mathbb{D}.$$

THEOREM 8.31. *Let* $f(z) = z+\sum_{n=2}^{\infty}a_nz^n \in S$. *Then* $|a_3|\leq 3$.

PROOF. Because of Theorem 7.32, it suffices to show the inequality for the case where f is a slit mapping. Furthermore, we can rotate the function such that the coefficient a_3 becomes a non-negative real number, and does not change its absolute value (see Example 3.33). So, in what follows, we assume $a_3\in[0,\infty)$ and we will show that $\operatorname{Re}(a_3)\leq 3$.

With the notation from above, we have $a_n = \lim_{t\to\infty} a_n(t)$ for all $n \geq 2$ due to Lemma 8.28. By using the expansion (8.10), and writing $\dot{x}(t)$ for $\frac{d}{dt}x(t)$, we obtain that $\dot{\varphi}_t$ satisfies[1]

$$\dot{\varphi}_t(z) = -e^{-t}\left(z + \sum_{n=2}^{\infty} a_n(t) z^n\right) + e^{-t}\left(\sum_{n=2}^{\infty} \dot{a}_n(t) z^n\right).$$

Equation (8.11) yields $\dot{\varphi}_t(z) = -\varphi_t(z) - 2\kappa(t)^{-1}\varphi_t(z)^2 - 2\kappa(t)^{-2}\varphi_t(z)^3 - \ldots$ and comparing coefficients leads to

$$\dot{a}_2(t) = -2e^{-t}\kappa(t)^{-1}, \tag{8.13}$$

$$\dot{a}_3(t) = -2e^{-2t}\kappa(t)^{-2} - 4e^{-t}\kappa(t)^{-1}a_2(t). \tag{8.14}$$

From (8.13), we get

$$a_2 = \int_0^\infty \dot{a}_2(t)\,dt = -2\int_0^\infty e^{-t}\kappa(t)^{-1}\,dt,$$

and equation (8.14) becomes

$$\dot{a}_3(t) = -2e^{-2t}\kappa(t)^{-2} + 2a_2(t)\dot{a}_2(t).$$

Thus, we obtain

$$\begin{aligned} a_3 = \int_0^\infty \dot{a}_3(t)\,dt &= \int_0^\infty -2e^{-2t}\kappa(t)^{-2} + 2a_2(t)\dot{a}_2(t)\,dt \\ &= \int_0^\infty -2e^{-2t}\kappa(t)^{-2}\,dt + 4\left(\int_0^\infty e^{-t}\kappa(t)^{-1}\,dt\right)^2. \end{aligned}$$

By writing $\kappa(t)^{-1} = e^{i\alpha(t)}$, we get

$$\begin{aligned} \operatorname{Re}(a_3) = &\int_0^\infty -2e^{-2t}(\cos^2(\alpha(t)) - \sin^2(\alpha(t)))\,dt \\ &+4\operatorname{Re}\left(\left(\int_0^\infty e^{-t}(\cos(\alpha(t) + i\sin(\alpha(t)))\,dt\right)^2\right), \end{aligned}$$

and with $\operatorname{Re}(z^2) \leq (\operatorname{Re} z)^2$, we obtain

$$\operatorname{Re}(a_3) \leq \int_0^\infty 2e^{-2t}(1 - 2\cos^2(\alpha(t)))\,dt + 4\left(\int_0^\infty e^{-t}\cos(\alpha(t))\,dt\right)^2.$$

Finally, we use the Cauchy–Schwarz inequality for square-integrable functions:

$$\begin{aligned} &\operatorname{Re}(a_3) \leq 1 - 4\int_0^\infty e^{-2t}\cos^2(\alpha(t))\,dt + 4\left(\int_0^\infty e^{-t/2}\cdot e^{-t/2}\cos(\alpha(t))\,dt\right)^2 \\ &\leq 1 - 4\int_0^\infty e^{-2t}\cos^2(\alpha(t))\,dt + 4\int_0^\infty e^{-t}\,dt \cdot \int_0^\infty e^{-t}\cos^2(\alpha(t))\,dt \\ &= 1 + 4\int_0^\infty (e^{-t} - e^{-2t})\cos^2(\alpha(t))\,dt \leq 1 + 4\int_0^\infty (e^{-t} - e^{-2t})\,dt = 3. \end{aligned}$$

□

[1] Here we can exchange differentiation and summation for the following reason: The function $\frac{\varphi_{t+h}-\varphi_t}{h}$ converges pointwise as $h \to 0$ due to Theorem 8.29. The proof shows that these functions are locally uniformly bounded, and Vitali's theorem implies that they converge in fact locally uniformly. The nth coefficient of $\frac{\varphi_{t+h}-\varphi_t}{h}$ can be represented as an integral (see (A.1)), and we see that $a_n(t)$ is differentiable and that its derivative is equal to the nth coefficient of the right-hand side of (8.11).

The Bieberbach conjecture had a significant impact on the development of complex analysis, and many partial results were obtained before its complete proof. These include:

- 1921: The case when f is starlike (Nevanlinna),
- 1923: The case $n = 3$ (Loewner),
- 1925: $|a_n| < en$ for all $n \geq 2$ (Littlewood),
- 1955: The case $n = 4$ (Garabedian and Schiffer),

and many more; see [**109**] for the historical development. Loewner proved his result in [**123**]. Note that the title of [**123**] ends with a "I", expressing hope that the new method might soon solve the complete conjecture, but there has never been a subsequent "II". The Bieberbach conjecture was finally proven in 1984 by Louis de Branges, and indeed, this final proof also uses Loewner's method.

THEOREM 8.32 ([**58**]). *Let* $f(z) = z + \sum_{n=2}^{\infty} a_n z^n \in S$. *Then* $|a_n| \leq n$ *for all* $n \geq 2$.

More on Loewner theory applied to the class S can be found in the books [**151**], [**62**], [**80**].

8.7. Exercises

E 8.7.1. The theory of Schwarz–Christoffel mappings allows us to construct a concrete conformal mapping from $\mathbb{H}$ onto $\mathbb{H}$ minus a straight line segment. Let $\alpha \in (0,1)$, and consider the function

$$G_\alpha(z) = (z+\alpha-1)^\alpha (z+\alpha)^{1-\alpha}$$

on $\mathbb{H}$. G_α maps $\mathbb{H}$ conformally onto $\mathbb{H} \setminus [0, K \cdot e^{i\alpha\pi}]$ with $K = (1-\alpha)^{1-\alpha}\alpha^\alpha$. The points $-\alpha$ and $1-\alpha$ are mapped onto 0, and $1-2\alpha$ is mapped onto the tip $K \cdot e^{i\alpha\pi}$. (See, e.g., [**72**, Section XI]. We note that $G'_\alpha(z)$ is given by $G'_\alpha(z) = (z+\alpha-1)^{\alpha-1}(z+\alpha)^{-\alpha}(z+2\alpha-1)$.) Compute the half-plane capacity of the hull $H = [0, e^{i\alpha\pi}]$.

E 8.7.2. Let $c \in \mathbb{R}$. Use Exercise 8.7.1 to find the solution of the Loewner equation

$$\frac{\partial}{\partial t} f_t(z) = \frac{\partial}{\partial z} f_t(z) \frac{1}{c\sqrt{t} - z}, \quad t \geq 0, z \in \mathbb{H}, f_0(z) = z.$$

E 8.7.3. Let $U : [0,1] \to \mathbb{R}$ be continuous such that the slit equation (8.2) generates a simple curve. Assume that this curve is contained within a sector, i.e., within the set

$$\{z \in \mathbb{H} \,|\, \alpha < \arg(z - U(0)) < \pi - \alpha\}$$

for some $\alpha \in (0, \pi/2)$. Prove that

$$\sup_{t \in (0,1]} \frac{|U(t) - U(0)|}{\sqrt{t}} < \infty.$$

8.8. Solutions

S 8.8.1. We first compute the half-plane capacity c of the hull $[0, K \cdot e^{i\alpha\pi}]$ by expanding $G_\alpha(z) = z - \frac{c}{z} + \dots$. Using the binomial series and $G_\alpha(1/z) = \frac{1}{z}(1+(\alpha-1)z)^\alpha(1+\alpha z)^{1-\alpha}$ gives

$$\begin{aligned} G_\alpha(1/z) &= \frac{1}{z}(1+\alpha(\alpha-1)z+\alpha(\alpha-1)^3z^2/2+\dots) \\ &\qquad \times (1+\alpha(1-\alpha)z+(\alpha-1)\alpha^3z^2/2+\dots) \\ &= \frac{1}{z} - \frac{1}{2}\alpha(1-\alpha)z+\dots. \end{aligned}$$

Hence, $c = \frac{1}{2}(1-\alpha)\alpha$.

The mapping $G_\alpha(K\cdot z)/K$ has hydrodynamic parametrization ($G_\alpha(K\cdot z)/K = z - \frac{c/K^2}{z} + \dots$) and maps $\mathbb{H}$ conformally onto $\mathbb{H}\backslash H$. We conclude that the half-plane capacity of H is given by

$$c/K^2 = \frac{(1-\alpha)\alpha}{2(1-\alpha)^{2-2\alpha}\alpha^{2\alpha}} = \frac{1}{2}(1-\alpha)^{2\alpha-1}\alpha^{1-2\alpha}.$$

S 8.8.2. The function $U(t) = c\sqrt{t}$ is invariant under the scaling $U(C^2\cdot t)/C$, $C > 0$, see Example 8.6. We can expect that U generates a straight line segment. So we scale G_α from Exercise 8.7.1 to produce a Loewner chain. We let f_0 be the identity, and for $t > 0$, we define f_t by

$$f_t(z) = \sqrt{\frac{2t}{(1-\alpha)\alpha}}G_\alpha\left(z\cdot\sqrt{\frac{1}{2t}(1-\alpha)\alpha}\right).$$

Then f_t is normalized by $f_t(z) = z - \frac{t}{z} + \dots$, and $(f_t)_{t\geq 0}$ is a Loewner chain satisfying the slit equation (8.2) for a continuous function U. As G_α maps $1-2\alpha$ onto the tip of the line segment, we conclude that f_t maps $(1-2\alpha)\sqrt{\frac{2t}{(1-\alpha)\alpha}}$ onto the tip of the generated slit. As $U(t) \in \mathbb{R}$ is exactly this point, it follows that

$$U(t) = \frac{\sqrt{2}(1-2\alpha)}{\sqrt{(1-\alpha)\alpha}}\sqrt{t}.$$

Define c_α as $c_\alpha = \frac{\sqrt{2}(1-2\alpha)}{\sqrt{(1-\alpha)\alpha}}$. The function $(0,1) \ni \alpha \mapsto c_\alpha \in \mathbb{R}$ can easily be shown to be bijective with inverse $\alpha = \frac{c_\alpha^2 - c_\alpha\sqrt{c_\alpha^2+8}+8}{2(c_\alpha^2+8)}$, and thus, we have found the solution for all $c \in \mathbb{R}$.

S 8.8.3. By translation, we can assume that $U(0) = 0$. Let $(K_t)_{t\in[0,1]}$ be the generated hulls. Fix some $t \in (0,1]$, and let $H_t = t^{-1/2}K_t$. Then the half-plane capacity of H_t is equal to 1, and also H_t is contained within the same sector. The hull H_t is generated by the scaled driving function $s \mapsto V(s) := U(s\cdot t)/\sqrt{t}$ at $s = 1$. From Exercise 6.8.3, it follows that

$$\sup_{z\in H_t} \operatorname{Im}(z) \leq \sqrt{2}.$$

Let Δ be the following Euclidean triangle with height $\sqrt{2}$:

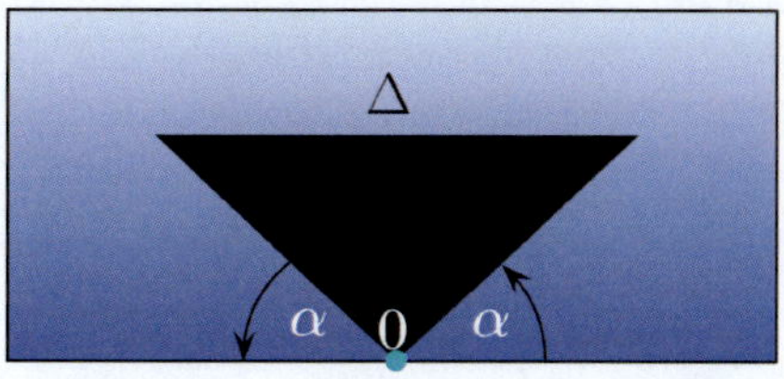

Then $H_t \subset \Delta$. Let F_Δ be the unique conformal mapping from $\mathbb{H}$ onto $\mathbb{H} \setminus \Delta$ with hydrodynamic normalization, and let $c = \lim_{x \searrow 0} F_\Delta^{-1}(x)$. Then F_Δ maps the interval $[-c, c]$ onto the boundary of Δ.

Furthermore, let F_t be the unique conformal mapping from $\mathbb{H}$ onto $\mathbb{H} \setminus H_t$ with hydrodynamic normalization. Then F_t maps an interval $[a_t, b_t]$ onto the slit H_t. The point that is mapped by F_t onto the tip of H_t is equal to $V(1) = U(t)/\sqrt{t}$. Thus $U(t)/\sqrt{t} \in [a_t, b_t]$.

As $H_t \subset \Delta$, we have $[a_t, b_t] \subset [-c, c]$ and, consequently,

$$\sup_{t \in (0,1]} \frac{|U(t)|}{\sqrt{t}} \leq c < \infty.$$

CHAPTER 9

Free Hemigroups, Loewner Chains, and Nonlinear Resolvents

We have seen that $\rhd$-hemigroups on $\mathbb{R}$ correspond to additive Loewner chains, and that $\rhd$-hemigroup distributions correspond to univalent F-transforms. In this chapter, we will show that the F-transforms of a $\boxplus$-hemigroup also form an additive Loewner chain, and that in the case of a $\boxplus$-semigroup, they correspond to the nonlinear resolvents of an infinitesimal generator.

The Loewner equation for a general additive Loewner chain (see Theorem 6.52) is given by

$$\frac{\partial}{\partial t} f_t(z) = \frac{\partial}{\partial z} f_t(z) \cdot G(z,t) \quad \text{for a.e. } t \geq 0, f_0(z) = z \in \mathbb{H}, \tag{9.1}$$

with

$$G(z,t) = a_t + \int_{\mathbb{R}} \frac{1+xz}{x-z} \rho_t(dx).$$

Now replace $G(z,t)$ by $G(f_t(z),t)$ to obtain the following modified equation:

$$\frac{\partial}{\partial t} f_t(z) = \frac{\partial}{\partial z} f_t(z) \cdot G(f_t(z),t) \quad \text{for a.e. } t \geq 0, f_0(z) = z \in \mathbb{H}. \tag{9.2}$$

At first glance, it seems unnatural to consider this equation. If G is an infinitesimal generator on $\mathbb{H}$ and $\varphi : \mathbb{H} \to \mathbb{H}$ is a holomorphic self-mapping, then $G \circ \varphi$ is not necessarily an infinitesimal generator as well. However, our generators have the form $G : \mathbb{H} \to \mathbb{H} \cup \mathbb{R}$, and in this case, $G \circ \varphi : \mathbb{H} \to \mathbb{H} \cup \mathbb{R}$ is also a generator on $\mathbb{H}$ (see Lemma 3.30).

In fact, the modified equation (9.2) is much simpler than (9.1). While the inverse functions $g_t = f_t^{-1}$ of the solution to (9.1) satisfy Loewner's ordinary differential equation

$$\frac{\partial}{\partial t} g_t(z) = -G(g_t(z),t),$$

the inverse functions g_t of a solution to (9.2), provided they exist, satisfy

$$\frac{\partial}{\partial t} g_t(z) = -G(z,t), \quad \text{i.e.,} \quad g_t(z) = z - \int_0^t G(z,s)ds.$$

Note that this implies that $g_t = f_t^{-1}$ extends holomorphically to the whole domain $\mathbb{H}$. We will see that (9.2) characterizes Loewner chains that arise as F-transforms of $\boxplus$-hemigroups.

9.1. Nonlinear resolvents

DEFINITION 9.1. Let G be an infinitesimal generator of a continuous semigroup on a simply connected domain $D \subsetneq \mathbb{C}$. Let $t \geq 0$. If there exists a holomorphic mapping $J_t : D \to D$ such that the equation

$$w = z - t \cdot G(z) \tag{9.3}$$

has a unique solution $z = J_t(w)$ for all $w \in D$, then we call J_t the *nonlinear resolvent* (at "time" t) of G.

If the nonlinear resolvent J_t exists for some $t \geq 0$, then J_t is a right inverse of the function $\varphi_t(z) := z - tG(z)$. Since $J_t(z)$ is the unique solution to the equation $\varphi_t(z) = w$, it is also a left inverse, i.e., if $\varphi_t(z) \in D$, then $J_t(\varphi_t(z)) = z$. Therefore, $J_t : D \to J_t(D)$ is the inverse of $\varphi_t|_{J_t(D)} : J_t(D) \to D$ and $\varphi_t(z) \notin D$ whenever $z \notin J_t(D)$.

EXAMPLE 9.2. Let $D = \mathbb{D}$ and consider the semigroup $F_t(z) = e^{-t}z$. Then $G(z) = -z$, and the solution to the equation

$$w = z + tz$$

is given by $J_t(w) = \frac{w}{1+t}$.

On unbounded domains, J_t may exist for some, but in general, not for all $t \geq 0$.

EXAMPLE 9.3. Let $D = \mathbb{H}$ and consider the semigroup $F_t(z) = e^t z$. Then $G(z) = z$, and the equation

$$w = z - tz$$

has the solution $J_t(w) = \frac{w}{1-t}$ for $t \in [0, 1)$, but for $t \geq 1$, it has no solution in $\mathbb{H}$.

The following theorem gives a geometric condition for the existence of nonlinear resolvents for all $t \geq 0$.

THEOREM 9.4. *Let $D \subset \mathbb{C}$ be a bounded and convex domain. If $G : D \to \mathbb{C}$ is an infinitesimal generator on D, then the nonlinear resolvents of G exist for all $t \geq 0$. Conversely, if $G : D \to \mathbb{C}$ is holomorphic and bounded such that the nonlinear resolvents $J_t : D \to D$ exist for all $t \geq 0$ in the sense of Definition* 9.1, *then G is an infinitesimal generator.*

PROOF. This result is shown in [**161**, Theorem 1.1] and [**161**, Corollary 1.2]. Uniqueness of the solution is shown in step 6 in the proof of [**161**, Theorem 1.1]. The proof utilizes the hyperbolic metric of bounded convex domains, which lies beyond the scope of this book. □

So let us first consider the case of bounded convex domains.

LEMMA 9.5. *Let $D \subset \mathbb{C}$ be a bounded and convex domain, and let G be an infinitesimal generator on D with nonlinear resolvents $J_t : D \to D$. Then we have the following properties:*

(a) *$t \mapsto J_t$ is continuous with respect to locally uniform convergence.*
(b) *Let $z \in D$. If $G(z) = 0$, then $J_t(z) = z$ for all $t \geq 0$. If $G(z) \neq 0$, then $J_s(z) \neq J_t(z)$ whenever $s \neq t$.*

PROOF. We let $\varphi_t(z) = z - tG(z)$.

(a) As D is bounded, $(J_t)_{t\geq 0}$ and $(J_t')_{t\geq 0}$ are both normal families and $(J_t')_{t\geq 0}$ is locally uniformly bounded.

Let $t \geq 0$ and $K \subset D$ be compact. We need to show that J_s converges uniformly on K to J_t as $s \to t$.

The set $J_t(K)$ is a compact subset of D, and we find a second compact set $L \subset D$ with $K \subset L$ and $\varepsilon > 0$ such that $\varphi_s(J_t(K)) \subset L$ for all $s \in [t-\varepsilon, t+\varepsilon] \cap [0,\infty)$ due to locally uniform convergence of φ_s to φ_t as $s \to t$. Let $z \in K$ and $s \in [t-\varepsilon, t+\varepsilon] \cap [0,\infty)$, and put $w = J_t(z)$. Then $\varphi_s(w) \in L$. We have

$$J_t(z) - J_s(z) = J_s(\varphi_s(J_t(z))) - J_s(z).$$

From $\varphi_s(J_t(z)) - z = \varphi_s(w) - \varphi_t(w)$, and the fact that $(J_t')_{t\geq 0}$ is locally uniformly bounded, we see that J_s converges to J_t uniformly on K as $s \to t$.

(b) If $G(z) = 0$, then $J_t(z) = z$ for all $t \geq 0$. Now assume $G(z) \neq 0$. Then $\varphi_t(z) \neq z$, and thus, $J_t(z) \neq z$ for all $t > 0$. Assume that $J_s(z) = J_t(z)$ for some $0 \leq s < t$. Then

$$\begin{aligned} z &= \varphi_s(J_t(z)) = J_t(z) - sG(J_t(z)) \\ &= J_t(z) - tG(J_t(z)) + (t-s)G(J_t(z)) = z + (t-s)G(J_t(z)). \end{aligned}$$

So $G(J_t(z)) = 0$ and $z = \varphi_t(J_t(z)) = J_t(z)$, a contradiction. □

We can now prove that nonlinear resolvents on a bounded and convex domain form a decreasing Loewner chain. This result has been shown in [**64**] for the unit disk and for higher dimensions in [**77**], [**96**].

THEOREM 9.6. *Let $D \subset \mathbb{C}$ be a bounded and convex domain, and let G be an infinitesimal generator on D with nonlinear resolvents $J_t : D \to D$. Then $(J_t)_{t\geq 0}$ is a decreasing Loewner chain on D satisfying the Loewner partial differential equation*

$$\frac{\partial}{\partial t} J_t(z) = \frac{\partial}{\partial z} J_t(z) \cdot G(J_t(z)), \quad J_0(z) = z \in D. \tag{9.4}$$

Furthermore, $\bigcap_{t\geq 0} J_t(D) = G^{-1}(\{0\})$.

PROOF. Let $\varphi_t(z) = z - tG(z)$. Since $(\varphi_t \circ J_t)(w) = w$, each J_t is clearly a univalent function. Furthermore, Lemma 9.5 (a) implies that $t \mapsto J_t$ is continuous with respect to locally uniform convergence.

Let $z \in D$. The statement $z \in J_t(D)$ is equivalent to $\varphi_t(z) \in D$. Hence, as D is convex, we see that $z \in J_t(D)$ implies $z \in J_s(D)$ for all $s \in [0,t]$. We conclude that $J_t(D) \subset J_s(D)$ whenever $s \leq t$, and that $(J_t)_{t\geq 0}$ is a decreasing Loewner chain.

From $z = J_t(z) - tG(J_t(z))$, we differentiate with respect to z to obtain the equation

$$J_t'(z) - tG'(J_t(z)) \cdot J_t'(z) = 1,$$

i.e., $J_t'(z) = (1 - tG'(J_t(z)))^{-1}$ for all $z \in D$.

Let $z \in D$. If $G(z) = 0$, then $J_t(z) = z$ for all $t \geq 0$ and (9.4) is clearly satisfied. Now assume that $G(z) \neq 0$. Then $J_s(z) \neq J_t(z)$ whenever $s \neq t$ by Lemma 9.5 (b). We can write

$$0 = \frac{(\varphi_{t+h} \circ J_{t+h})(z) - (\varphi_t \circ J_t)(z)}{h} = \frac{(\varphi_{t+h} \circ J_t)(z) - (\varphi_t \circ J_t)(z)}{h}$$
$$+ \frac{(\varphi_{t+h} \circ J_{t+h})(z) - (\varphi_{t+h} \circ J_t)(z)}{h} = \frac{(\varphi_{t+h} \circ J_t)(z) - (\varphi_t \circ J_t)(z)}{h}$$
$$+ \frac{(\varphi_{t+h} \circ J_{t+h})(z) - (\varphi_{t+h} \circ J_t)(z)}{J_{t+h}(z) - J_t(z)} \cdot \frac{J_{t+h}(z) - J_t(z)}{h}$$

for all $h \neq 0$ small enough, which shows that $t \mapsto J_t(z)$ is differentiable with

$$0 = (1 - tG'(J_t(z)))\frac{\partial}{\partial t} J_t(z) - G(J_t(z)),$$

which gives us

$$\frac{\partial}{\partial t} J_t(z) = \frac{\partial}{\partial z} J_t(z) \cdot G(J_t(z)).$$

If $G(z) = 0$, then $J_t(z) = z$ for all $t \geq 0$. If $G(z) \neq 0$, then there exists $T > 0$ such that $z - tG(z) \notin D$ for all $t \geq T$ as D is bounded. Hence, $z \notin J_t(D)$ for all $t \geq T$, and we conclude $\bigcap_{t\geq 0} J_t(D) = G^{-1}(\{0\})$. □

REMARK 9.7. Theorem 6.50 implies that $(z,t) \mapsto (G \circ J_t)(z)$ is a Herglotz vector field. This can also be seen as follows: We need to verify that $G \circ J_t$ is an infinitesimal generator on D for all $t \geq 0$. This is clearly true for $t = 0$. So let $t > 0$. Then $(G \circ J_t)(w) = (J_t(w) - w)/t$. It is a general fact that $f(z) - z$, $f : D \to D$ holomorphic, is always an infinitesimal generator on bounded convex domains, see [**160**, Proposition 4.3], and $r \cdot G$ is an infinitesimal generator for every infinitesimal generator G and $r > 0$.

Next, we prove a converse result.

THEOREM 9.8. *Let $D \subsetneq \mathbb{C}$ be a (possibly unbounded) convex domain, and let G be an infinitesimal generator on D. Furthermore, let $J_t : D \to D$ be a family of holomorphic functions satisfying* (9.4). *Then $(J_t)_{t\geq 0}$ are the nonlinear resolvents of G.*

PROOF. Consider the differential equation

$$\frac{\partial}{\partial t}\varphi_t(z) = -G(J_t(\varphi_t(z))), \quad \varphi_0(z) = z. \tag{9.5}$$

Fix $z \in D$. We can solve this equation at least for t small enough. We have that

$$\frac{d}{dt}[J_t(\varphi_t(z))] = J_t'(\varphi_t(z)) \cdot G(J_t(\varphi_t(z))) + J_t'(\varphi_t(z)) \cdot [-G(J_t(\varphi_t(z)))] = 0,$$

i.e., $J_t(\varphi_t(z))$ does not depend on t, and $J_t(\varphi_t(z)) = J_0(\varphi_0(z)) = z$. Hence, $\frac{\partial}{\partial t}\varphi_t(z) = -G(z)$. We conclude that $t \mapsto \varphi_t(z)$ simply describes a straight line:

$$\varphi_t(z) = z - tG(z).$$

In particular, we can now define $\varphi_t(z)$ for all $z \in D$ and all $t \geq 0$ by $\varphi_t(z) = z - tG(z)$. (But $\varphi_t(z)$ may not solve (9.5) for all $t \geq 0$.)

Let $D_t = \{z \in D \,|\, \varphi_t(z) \in D\}$. The convexity of D implies that $D_t \subset D_s$ whenever $s \leq t$.

Let $T > 0$ and assume that D_T contains a non-empty open subset. We find such a T as $D_0 = D$ and by using the continuity of $t \mapsto \varphi_t(z)$. Let $z \in D_T$. Then $z \in D_t$ for all $t \in [0,T]$ and a local solution to (9.5) can be extended up to $t = T$: If the solution, denoted by $\hat{\varphi}_t$, cannot be extended beyond $[0,T]$, then the maximal solution exists for $t \in [0,T_0)$ with some $T_0 \leq T$, and either $\lim_{t \nearrow T_0} \hat{\varphi}_t$ does not exist, which is impossible since $\hat{\varphi}_t = z - tG(z)$, or $\lim_{t \nearrow T_0} \hat{\varphi}_t$ exists and belongs to ∂D, which is excluded as $z \in D_T$. Hence, $J_T(\varphi_T(z)) = z$ for all $z \in D_T$, and φ_T is univalent on D_T.

Applying φ_T gives $\varphi_T(J_T(w)) = w$ for all $w \in \varphi_T(D_T)$, and $\varphi_T(D_T)$ contains a non-empty open subset. Since the left-hand side extends holomorphically to D, we can apply the identity theorem to conclude that $\varphi_T(J_T(w)) = w$ for all $w \in D$. Thus, J_T is univalent, and $J_T(D) \subset D_T$. Furthermore, φ_T is univalent on D_T and maps $J_T(D)$ onto D. Therefore, $D_T = J_T(D)$, and $J_T : D \to J_T(D)$ is the inverse of $\varphi_T|_{J_T(D)} : J_T(D) \to D$. If $z \in D \setminus J_T(D)$, then $\varphi_T(z) \notin D$. Hence, J_T is the nonlinear resolvent of G at time T.

Let $T_0 > 0$ be the infimum of all $T > 0$ such that D_T does not contain a non-empty open subset, and assume that $T_0 < \infty$. If D_{T_0} contains a non-empty open subset, then this property also holds for D_t with $t \in [T_0, T_0 + \varepsilon]$ for some $\varepsilon > 0$, due to the continuity of $t \mapsto \varphi_t(z)$. Hence, D_{T_0} does not contain a non-empty open subset.

Choose some $z_0 \in D$ and let $t \in [0,T_0)$. Then $\varphi_t(J_t(z_0)) = z_0$. Let $(t_n)_n \subset [0,T_0)$ be a sequence converging to T_0. Then $w_n := J_{t_n}(z_0)$ converges to $J_{T_0}(z_0)$ as $n \to \infty$. Furthermore, $\varphi_{t_n}(w_n) = z_0$, i.e., $w_n - t_n G(w_n) = z_0$. By taking the limit as $n \to \infty$, we get $J_{T_0}(z_0) - T_0 G(J_{T_0}(z_0)) = z_0$. Therefore, $J_{T_0}(D) \subset D_{T_0}$.

If J_{T_0} is non-constant, then D_{T_0} contains a non-empty open subset. So J_{T_0} must be constant, say $c \in D$. For each $z \in D$, we have $\varphi_{t_n}(J_{t_n}(z)) \to \varphi_{T_0}(c)$ as $n \to \infty$, but $\varphi_{t_n}(J_{t_n}(z)) = z$, which is a contradiction.

We conclude that $(J_t)_{t \geq 0}$ is the family of nonlinear resolvents of G. □

9.2. Free hemigroups and Loewner chains

We now show that the F-transforms of a $\boxplus$-semigroup are nonlinear resolvents that satisfy (9.4) for an infinitesimal generator on $\mathbb{H}$.

THEOREM 9.9 ([**96**]). *Let $G : \mathbb{H} \to \mathbb{H} \cup \mathbb{R}$ be holomorphic with $\lim_{y \to \infty} G(iy)/y = 0$. There exists a unique solution $(f_t)_{t \geq 0}$ of holomorphic mappings $f_t : \mathbb{H} \to \mathbb{H}$ to*

$$\frac{\partial}{\partial t} f_t(z) = \frac{\partial}{\partial z} f_t(z) \cdot G(f_t(z)), \quad t \geq 0, \quad f_0(z) = z \in \mathbb{H}. \tag{9.6}$$

The solution can be expressed as $(f_t)_{t \geq 0} = (F_{\mu_t})_{t \geq 0}$ for a $\boxplus$-semigroup $(\mu_t)_{t \geq 0}$, and $(f_t)_{t \geq 0}$ is a decreasing Loewner chain. Furthermore, $(f_t)_{t \geq 0}$ is the family of nonlinear resolvents of G, and

$$G(z) = -R_{\mu_1}(1/z).$$

PROOF. For $t \geq 0$, we consider $H_t(z) := -tG(z)$. This function is of the form (5.11), and Theorem 5.41 implies that we find a free $\boxplus$-semigroup $(\mu_t)_{t \geq 0}$ such that $R_{\mu_t}(1/z) = H_t(z)$. Let $f_t = F_{\mu_t}$. Then each f_t is univalent, see Corollary 5.42, and with $\Phi_t(z) := z + H_t(z)$ we have $\Phi_t(f_t(z)) = z$ for all $z \in \mathbb{H}$, due to the definition of the R-transform.

Let $z \in \mathbb{H}$ and $t \geq 0$. Then

$$0 = \frac{(\Phi_{t+h} \circ f_{t+h})(z) - (\Phi_t \circ f_t)(z)}{h} = \frac{(\Phi_{t+h} \circ f_t)(z) - (\Phi_t \circ f_t)(z)}{h}$$

$$+ \frac{(\Phi_{t+h} \circ f_{t+h})(z) - (\Phi_{t+h} \circ f_t)(z)}{h} = \frac{(\Phi_{t+h} \circ f_t)(z) - (\Phi_t \circ f_t)(z)}{h}$$

$$+ \frac{(\Phi_{t+h} \circ f_{t+h})(z) - (\Phi_{t+h} \circ f_t)(z)}{f_{t+h}(z) - f_t(z)} \cdot \frac{f_{t+h}(z) - f_t(z)}{h},$$

for all $h \neq 0$ small enough, and provided that $f_{t+h}(z) \neq f_t(z)$. Let h_n be a null-sequence such that $f_{t+h_n}(z) \neq f_t(z)$ for all n. Then

$$0 = \Phi_t'(f_t(z)) \cdot \lim_{n\to\infty} \frac{f_{t+h_n}(z) - f_t(z)}{h_n} - G(f_t(z)),$$

and as $\Phi_t'(f_t(z)) = 1/f_t'(z)$, we have

$$\lim_{n\to\infty} \frac{f_{t+h_n}(z) - f_t(z)}{h_n} = f_t'(z) \cdot G(f_t(z)).$$

Now let h_n be a null-sequence such that $f_{t+h_n}(z) = f_t(z)$ for all n. Then

$$\Phi_{t+h_n}(f_t(z)) = \Phi_t(f_t(z)),$$

and thus, $0 = \lim_{n\to\infty}(\Phi_{t+h_n}(f_t(z)) - \Phi_t(f_t(z)))/h_n \to -G(f_t(z))$. Together, we conclude that $t \mapsto f_t(z)$ is differentiable for all $z \in \mathbb{H}$ with

$$\frac{\partial}{\partial t} f_t(z) = \frac{\partial}{\partial z} f_t(z) \cdot G(f_t(z)).$$

Uniqueness of the solution follows from Theorem 9.8, which also implies that the functions $(f_t)_{t\geq 0}$ are the nonlinear resolvents of the generator $G(z) = -R_{\mu_1}(1/z)$.

The continuity of $t \mapsto f_t$ with respect to locally uniform convergence can be shown as in Lemma 9.5 (a). The statement $z \in f_t(\mathbb{H})$ is equivalent to $\Phi_t(z) \in \mathbb{H}$. Hence, as $\mathbb{H}$ is convex, $z \in f_t(\mathbb{H})$ implies $z \in f_s(\mathbb{H})$ for all $s \in [0, t]$. We conclude that $f_t(\mathbb{H}) \subset f_s(\mathbb{H})$ whenever $s \leq t$, and that $(f_t)_{t\geq 0}$ is a decreasing Loewner chain.

Alternatively, we have that $z \mapsto G(f_t(z))$ is an infinitesimal generator on $\mathbb{H}$ with $\lim_{y\to\infty} G(f_t(iy))/y = 0$ for every $t \geq 0$. So $(z, t) \mapsto G(f_t(z))$ is a Herglotz vector field of the form (6.20), and Theorem 6.52 implies that $(f_t)_{t\geq 0}$ is a decreasing Loewner chain. □

REMARK 9.10. We note that Theorem 9.9 can also be seen as an embedding theorem. By Theorem 5.41, for any $\boxplus$-infinitely divisible distribution μ, $G(z) = -R_\mu(1/z)$ defines a holomorphic function on $\mathbb{H}$, which maps $\mathbb{H}$ into $\mathbb{H} \cup \mathbb{R}$ and satisfies $\lim_{y\to\infty} G(iy)/y = 0$. Thus, Theorem 9.9 produces an additive Loewner chain $(F_{\mu_t})_{t\geq 0}$ with $R_\mu = R_{\mu_1}$, and therefore $\mu_1 = \mu$.

EXAMPLE 9.11. Consider the semicircle distributions $W(0, t)$, which form a $\boxplus$-semigroup. Then $F_{W(0,t)}(z) = \frac{2t}{z - \sqrt{z^2 - 4t}}$ and $R_{W(0,1)}(z) = z$, see Example 5.35, and use the scaling rule $F_{W(0,t)}(z) = \sqrt{t}F_{W(0,1)}(z/\sqrt{t})$. A simple calculation yields

$$\frac{\partial}{\partial t}F_{W(0,t)}(z) = F'_{W(0,t)}(z) \cdot \frac{\sqrt{z^2-4t}-z}{2t},$$

which is (9.6) with $G(F_{W(0,t)}) = -1/F_{W(0,t)} = -R_{W(0,1)}(1/F_{W(0,t)})$.

A special example of a generator as in Theorem 9.9 is the function

$$G_S(z) = i\frac{1+e^{iz}}{1-e^{iz}}.$$

Since e^{iz} maps $\mathbb{H}$ onto $\mathbb{D}\setminus\{0\}$, G_S maps $\mathbb{H}$ onto $\mathbb{H}\setminus\{i\}$, and

$$\lim_{y\to\infty} G_S(iy)/y = \lim_{y\to\infty} i\frac{1+e^{-y}}{1-e^{-y}}/y = 0.$$

Let $(\mu_t)_{t\geq 0}$ be a free $\boxtimes$-semigroup of probability measures from $\mathcal{P}_\times(\partial\mathbb{D})$. By Theorem 7.24, we have $\Sigma_{\mu_t}(z) = \exp(tu(z))$ for a holomorphic function $u:\mathbb{D}\to\mathbb{C}$ with $\mathrm{Re}(u)\geq 0$.

Let $M_t(z) = \int_{\partial\mathbb{D}} \frac{1+xz}{1-xz}\mu_t(dx)$. Then $M_t(z) = 1+2\psi_{\mu_t}(z) = \frac{1+\eta_t(z)}{1-\eta_t(z)}$ is a univalent function and $iM_t(e^{iz})$ maps $\mathbb{H}$ into $\mathbb{H}\setminus\{i\}$.

Next, let $F_S(z) = i\frac{1+z}{1-z}$. Then $G_S = F_S(e^{iz})$. The function $F_S^{-1}(iM_t(e^{iz}))$ maps $\mathbb{H}$ into $\mathbb{D}\setminus\{0\}$. Since $\mathbb{H}$ is simply connected and this function has no zeros, we can take the logarithm:

$$J_t(z) = -i\log(F_S^{-1}(iM_t(e^{iz}))), \tag{9.7}$$

and J_t is well-defined modulo shifts by $2k\pi$, $k\in\mathbb{Z}$. For $t=0$, we have $J_0(z) = z+2k\pi$. We choose $k=0$ such that J_0 is the identity. Furthermore, we require continuity of $t\mapsto J_t$, and then each J_t is uniquely defined.

THEOREM 9.12 ([**96**]). *Let $(\mu_t)_{t\geq 0}$ be a free $\boxtimes$-semigroup of probability measures from $\mathcal{P}_\times(\partial\mathbb{D})$, and let u and $(J_t)_{t\geq 0}$ be defined as above. Then $(J_t)_{t\geq 0}$ is a decreasing Loewner chain on $\mathbb{H}$ and $(J_t)_{t\geq 0}$ are the nonlinear resolvents of the infinitesimal generator $G(z) = iu(-e^{iz})$.*

PROOF. Let $\eta_t = \eta_{\mu_t}$. Then, by the definition of the Σ-transform, we have

$$\eta_t^{-1}(z) = z\exp(tu(z))$$

for all $z\in\mathbb{D}$. As $(\eta_t^{-1}\circ\eta_t)(z) = z$ for all $z\in\mathbb{D}$, we have, similar to the proof of Theorem 9.6, that $t\mapsto\eta_t(z)$ is differentiable for all $z\in\mathbb{D}$ and $t\geq 0$. We obtain

$$0 = \frac{\partial}{\partial s}\eta_s^{-1}(\eta_t(z))\bigg|_{s=t} + (\eta_t^{-1})'(\eta_t(z))\frac{\partial}{\partial t}\eta_t(z) = \eta_t(z)u(\eta_t(z))\exp(tu(\eta_t(z)))$$
$$+ (\eta_t^{-1})'(\eta_t(z))\frac{\partial}{\partial t}\eta_t(z) = zu(\eta_t(z)) + (\eta_t^{-1})'(\eta_t(z))\frac{\partial}{\partial t}\eta_t(z),$$

which yields the differential equation

$$\frac{\partial}{\partial t}\eta_t(z) = -zu(\eta_t(z))\cdot\eta_t'(z), \quad \eta_0(z) = z.$$

From $M_t(z) = \frac{1+\eta_t(z)}{1-\eta_t(z)}$, we obtain the partial differential equation

$$\frac{\partial}{\partial t}M_t(z) = -zS(M_t(z))\cdot M_t'(z), \quad M_0(z) = \frac{1+z}{1-z},$$

with $S(z) = u(\frac{1-z}{1+z})$. Note that S maps the right half-plane $-i\mathbb{H}$ holomorphically into $-i\mathbb{H} \cup i\mathbb{R}$.

For $\mathrm{Im}(z) > 0$, define $V_t(z) = iM_t(e^{iz})$. Then we obtain

$$\frac{\partial V_t(z)}{\partial t} = iS(-iV_t(z))V_t'(z), \quad V_0(z) = i\frac{1+e^{iz}}{1-e^{iz}}.$$

From (9.7), we have $J_t(z) = -i\log(F_S^{-1}(V_t(z)))$, and thus,

$$\begin{aligned}\frac{\partial J_t(z)}{\partial t} &= iS(-iV_t(z))J_t'(z) = iS(-i(V_0 \circ J_t)(z))J_t'(z) \\ &= G(J_t(z))J_t'(z), \quad J_0(z) = z \in \mathbb{H},\end{aligned}$$

with $G(z) = iS(-iV_0(z)) = iu(-e^{iz})$, which maps $\mathbb{H}$ into $\mathbb{H} \cup \mathbb{R}$ and

$$\lim_{y\to\infty} G(iy)/y = \lim_{y\to\infty} iu(-e^{-y})/y = 0.$$

By Theorem 9.9, we conclude that $(J_t)_{t\geq 0}$ are the nonlinear resolvents of the infinitesimal generator $iu(-e^{iz})$. □

Next we consider the non-autonomous equation (9.2). Let $G : \mathbb{H} \times [0,\infty) \to \mathbb{H} \cup \mathbb{R}$ be a Herglotz vector field as in Theorem 6.52, i.e., it has the form

$$G(z,t) = a_t + \int_{\mathbb{R}} \frac{1+xz}{x-z}\rho_t(dx),$$

where $a_t \in \mathbb{R}$ and ρ_t is a finite, non-negative Borel measure on $\mathbb{R}$ for a.e. $t \geq 0$.

Let us assume that G is of order ∞, i.e., for any compact subset $K \subset \mathbb{H}$ and $T > 0$, $|G(z,t)|$ is uniformly bounded for all $z \in K$ and a.e. $t \in [0,T]$. Then (see the proof of Theorem 6.52), for each $T > 0$, there exists $C_T > 0$ such that

$$|a_t| \leq C_T \quad \text{and} \quad \rho_t(\mathbb{R}) \leq C_T \quad \text{for a.e. } t \in [0,T].$$

We can integrate G with respect to t, and so we consider

$$H(z,t) := \int_0^t G(z,s)ds.$$

LEMMA 9.13. *For every $t \geq 0$, the function $z \mapsto H(z,t)$ maps $\mathbb{H}$ holomorphically into $\mathbb{H} \cup \mathbb{R}$ and $H(iy,t)/y \to 0$ as $y \to \infty$. Furthermore, there exists a set $\mathcal{N} \subset [0,\infty)$ of Lebesgue measure 0 such that $t \mapsto H(z,t)$ is differentiable for all $t \in [0,\infty) \setminus \mathcal{N}$ and all $z \in \mathbb{H}$.*

PROOF. Clearly, each $H(\cdot,t)$ maps $\mathbb{H}$ holomorphically into $\mathbb{H} \cup \mathbb{R}$ and from

$$\begin{aligned}|H(iy,t)/y| &\leq \int_0^t |a_s/y|ds + \int_0^t \left|\int_{\mathbb{R}} \frac{1/y+ix}{x-iy}\rho_s(dx)\right| ds \\ &\leq tC_t/|y| + \int_0^t \int_{\mathbb{R}} \frac{|x/y+ix^2+i-xy|}{x^2+y^2}\rho_s(dx)ds\end{aligned}$$

and the dominated convergence theorem, we obtain that $H(iy,t)/y \to 0$ as $y \to \infty$. The existence of $\mathcal{N}$ follows as in the proof of Theorem 6.12. □

THEOREM 9.14. *Let G and H be defined as above, and let $f_t : \mathbb{H} \to \mathbb{H}$ be the nonlinear resolvent of $H(\cdot,t)$ at time 1, i.e., the inverse of $z \mapsto z - H(z,t)$. Then $t \mapsto f_t(z)$ is locally absolutely continuous for every $z \in \mathbb{H}$, $t \mapsto f_t$ is continuous* *with respect to locally uniform convergence, and there exists a set* $\mathcal{N} \subset [0, \infty)$ *of Lebesgue measure* 0 *such that*

$$\frac{\partial}{\partial t} f_t(z) = \frac{\partial}{\partial z} f_t(z) \cdot G(f_t(z), t) \quad \text{for all } t \in [0, \infty) \setminus \mathcal{N} \text{ and all } z \in \mathbb{H}. \tag{9.8}$$

Furthermore, we have $(f_t)_{t\geq 0} = (F_{\mu_{0,t}})_{t\geq 0}$ *for a* $\boxplus$*-hemigroup* $(\mu_{s,t})_{0\leq s\leq t}$*, and* $(f_t)_{t\geq 0}$ *is a decreasing Loewner chain.*

PROOF. For $0 \leq s \leq t$, Theorem 5.41 implies that we find $\boxplus$-infinitely divisible probability measures $(\mu_{s,t})_{0\leq s\leq t}$ such that $R_{\mu_{s,t}}(1/z) = -(H(z,t) - H(z,s)) = -\int_s^t G(z,\sigma)d\sigma$. Then $(\mu_{s,t})_{0\leq s\leq t}$ is a $\boxplus$-hemigroup.

We let $\Phi_t(z) = z - H(z,t) = z + R_{\mu_{0,t}}(1/z)$. Then $F_{\mu_{0,t}}$ is the nonlinear resolvent of $H(z,t)$ at time 1, so $F_{\mu_{0,t}} = f_t$.

The set $\{f_\tau \,|\, \tau \geq 0\}$ is a normal family, and thus, $\{f'_\tau \,|\, \tau \geq 0\}$ is locally uniformly bounded. Let $s, t \geq 0$ and $z \in \mathbb{H}$. Let $w = f_t(z)$. If s is close enough to t, then $\Phi_s(w) \in \mathbb{H}$. We have

$$f_t(z) - f_s(z) = f_s(\Phi_s(w)) - f_s(z).$$

From $\Phi_s(w) - z = \Phi_s(w) - \Phi_t(w)$, we see that $\tau \mapsto f_\tau(z)$ is locally absolutely continuous for any $z \in \mathbb{H}$, and as $t \mapsto \Phi_t$ is continuous with respect to locally uniform convergence, we can argue as in Lemma 9.5 to show that also $t \mapsto f_t$ is continuous with respect to locally uniform convergence.

Let $t \geq 0$, and assume that $u \mapsto \Phi_u(z)$ is differentiable at $u = t$ with

$$\left.\frac{\partial}{\partial u}\Phi_u(z)\right|_{u=t} = -G(z,t)$$

for all $z \in \mathbb{H}$. This is possible for almost all $t \geq 0$ due to Lemma 9.13. As in the proof of Theorem 9.9, we can show that $u \mapsto f_u(z)$ is differentiable at $u = t$ for all $z \in \mathbb{H}$ with

$$\frac{\partial}{\partial t} f_t(z) = \frac{\partial}{\partial z} f_t(z) \cdot G(f_t(z), t).$$

For a.e. $t \geq 0$, $z \mapsto G(f_t(z), t)$ maps $\mathbb{H}$ into $\mathbb{H} \cup \mathbb{R}$ with $\lim_{y\to\infty} G(f_t(iy), t)/y = 0$, which we show as in (5.7). Thus, $(z,t) \mapsto G(f_t(z), t)$ is again a Herglotz vector field of the form (6.20), and Theorem 6.52 implies that $(f_t)_{t\geq 0}$ is a decreasing Loewner chain. □

COROLLARY 9.15. *Let* $(\mu_{s,t})_{0\leq s\leq t}$ *be a* $\boxplus$*-hemigroup, and let* $F_t = F_{\mu_{0,t}}$*. Then* $(F_t)_{t\geq 0}$ *is an additive Loewner chain on* $\mathbb{H}$*.*

PROOF. Each $\mu_{0,t}$ is $\boxplus$-infinitely divisible due to Theorem 5.50, and each F_t is univalent. Clearly, $t \mapsto F_t$ is continuous. It only remains to show that we can write $F_t = F_s \circ G$ for some univalent $G : \mathbb{H} \to \mathbb{H}$ when $s \leq t$. (Thus, $F_{s,t} := F_s^{-1} \circ F_t$ defines the transition mappings of the Loewner chain.)

Fix $0 \leq s \leq t$. We embed $\mu_{0,s}$ into a $\boxplus$-semigroup $(\alpha_t)_{t\geq 0}$ with $\alpha_1 = \mu_{0,s}$, and we embed $\mu_{s,t}$ into a $\boxplus$-semigroup $(\beta_t)_{t\geq 0}$ with $\beta_1 = \mu_{s,t}$. Define $R_t = R_{\alpha_t} = tR_{\alpha_1}$ for $0 \leq t \leq 1$, and

$$R_t = R_{\alpha_1} + R_{\beta_{t-1}} = R_{\alpha_1} + (t-1)R_{\beta_1}$$

for $t > 1$. We obtain probability measures $(\gamma_t)_{t\geq 0}$, defined via $R_{\gamma_t} = R_t$, which corresponds to taking $G(z,t) = -R_{\alpha_1}(1/z)$ for $t < 1$, and $G(z,t) = -R_{\beta_1}(1/z)$ for $t > 1$ in Theorem 9.14. Thus, F_{γ_t} is a decreasing Loewner chain. We have $F_{\gamma_1} = F_{\mu_{0,s}} = F_s$, and $F_{\gamma_2} = F_{\mu_{0,s} \boxplus \mu_{s,t}} = F_{\mu_{0,t}} = F_t$. Therefore, $F_t = F_s \circ G$ for some univalent $G : \mathbb{H} \to \mathbb{H}$. $\square$

REMARK 9.16. We see that for every $\boxplus$-hemigroup $(\mu_{s,t})_{0\leq s\leq t}$, there exists a $\rhd$-hemigroup $(\nu_{s,t})_{0\leq s\leq t}$ with $\mu_{0,t} = \nu_{0,t}$ for all $t \geq 0$. The special case $\mu_{s,t} = \mu_{0,t-s}$ corresponds to a free $\boxplus$-semigroup and to free Lévy processes. If we instead require $\nu_{s,t} = \nu_{0,t-s}$, we obtain another class of processes, which are called Lévy processes of the second kind in [**37**].

9.3. The nonlinear resolvent equation on the upper half-plane

THEOREM 9.17 ([**96**]). *Let $G : \mathbb{H} \to \mathbb{C}$ be an infinitesimal generator.*

(1) *Assume that $G(\mathbb{H}) \subset \mathbb{H} \cup \mathbb{R}$. Let $b = \lim_{y\to\infty} G(iy)/(iy)$. Then the nonlinear resolvent J_t of G exists if and only if $t \in [0, 1/b)$.*
If $b = 0$, then the nonlinear resolvents J_t satisfy
$$J_t = F_{\mu_t},$$
where $(\mu_t)_{t\geq 0}$ is a $\boxplus$-semigroup of probability measures on $\mathbb{R}$.
(2) *In all other cases, the nonlinear resolvent of G exists for every $t \geq 0$.*

PROOF.

(1) If $b = 0$, then the statement follows from Theorem 9.9. Now assume that $b \neq 0$. The Nevanlinna representation formula, Theorem 3.12, shows that $\hat{G}(z) = G(z) - bz$ also maps $\mathbb{H}$ into $\mathbb{H} \cup \mathbb{R}$. The equation $w = z - tG(z)$ becomes $w = (1 - tb)z - t\hat{G}(z)$. For $t \geq 1/b$ and $z \in \mathbb{H}$, the right side belongs to $\mathbb{H}^- \cup \mathbb{R}$. Thus, there is no solution $z \in \mathbb{H}$ for any $w \in \mathbb{H}$. For $t < 1/b$, we can rewrite the equation as
$$w/(1 - tb) = z - t\hat{G}(z)/(1 - tb),$$
and the previous case applied to the generator $\hat{G}(z)/(1 - tb)$ shows the existence of the nonlinear resolvent J_t of G.
(2) Now we assume that $G(\mathbb{H})$ is not a subset of $\mathbb{H}\cup\mathbb{R}$. We define $p \in \mathbb{H}\cup\mathbb{R}$ as follows. By Lemma 3.30, the associated semigroup $(F_t)_{t\geq 0}$ either consists of elliptic automorphisms with fixed point $p \in \mathbb{H}$, or the Denjoy–Wolff point p of $(F_t)_{t\geq 0}$ is different from ∞ and $p \in \mathbb{H} \cup \mathbb{R}$.

Case 1: $p \in \mathbb{H}$. We find a Möbius transformation M such that $\tilde{F}_t = M^{-1} \circ F_t \circ M$ is a semigroup on $\mathbb{D}$ with $\tilde{F}_t(0) = 0$ for all $t \geq 0$. For $r \in (0, 1)$, let $D_r = M(\{|z| < r\})$. Due to the Schwarz lemma (Theorem 3.4), we have $|\tilde{F}_t(z)| \leq |z|$, and thus, $F_t(D_r) \subset D_r$ for every $t \geq 0$ and $r \in (0, 1)$. Therefore, $(F_t)_{t\geq 0}$ restricted to D_r is a semigroup on the bounded convex domain D_r. (D_r is again a disk within $\mathbb{H}$.) By Theorem 9.4, we obtain nonlinear resolvents J_t of G restricted to D_r for all $t \geq 0$. Since $\mathbb{H} = \cup_{0<r<1} D_r$, each J_t extends to $\mathbb{H}$ and is the nonlinear resolvent of G.

Case 2: $p \in \mathbb{R}$. In this case, we find a Möbius transformation M such that $\tilde{F}_t = M^{-1} \circ F_t \circ M$ is a semigroup on $\mathbb{D}$ with Denjoy–Wolff point equal to 1. For $r \in (0,1)$, we let H_r be the horodisk at the boundary point 1 with Euclidean radius r, see Section 3.5. Let $D_r = M(H_r)$. Due to the boundary version of the Schwarz lemma, Theorem 3.19 (d), we have $\tilde{F}_t(H_r) \subset H_r$, and thus, $F_t(D_r) \subset D_r$ for every $t \geq 0$ and $r \in (0,1)$. Therefore, $(F_t)_{t\geq 0}$ restricted to D_r is a semigroup on the bounded convex domain D_r. (Again, D_r is a disk within $\mathbb{H}$.) We now finish the proof just as in case 1. □

Part (2) of the proof can be generalized to obtain that the nonlinear resolvents of an infinitesimal generator $G : D \to \mathbb{C}$ exist for all $t \geq 0$ also if D is unbounded and convex, provided that the Denjoy–Wolff point of the semigroup associated with G is finite, see [**96**].

QUESTION 9.18. Let $D \subsetneq \mathbb{C}$ be an unbounded convex domain and let $G : D \to \mathbb{C}$ be an infinitesimal generator. Under which conditions do the nonlinear resolvents of G exist?

EXAMPLE 9.19. Let $D \subsetneq \mathbb{C}$ be an unbounded convex domain such that $z \in D$ implies $rz \in D$ for all $r \geq 1$. In this case, the family $(e^t z)_{t\geq 0}$ is a semigroup on D and its generator is given by $G(z) = z$. The equation $w = z - tz$ cannot be solved for $t = 1$.

9.4. Exercises

E 9.4.1. Let $(J_t)_{t\geq 0}$ be the nonlinear resolvents of the generator

$$G(z) = -z(1+z) \quad \text{on } \mathbb{D}.$$

(a) Is $\overline{J_t(\mathbb{D})} \subset \mathbb{D}$ for all $t > 0$?
(b) Is $J_t(\mathbb{D})$ convex for all $t \geq 0$?

E 9.4.2. Consider the unbounded convex domain $Q = \{z \in \mathbb{H} \,|\, \text{Re}(z) > 0\}$, the first quadrant of the complex plane. Prove that

$$H(z) := z - \frac{1}{z^3}$$

is an infinitesimal generator on Q. Can the nonlinear resolvent equation for H be solved for all $t \geq 0$?

E 9.4.3. Find a simply connected domain $D \subsetneq \mathbb{C}$ and an infinitesimal generator G on D such that the equation $w = z - tG(z)$ has more than one solution in D for some $t > 0$, $w \in D$.

9.5. Solutions

S 9.5.1.

(a) For $t > 0$, the equation $w = z + tz(1+z)$ leads to

$$J_t(w) = \frac{\sqrt{(1+t)^2 + 4tw} - 1 - t}{2t},$$

where the square root is chosen such that $\sqrt{(1+t)^2} = 1 + t$. Note that $(1+t)^2 + 4tw$ maps $\mathbb{D}$ onto a disk that is symmetric with respect to $\mathbb{R}$, and whose left boundary point is equal to $(1+t)^2 - 4t = (1-t)^2 \geq 0$.

J_t extends continuously to $\overline{\mathbb{D}}$, and for $0 < t \leq 1$, we have $J_t(-1) = \frac{1-t-1-t}{2t} = -1$, and thus, $\overline{J_t(\mathbb{D})}$ is not contained in $\mathbb{D}$. (But for $t > 1$, $J_t(-1) = \frac{t-1-1-t}{2t} = \frac{-1}{t} > -1$.)

(b) We use Theorem 3.43 and check whether $1 + w\frac{J_t''(w)}{J_t'(w)} \in \mathcal{P}$. Let $t > 0$. We have

$$1 + w\frac{J_t''(w)}{J_t'(w)} = \frac{(1+t)^2 + 2tw}{(1+t)^2 + 4tw} =: M(w).$$

The pole of the Möbius transformation M is at $w = \frac{-(1+t)^2}{4t}$, which is $= -1$ for $t = 1$ and < -1 otherwise.

For $t \neq 1$, M maps $\mathbb{D}$ onto a disk that is symmetric with respect to $\mathbb{R}$, and whose boundary points are given by

$$M(-1) = \frac{(1+t)^2 - 2t}{(1+t)^2 - 4t} = \frac{1+t^2}{(1-t)^2} > 1 \quad \text{and} \quad M(1) = \frac{(1+t)^2 + 2t}{(1+t)^2 + 4t} > 0.$$

For $t = 1$, M maps $(-1, 1)$ onto $(M(1), \infty) = (\frac{3}{4}, \infty)$. So M maps $\mathbb{D}$ in this case onto the half-plane $\mathrm{Re}(z) > \frac{3}{4}$.

Hence, M belongs to $\mathcal{P}$ for every $t > 0$, and thus, $J_t(\mathbb{D})$ is convex for all $t \geq 0$.

S 9.5.2. The function $G(z) = 2z - \frac{2}{z}$ maps $\mathbb{H}$ into $\mathbb{H}$, and thus, it is an infinitesimal generator on $\mathbb{H}$. Consider the conformal mapping $\varphi : Q \to \mathbb{H}$, $\varphi(z) = z^2$. With the calculation from Section 3.6, we obtain that the function

$$\frac{G(\varphi(z))}{\varphi'(z)} = \frac{G(z^2)}{2z} = z - \frac{1}{z^3} = H(z)$$

is an infinitesimal generator on Q. The equation $w = z - tH(z)$ leads for $t = 1$ to $z^3 = \frac{1}{w}$, which has 3 solutions in $\mathbb{C}$ for fixed $w \in Q$. But as $\mathrm{Re}(1/w) > 0$ and $\mathrm{Im}(1/w) < 0$, no solution lies within Q, since z^3 cannot belong to the fourth quadrant of the complex plane for $z \in Q$.

S 9.5.3. The map $f(z) = \sqrt{z}$ maps $D := \mathbb{C} \setminus [0, \infty)$ onto $\mathbb{H}$ with inverse $f^{-1}(z) = z^2$. The function $H(z) = \frac{1}{2}(z + i)$ maps $\mathbb{H}$ into itself, and thus, it is an infinitesimal generator on $\mathbb{H}$. Therefore,

$$G(z) := \frac{H(f(z))}{f'(z)} = \frac{\frac{1}{2}(\sqrt{z} + i)}{1/(2\sqrt{z})} = z + i\sqrt{z}$$

is an infinitesimal generator on D. For $t = 1/2$, we have

$$z - tG(z) = z/2 - i\sqrt{z}/2.$$

For $w = 1/8 + i/8 \in D$, we obtain the equation

$$1/8 + i/8 = z/2 - i\sqrt{z}/2.$$

Let $u = \sqrt{z}$. Then

$$1/8 + i/8 = u^2/2 - iu/2, \quad \text{i.e.,} \quad u^2 - iu - 1/4 - i/4 = 0,$$

which leads to the solutions $u_1 = 1/(2\sqrt{2}) + i(1/2 + 1/(2\sqrt{2}))$ and $u_2 = -1/(2\sqrt{2}) + i(1/2 - 1/(2\sqrt{2}))$, which are both in $\mathbb{H}$. Furthermore, $u_1^2 \neq u_2^2$, and the nonlinear resolvent equation has two solutions.

CHAPTER 10

Graph Products as Quantum Random Walks

In this chapter, we take a quantum probabilistic look at graph theory. Undirected graphs can be represented by their adjacency matrices, which are symmetric and real-valued and thus self-adjoint quantum random variables. The five independences now arise as five products of graphs. We will see that certain independent additive processes can be regarded as growing graphs.

A survey on spectra of infinite graphs is given in [**134**], see also [**133**]. The quantum probabilistic view on graphs is covered in the books [**93**] and [**145**], which also treat several further aspects such as the quantum decomposition of graphs.

10.1. Graphs as non-commutative random variables

Let V be a vertex set, finite or countably infinite, with a distinguished vertex $o \in V$, and let $A : V \times V \to \{0, 1\}$ be a symmetric matrix with $A_{x,x} = 0$ for all $x \in V$. We can interpret A as the adjacency matrix of an undirected, loop-free graph with vertex set V, where $A_{x,y} = 1$ if and only if $x \sim y$, i.e., x and y are connected by an edge.

DEFINITION 10.1. We define a *(rooted) graph* as such a triple $G = (V, A, o)$. For $x \in V$, the degree $\deg(x)$ of x is defined as the number of vertices that x is connected with, which can be written as $\sum_{y \in V} A_{x,y}$. The degree of the graph is defined as

$$\deg(G) := \deg(A) := \sup_{x \in V} \deg(x).$$

If $\deg(A) < \infty$, then A can be regarded as a bounded self-adjoint operator on the Hilbert space

$$l^2(V) = \left\{ (x_v)_{v \in V} \mid x_v \in \mathbb{C}, \sum_{v \in V} |x_v|^2 < \infty \right\}.$$

The distinguished vertex $o \in V$ enables us to regard A as a quantum random variable in the quantum probability space

$$(B(l^2(V)), \langle \delta_o, \cdot \delta_o \rangle),$$

where $\delta_o \in l^2(V)$ with $(\delta_o)(o) = 1, (\delta_o)(x) = 0$ for $x \neq o$.

It can be shown by induction that the entries of the n-th power A^n of the adjacency matrix represent the numbers of n-step walks between two vertices. As the n-th moment of A is the (o, o)-entry of A^n, we have

$$\langle \delta_o, A^n \delta_o \rangle = \text{ the number of } n\text{-step walks from } o \text{ to itself.}$$

EXAMPLE 10.2. Let $V = \{0, 1, 2\}$ and connect 0 with 1 and 2.

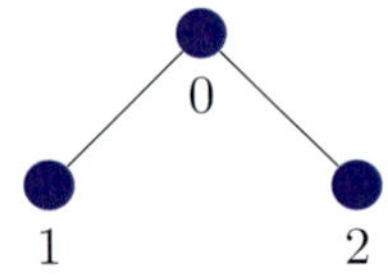

We choose $o = 0$. The number of n-step walks from o to itself is 0 for odd n, and if $n = 2k$ is even, then we have 2^k choices (k times we choose between going to 1 or to 2). The distribution μ of A is thus given by $\frac{1}{2}\delta_{\sqrt{2}} + \frac{1}{2}\delta_{-\sqrt{2}}$.

We can calculate μ also directly. The matrix

$$A = \begin{pmatrix} 0 & 1 & 1 \\ 1 & 0 & 0 \\ 1 & 0 & 0 \end{pmatrix}$$

has eigenvalues $\sqrt{2}$ with eigenvector $v_1 = (\sqrt{2}, 1, 1)^T/2$, $-\sqrt{2}$ with eigenvector $v_2 = (-\sqrt{2}, 1, 1)^T/2$, and 0 with eigenvector $v_3 = (0, -1, 1)^T/\sqrt{2}$. Hence, the distribution μ of A is given by

$$\mu = \langle \delta_o, v_1 \rangle^2 \delta_{\sqrt{2}} + \langle \delta_o, v_2 \rangle^2 \delta_{-\sqrt{2}} + \langle \delta_o, v_3 \rangle^2 \delta_0 = \frac{1}{2}\delta_{\sqrt{2}} + \frac{1}{2}\delta_{-\sqrt{2}}.$$

More generally, if we consider $V = \{0, 1, 2, \ldots, n\}$ and only connect 0 with 1, 2, ..., n, then the distribution of this graph is given by $\frac{1}{2}\delta_{\sqrt{n}} + \frac{1}{2}\delta_{-\sqrt{n}}$.

EXAMPLE 10.3. Let $V = \mathbb{Z}$ with $A_{j,k} = 1$ if and only if $|j - k| = 1$ and choose $o = 0$.

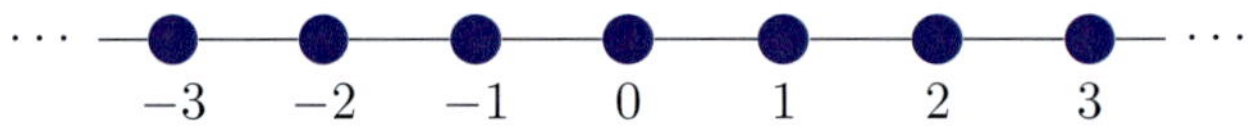

The number of n-step walks from o to itself is 0 for odd n. If $n = 2k$ is even, then a $2k$-step walk from o to itself consists of k steps to the right and k steps to the left. So we have $\binom{2k}{k}$ different choices. By Remark 5.4, the arcsine distribution $Arc(0, 1)$ has odd moments equal to 0, and its $2k$-th moment is given by $2^{-k}\binom{2k}{k}$. Thus, the $2k$-th moment of $Arc(0, 2)$ is equal to

$$(\sqrt{2})^{2k} \cdot 2^{-k}\binom{2k}{k} = \binom{2k}{k}.$$

Therefore, the distribution of A within the probability space $(B(l^2(\mathbb{Z})), \langle \delta_o, \cdot \delta_o \rangle)$ is given by $Arc(0, 2)$.

EXAMPLE 10.4. Consider the graph with vertex set $V = \mathbb{N}_0$, $o = 0$, and $j \sim k$ if and only if $|j - k| = 1$.

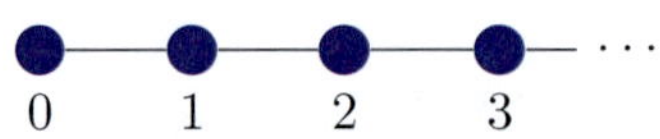

The distribution of this rooted graph is given by $W(0, 1)$, see Exercise 10.5.4.

Note that the previous three examples yield graphs whose distributions are (scalings of) the normal distributions from Boolean, monotone, and free probability theory.

Obviously, only special distributions arise from adjacency matrices.

THEOREM 10.5. *Let μ be the distribution of a rooted graph (V, A, o), where A is interpreted as a quantum random variable from $(B(l^2(V)), \langle \delta_o, \cdot \delta_o \rangle)$. Then*

$$\int_{\mathbb{R}} x\mu(dx) = 0 \quad \text{and} \quad \int_{\mathbb{R}} x^n \mu(dx) \in \mathbb{N}_0 \quad \text{for all} \quad n \geq 2.$$

PROOF. Clearly, $\int_{\mathbb{R}} x\mu(dx) = \langle \delta_o, A\delta_o \rangle = A_{o,o} = 0$ and for $n \geq 2$, $\int_{\mathbb{R}} x^n \mu(dx) = \langle \delta_o, A^n \delta_o \rangle$, which is a sum of zeros and ones, and thus a non-negative integer. □

Theorem 10.5 suggests the following inverse problem: Let $\mu \in \mathcal{P}_c(\mathbb{R})$ with $\int_{\mathbb{R}} x\mu(dx) = 0$ and $\int_{\mathbb{R}} x^n \mu(dx) \in \mathbb{N}_0$ for all $n \geq 2$. Is there a graph (V, A, o) whose distribution is μ?

REMARK 10.6. We note that our setting can be generalized by allowing loops and weighted edges. In this way, every $A \in B(l^2(V))$ with $A_{j,k} = A_{k,j} \in \mathbb{R}$ can be interpreted as a graph.

10.2. Graph products and independence

Let $G_1 = (V_1, A^1, o_1), G_2 = (V_2, A^2, o_2)$ be two graphs. (We use upper indices for the matrices for better readability.) Following [**93, 145**], we now construct new graphs with vertex set $V_3 = V_1 \times V_2$ and distinguished vertex $o_3 = (o_1, o_2)$.

The comb product. The comb product $G_1 \rhd G_2 = (V_3, A^3, o_3)$, see Figure 10.1, is defined via

$$A^3_{(x,y),(x',y')} = A^1_{x,x'} \delta_{y,o_2} \delta_{y',o_2} + \delta_{x,x'} A^2_{y,y'}. \tag{10.1}$$

Here we use the symbol $\delta_{x,y} = 1$ if $x = y$, $\delta_{x,y} = 0$ if $x \neq y$. So we have $(x, y) \sim (x', y')$ if and only if

- $x \sim x'$ and $y = y' = o_2$, or
- $x = x'$ and $y \sim y'$.

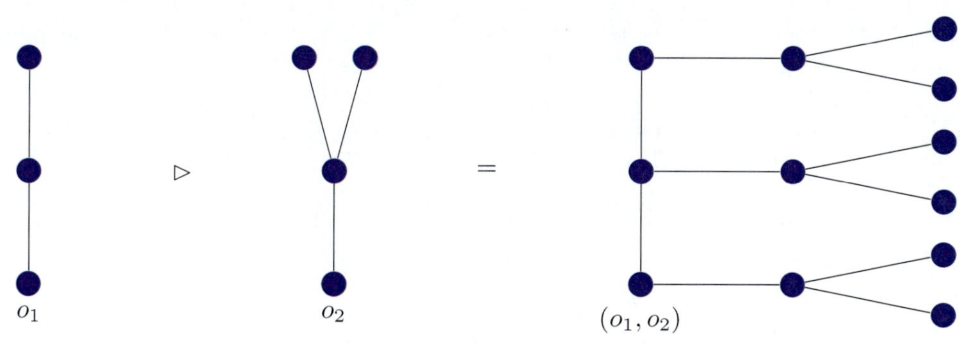

FIGURE 10.1. The comb product of two graphs.

If $\deg(G_1), \deg(G_2) < \infty$, then the adjacency matrix A^3 of $G_1 \rhd G_2$ acts on $l^2(V_1 \times V_2) \simeq l^2(V_1) \otimes l^2(V_2)$. Denote by I^1 the identity on $l^2(V_1)$ and by P^2 the projection from $l^2(V_2)$ onto the subspace spanned by δ_{o_2}, i.e., $(P^2(\psi))(y) = \delta_{y,o_2}\psi(o_2)$. It can easily be verified that

$$A^3 = A^1 \otimes P^2 + I^1 \otimes A^2.$$

REMARK 10.7. Note that both $A^1 \otimes P^2$ and $I^1 \otimes A^2$ are again adjacency matrices. While $A^1 \otimes P^2$ represents a natural embedding of G_1 into the larger vertex set $V_1 \times V_2$, $I^1 \otimes A^2$ represents $|V_1|$ many copies of G_2.

More generally, we have the following decomposition.

LEMMA 10.8. *Let $G_1 = (V_1, A^1, o_1), \dots, G_n = (V_n, A^n, o_n)$ be graphs. Denote by I^k the identity on $l^2(V_k)$ and by P^k the projection from $l^2(V_k)$ onto the subspace spanned by δ_{o_k}, i.e., $(P^k(\psi))(y) = \delta_{y,o_k}\psi(o_k)$. Denote by B the adjacency matrix of the graph $G_1 \rhd G_2 \rhd \cdots \rhd G_n$. Then*

$$B = \sum_{j=1}^{n} I^1 \otimes \cdots \otimes I^{j-1} \otimes A^j \otimes P^{j+1} \otimes \cdots \otimes P^n. \tag{10.2}$$

The lemma is a slightly more general version of [**4**, Theorem 3.1]. Its proof follows from definition (10.1) and by induction.

Assume that $\deg(G_j) < \infty$ for all $j = 1, \dots, n$. Then the adjacency matrix B can be regarded as a quantum random variable in

$$(B(l^2(V_1 \times \cdots \times V_n)), \langle \delta_{o_1} \otimes \cdots \otimes \delta_{o_n}, \cdot(\delta_{o_1} \otimes \cdots \otimes \delta_{o_n})\rangle).$$

The decomposition (10.2) is a generalization of the model of monotone independence from Theorem 4.26. So the tuple

$$(I^1 \otimes \cdots \otimes I^{j-1} \otimes A^j \otimes P^{j+1} \otimes \cdots \otimes P^n)_{j=1,\dots,n}$$

is monotonically independent (see also [**4**, Proposition 4.1]). Thus, the distribution of B is given by the monotone convolution of the distributions of the summands in (10.2). Furthermore, it is easy to see that the moments of $I^1 \otimes \cdots \otimes I^{j-1} \otimes A^j \otimes P^{j+1} \otimes \cdots \otimes P^n$ agree with the moments of A^j within $(B(l^2(V_j)), \langle \delta_{o_j}, \cdot \delta_{o_j}\rangle)$. Thus, we obtain:

LEMMA 10.9. *Assume that $\deg(G_j) < \infty$ for all $j = 1, \dots, n$. Then the random variables $(I^1 \otimes \cdots \otimes I^{j-1} \otimes A^j \otimes P^{j+1} \otimes \cdots \otimes P^n)_{j=1,\dots,n}$ are monotonically independent (ordered by j from $j = 1$ to $j = n$) in the quantum probability space $(B(l^2(V_1 \times \cdots \times V_n)), \langle \delta_{o_1} \otimes \cdots \otimes \delta_{o_n}, \cdot(\delta_{o_1} \otimes \cdots \otimes \delta_{o_n})\rangle)$. Let μ_j be the distribution of A_j within $(B(l^2(V_j)), \langle \delta_{o_j}, \cdot \delta_{o_j}\rangle)$. Then B has the distribution*

$$\mu_1 \rhd \mu_2 \rhd \cdots \rhd \mu_n.$$

The direct product. The direct product $G_1 \times G_2 = (V_3, A^3, o_3)$, see Figure 10.2, is defined by $(x, y) \sim (x', y')$ if and only if $x = x'$ and $y \sim y'$ or $x \sim x'$ and $y = y'$.

LEMMA 10.10. *Let $G_1 = (V_1, A^1, o_1), \dots, G_n = (V_n, A^n, o_n)$ be graphs. Denote by I^k the identity on $l^2(V_k)$ and by B the adjacency matrix of the graph $G_1 \times G_2 \times \cdots \times G_n$. Then*

$$B = \sum_{j=1}^{n} I^1 \otimes \cdots \otimes I^{j-1} \otimes A^j \otimes I^{j+1} \otimes \cdots \otimes I^n.$$

LEMMA 10.11. *Assume that $\deg(G_j) < \infty$ for all $j = 1, \dots, n$. Then the random variables $(I^1 \otimes \cdots \otimes I^{j-1} \otimes A^j \otimes I^{j+1} \otimes \cdots \otimes I^n)_{j=1,\dots,n}$ are tensor independent in the quantum probability space $(B(l^2(V_1 \times \cdots \times V_n)), \langle \delta_{o_1} \otimes \cdots \otimes \delta_{o_n}, \cdot(\delta_{o_1} \otimes \cdots \otimes \delta_{o_n})\rangle)$.*

Let μ_j be the distribution of A_j within $(B(l^2(V_j)), \langle \delta_{o_j}, \cdot \delta_{o_j} \rangle)$. Then B has the distribution

$$\mu_1 * \mu_2 * \cdots * \mu_n.$$

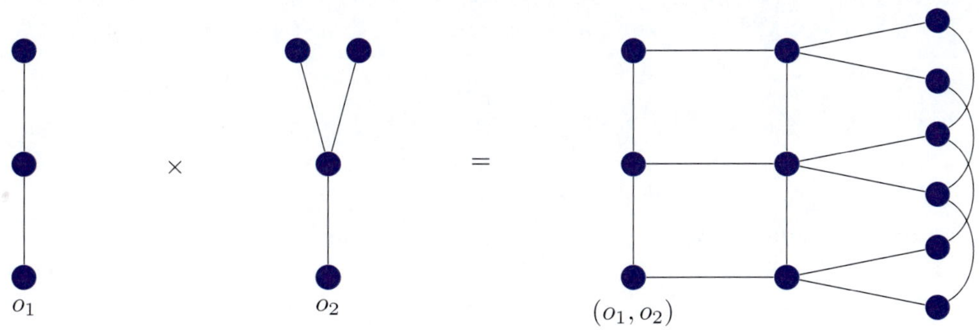

FIGURE 10.2. The direct product of two graphs.

The star product. The star product $G_1 \star G_2 = (V_3, A^3, o_3)$, see Figure 10.3, is defined by $(x, y) \sim (x', y')$ if and only if $x = x' = o_1$ and $y \sim y'$ or $x \sim x'$ and $y = y' = o_2$.

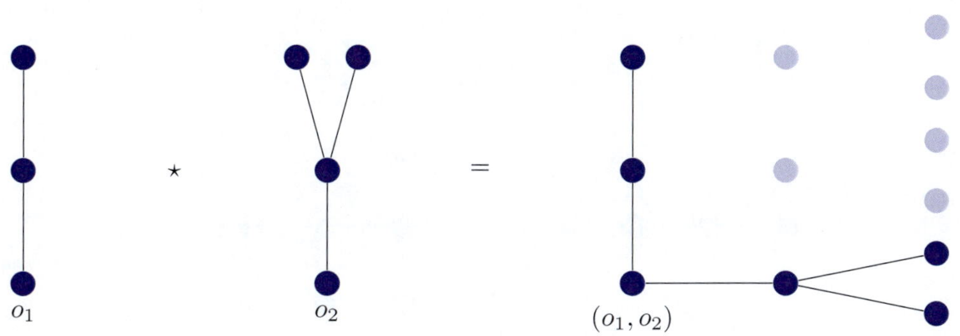

FIGURE 10.3. The star product of two graphs.

LEMMA 10.12. *Let $G_1 = (V_1, A^1, o_1), \ldots, G_n = (V_n, A^n, o_n)$ be graphs. Denote by P^k the projection from $l^2(V_k)$ onto the subspace spanned by δ_{o_k}. Denote by B the adjacency matrix of the graph $G_1 \star G_2 \star \cdots \star G_n$. Then*

$$B = \sum_{j=1}^{n} P^1 \otimes \cdots \otimes P^{j-1} \otimes A^j \otimes P^{j+1} \otimes \cdots \otimes P^n.$$

LEMMA 10.13. *Assume that $\deg(G_j) < \infty$ for all $j = 1, \ldots, n$. Then the random variables $(P^1 \otimes \cdots \otimes P^{j-1} \otimes A^j \otimes P^{j+1} \otimes \cdots \otimes P^n)_{j=1,\ldots,n}$ are Boolean independent in $(B(l^2(V_1 \times \cdots \times V_n)), \langle \delta_{o_1} \otimes \cdots \otimes \delta_{o_n}, \cdot (\delta_{o_1} \otimes \cdots \otimes \delta_{o_n}) \rangle)$. Let μ_j be the distribution of A_j within $(B(l^2(V_j)), \langle \delta_{o_j}, \cdot \delta_{o_j} \rangle)$. Then B has the distribution*

$$\mu_1 \uplus \mu_2 \uplus \cdots \uplus \mu_n.$$

The free product. Free independence can be realized via the free product of graphs, which is more complicated than the other cases. We refer to [**5**].

Approximation of additive processes. Fix a type of independence and its convolution $\star \in \{*, \boxplus, \uplus, \rhd\}$ and let $\star_G$ be the corresponding graph product. Let $T > 0$ and let $(X_t)_{t\in[0,T]}$ be an additive process with distributions $(\mu_t)_{t\in[0,T]}$.

Next let $G_{n,k}$, $n \in \mathbb{N}$, $k = 1, \dots, n$, be graphs and consider the discrete quantum process $(Y_k^n)_{k=1,\dots,n}$ given by the adjacency matrices of the n graphs

$$G_{n,1}, \quad G_{n,1} \star_G G_{n,2}, \quad \dots, \quad G_{n,1} \star_G G_{n,2} \star_G \cdots \star_G G_{n,n}.$$

Can we approximate X_t by Y_k^n, such that the adjacency matrix of $G_{n,k}$ approximates the increment $X_{kT/n} - X_{(k-1)T/n}$? If $\mu_{n,k}$ is the distribution of $G_{n,1} \star_G \cdots \star_G G_{n,k}$, then we would like that the scaled distribution $\mu_{n,\lfloor tn/T \rfloor}(c(n)\,\cdot)$ converges to μ_t as $n \to \infty$, where $c(n) > 0$ is some scaling factor.

We already know a necessary condition. By Theorem 10.5, μ_t, as well as the distributions of the increments, must have compact support with mean 0 and non-negative nth moments for all $n \geq 2$.

For example, as observed in [**4**], taking comb products of $\mathbb{Z}$, regarded as the graph from Example 10.3, can be interpreted as a discrete approximation of a monotone Brownian motion, a "monotone quantum random walk"; see Figure 10.4. With Examples 10.2 and 10.4, we obtain discrete approximations of Boolean (Figure 10.5) and free Brownian motions. The scalings will be different due to the different variances of the graphs.

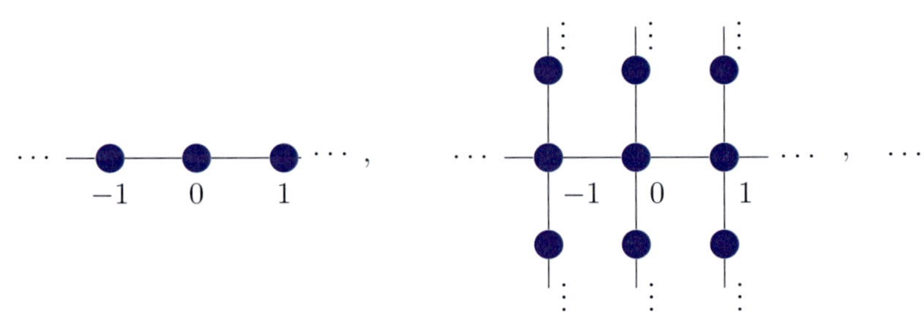

FIGURE 10.4. Approximation of a monotone Brownian motion.

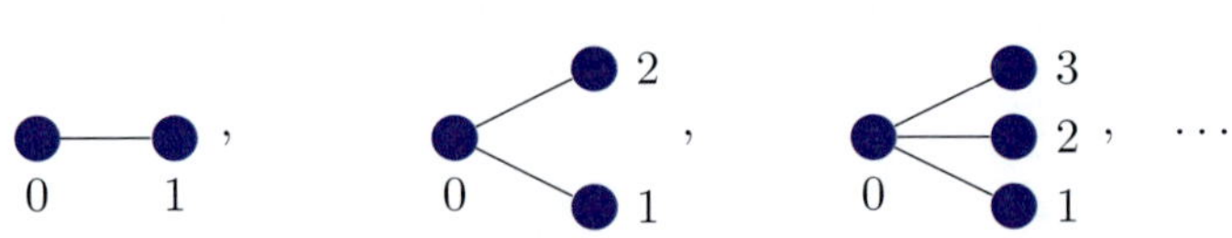

FIGURE 10.5. Approximation of a Boolean Brownian motion.

Next, we will see that in the case of monotone independence, an additional assumption on $(\mu_t)_{t\in[0,T]}$ guarantees that we can approximate $(X_t)_{t\in[0,T]}$ with $G_{n,k}$ being spidernets.

10.3. Spidernets

We now follow [**170**], which extends the idea from [**4**] to more general processes.

First, we construct special graphs whose distributions will be related to the Loewner equation. We denote by $d(x,y)$ the length of the shortest walk within a graph connecting x and y. For $\varepsilon \in \{-1,0,+1\}$, we define, for any $x \in V$,

$$\omega_\varepsilon(x) = |\{y \in V \mid y \sim x, d(o,y) = d(o,x) + \varepsilon\}|.$$

Let $a \in \mathbb{N}, b \in \mathbb{N} \setminus \{1\}$ and $c \in \mathbb{N}$ with $c \le b-1$. A *spidernet with data* (a,b,c), see [**93**, Def. 4.25]), is a graph (V,A,o) with root $o \in V$ such that

$$\omega_{+1}(o) = a, \quad \omega_{-1}(o) = \omega_0(o) = 0, \quad \text{and}$$
$$\omega_{+1}(x) = c, \quad \omega_{-1}(x) = 1, \quad \omega_0(x) = b-1-c$$

for all $x \in V \setminus \{o\}$ (and $A_{xy} \in \{0,1\}$ for all $x,y \in V$).

EXAMPLE 10.14. Example 10.4 with $V = \mathbb{N}_0$ is a spidernet with data $(1,2,1)$. Figure 10.6 shows two spidernets with data $(4,4,2)$.

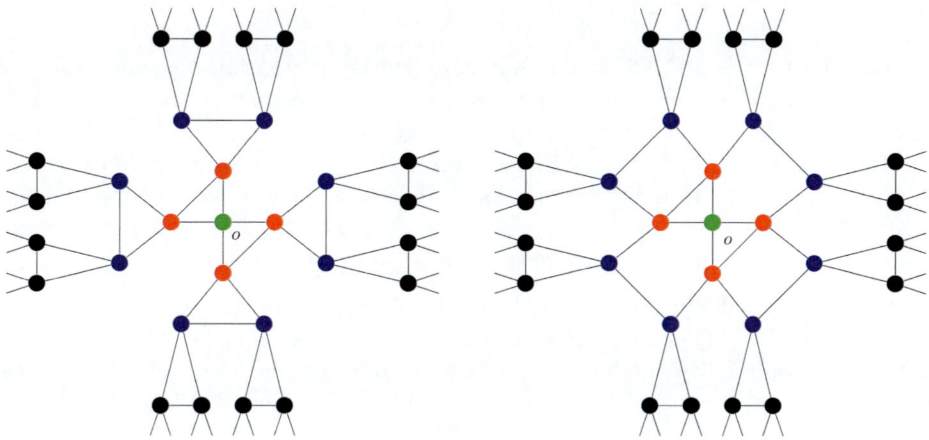

FIGURE 10.6. Two spidernets with data $(4,4,2)$.

The spectrum of the adjacency matrix of a spidernet with respect to the quantum probability space $(B(l^2(V)), \langle\delta_o, \cdot\delta_o\rangle)$ is the free Meixner law $m_{a,c,b-1-c}$, see [**98**, Thm. 7.3]. This distribution is described explicitly in [**98**, Section B]. For our purpose, it is sufficient to know its F-transform for the following special case.

LEMMA 10.15. *Let $n \in \mathbb{N}$ and $u \in \{0,\dots,2n-1\}$. Then there exists a spidernet $S_{n,u}$ with data $(2n, n+1+u, n)$. The distribution $\psi_{n,u}$ of $S_{n,u}$ (the free Meixner law $m_{2n,n,u}$) has mean 0 and variance $2n$, and its F-transform is given by*

$$F_{\psi_{n,u}}(z) = \sqrt{(z-u)^2 - 4n} + u.$$

PROOF. From looking at the $2n$ vertices with $d(o,x) = 1$, we get the necessary condition $b-1-c = u \le 2n-1$ for the existence of a spidernet with data $(2n, n+1+u, n)$. Conversely, one can verify by induction that for each $n \in \mathbb{N}$ and every $u \in \{0,\dots,2n-1\}$ there exists a spidernet with data $(2n, n+1+u, n)$.

The number of m-step walks from o to itself is equal to 0 for $m = 1$ and $2n$ for $m = 2$, so $\psi_{n,u}$ has mean 0 and variance $2n$. The statement that $\psi_{n,u}$ is given by the F-transform above is shown in [**98**] via a decomposition of the adjacency matrix A and the method of Jacobi coefficients. It is shown that $F_{\psi_{n,u}}$ can be written as a continued fraction

$$F_{\psi_{n,u}}(z) = z - \cfrac{2n}{z - u - \frac{n}{z-u-\frac{n}{z-u-\dots}}}.$$

See [**98**, Theorem 7.3] and the formula for the Cauchy transform of the free Meixner law in [**98**, Equation (B.1)]; see also [**145**, Section 6.6]. □

In the following, we denote by $S_{n,u}$ a fixed spidernet with data $(2n, n+1+u, n)$, $n \in \mathbb{N}$ and $u \in \{0, \dots, 2n-1\}$.

Lemma 10.15 is a fortunate result for us. The spidernets $S_{n,u}$ have a distribution whose F-transform is univalent. Moreover, $\sqrt{(z-u)^2 - 4n} + u$ is a simple slit mapping of a vertical line segment in the upper half-plane from u to $u + i2\sqrt{n}$. In other words, this mapping is the solution of the slit Loewner equation (8.2) with $U(t) \equiv u$ at $t = 2n$, see Exercise 5.9.7.

Therefore, approximating a non-negative driving function by piecewise constant, non-negative driving functions is related to approximating the corresponding measures by distributions of spidernets. Together with Lemma 10.9, we expect that the monotone additive process associated with the measures driven by a non-negative driving function can be approximated by a sequence of growing graphs.

10.4. Approximation via spidernets

We now consider a driving function $U : [0, \infty) \to \mathbb{R}$ that is continuous and non-negative. Let $(f_t)_{t\geq 0}$ be the solution to the slit equation (8.2) and denote by $(\mu_t)_{t\geq 0}$ the probability measures with $F_{\mu_t} = f_t$. Furthermore, let $(X_t)_{t\geq 0}$ be a corresponding monotone additive process given by Theorems 5.52, 5.57.

Fix some $T > 0$. We aim to approximate $(X_t)_{t\in[0,T]}$ by a discrete quantum process, where each random variable is the adjacency matrix of a graph. Using the lemmas above, we proceed as follows.

Choose $n_0 \in \mathbb{N}$ such that

$$0 \leq U(t) \leq \sqrt{\frac{T}{2}}\left(2\sqrt{n} - \frac{1}{\sqrt{n^3}}\right) \quad \text{on } [0, T] \tag{10.3}$$

for all $n \geq n_0$.

Now assume that $n \geq n_0$. For $k = 1, \dots, n$, we define

$$u_{n,k} = \lfloor \sqrt{2T}\sqrt{n} \cdot \frac{U(k/n \cdot T)}{\frac{T}{n}} \rfloor \in \{0, \dots, 2n^2 - 1\},$$

where $\lfloor x \rfloor$ denotes the largest $m \in \mathbb{N}_0$ with $m \leq x$. With Lemma 10.15 we see that (10.3) implies that the spidernet $S_{n^2, u_{n,k}}$ exists for all $k = 1, \dots, n$. We denote by $V_{n,k}$ the vertex set and by $o_{n,k}$ the root of $S_{n^2, u_{n,k}}$.

THEOREM 10.16. *For $k = 1, \ldots, n$, let $\mathcal{C}_{n,k}$ be the graph*

$$\mathcal{C}_{n,k} := S_{n^2,u_{n,1}} \rhd S_{n^2,u_{n,2}} \rhd \cdots \rhd S_{n^2,u_{n,k}}.$$

Then $(\mathcal{C}_{n,k})_{k=1,\ldots,n}$ is an approximation of the quantum process $(X_t)_{t\in[0,T]}$ in the following sense:

(a) *Let $A_{n,k}$ be the adjacency matrix of $\mathcal{C}_{n,k}$. Denote by $\mu_{n,k}$ the distribution of $A_{n,k}$ with respect to the quantum probability space*

$$(B(l^2(V_{n,1} \times \cdots \times V_{n,k})), \langle \delta_{o_{n,1}} \otimes \cdots \otimes \delta_{o_{n,k}}, \cdot (\delta_{o_{n,1}} \otimes \cdots \otimes \delta_{o_{n,k}})\rangle).$$

Then

$$\lim_{n\to\infty} \mu_{n,\lfloor tn/T \rfloor}(\sqrt{2n^3/T} \cdot) = \mu_t(\cdot)$$

with respect to weak convergence for all $t \in [0, T]$.

(b) *Embed $A_{n,k}$ into $B(l^2(V_{n,1} \times \cdots \times V_{n,n}))$ by $\mathcal{A}_{n,k} := A_{n,k} \otimes P^{n,k+1} \otimes \cdots \otimes P^{n,n}$, where $P^{n,j}$ denotes the projection in $l^2(V_{n,j})$ onto $\delta_{o_{n,j}}$, see Remark 10.7. Then the increments $(\mathcal{A}_{n,1}, \mathcal{A}_{n,2} - \mathcal{A}_{n,1}, \ldots, \mathcal{A}_{n,n} - \mathcal{A}_{n,n-1})$ are monotonically independent within the quantum probability space*

$$(B(l^2(V_{n,1} \times \cdots \times V_{n,n})), \langle \delta_{o_{n,1}} \otimes \cdots \otimes \delta_{o_{n,n}}, \cdot (\delta_{o_{n,1}} \otimes \cdots \otimes \delta_{o_{n,n}})\rangle).$$

PROOF. Statement (b) follows directly from Lemmas 10.8 and 10.9.

In order to show (a), we define the driving function $U_n : [0, 2n^3] \to \mathbb{R}$ as the function that is equal to $u_{n,1}$ on $[0, 2n^2]$, equal to $u_{n,2}$ on $(2n^2, 4n^2]$, etc.

Let $f_{n,t}$ be the solution to (8.2) with driving function U_n and define the measures $\alpha_{n,t}$ by $F_{\alpha_{n,t}} = f_{n,t}$. By Lemma 10.15, we have

$$\alpha_{n,2n^2} = \psi_{n^2,u_{n,1}}.$$

Solving the Loewner equation (8.2) with the initial condition $h_{2n^2}(z) = z$ and driving function $U_n(t)$ yields the mappings (h_t) that satisfy $f_{n,t} = f_{n,2n^2} \circ h_t$. Obviously,

$$h_{4n^2} = F_{\psi_{n^2,u_{n,2}}},$$

and thus, $\alpha_{n,4n^2} = \psi_{n^2,u_{n,1}} \rhd \psi_{n^2,u_{n,2}}$. By induction, we obtain

$$\alpha_{n,2kn^2} = \rhd_{j=1}^{k} \psi_{n^2,u_{n,j}}.$$

On the other hand, Lemmas 10.9 and 10.15 imply

$$\mu_{n,k} = \rhd_{j=1}^{k} \psi_{n^2,u_{n,j}} \tag{10.4}$$

for all $k = 1, \ldots, n$.

The function $V_n : [0, T] \to \mathbb{R}, V_n(t) := \sqrt{\frac{T}{2n^3}} \cdot U_n(t/T \cdot 2n^3)$, is constant on the intervals $\left(\frac{(k-1)T}{n}, \frac{kT}{n}\right]$, $k = 1, \ldots, n$. We have

$$U(k/n \cdot T) - V_n(k/n \cdot T) = U(k/n \cdot T) - \sqrt{\frac{T}{2n^3}} \cdot U_n(k \cdot 2n^2) =$$

$$U(k/n \cdot T) - \sqrt{\frac{T}{2n^3}} \cdot \lfloor \sqrt{2T}\sqrt{n} \cdot \frac{U(k/n \cdot T)}{\frac{T}{n}} \rfloor \leq \sqrt{\frac{T}{2n^3}}.$$

Now let $t \in \left(\frac{(k-1)T}{n}, \frac{kT}{n}\right)$ and denote by $\omega : [0,T] \to [0,\infty)$ a modulus of continuity of U for the interval $[0,T]$, i.e., $|U(x) - U(y)| \le \omega(|x-y|)$ for all $x, y \in [0,T]$, ω is increasing, vanishes at 0, and is continuous at 0. We have

$$|U(t) - V_n(t)| = |U(t) - V_n(kT/n)| \le$$
$$|U(t) - U(kT/n)| + |U(kT/n) - V_n(kT/n)| \le \omega\left(\frac{T}{n}\right) + \sqrt{\frac{T}{2n^3}}.$$

Finally, for $t = 0$, we have $V_n(0) = V_n(T/n)$, and thus,

$$|U(0) - V_n(0)| = |U(0) - U(T/n)| + |U(T/n) - V_n(T/n)| \le \omega\left(\frac{T}{n}\right) + \sqrt{\frac{T}{2n^3}}.$$

Hence, we obtain

$$\sup_{t\in[0,T]} |U(t) - V_n(t)| \to 0 \quad \text{as } n \to \infty. \tag{10.5}$$

Let $(h_{n,t})_{t\in[0,T]}$ be the Loewner chain that corresponds to V_n. Define the measures $\nu_{n,t}$ by $h_{n,t} = F_{\nu_{n,t}}$. Note that V_n has the form $V_n = U_n(d \cdot t)/c$ with $d = c^2$. Hence, by Lemma 8.4, we have

$$\nu_{n,t}(M) = \alpha_{n,t/T\cdot 2n^3}(\sqrt{2n^3/T} \cdot M)$$

for all $t \in [0,T]$ and all Borel subsets $M \subset \mathbb{R}$. If t has the form $t = kT/n, k = 1, \dots, n$, then (10.4) gives

$$\begin{aligned}\nu_{n,t}(M) &= \mu_{n,k}(\sqrt{2n^3/T} \cdot M) = (\rhd_{j=1}^{k} \psi_{n^2,u_{n,j}})(\sqrt{2n^3/T} \cdot M)\\ &= (\rhd_{j=1}^{tn/T} \psi_{n^2,u_{n,j}})(\sqrt{2n^3/T} \cdot M).\end{aligned}$$

For every $t \in [0,T]$, we have $h_{n,t} \to f_t$ locally uniformly because of (10.5) and Lemma 8.25. By Theorem 3.18, we have $\nu_{n,t} \to \mu_t$ with respect to weak convergence, or

$$\mu_{n,\lfloor tn/T\rfloor}(\sqrt{2n^3/T}\cdot) = (\rhd_{j=1}^{\lfloor tn/T\rfloor} \psi_{n^2,u_{n,j}})(\sqrt{2n^3/T}\,\cdot) \to \mu_t(\cdot). \qquad \square$$

From Section 8.5, we know that solutions to the slit equation can approximate solutions of the (bounded) normalized Loewner equation for additive Loewner chains. Therefore, we now consider equation (8.9)

$$\frac{\partial}{\partial t} f_t(z) = \frac{\partial}{\partial z} f_t(z) \cdot H(z,t) \quad \text{for a.e. } t \ge 0, f_0(z) = z \in \mathbb{H}, \tag{10.6}$$

where $t \mapsto H(z,t) = \int_{\mathbb{R}} \frac{\nu_t(du)}{u-z}$ is measurable for every $z \in \mathbb{H}$, ν_t is a probability measure on $\mathbb{R}$ for every $t \ge 0$, and, in addition, we assume that there exists $M > 0$ such that for all $t \ge 0$:

$$\operatorname{supp} \nu_t \subset [0, M].$$

Let $(f_t)_{t\geq 0}$ be the solution to the corresponding Loewner equation and denote by $(\mu_t)_{t\geq 0}$ the probability measures with $F_{\mu_t} = f_t$. Furthermore, let $(X_t)_{t\geq 0}$ be a corresponding additive process. Let $T > 0$. Then the process $(X_t)_{t\in[0,T]}$ can be approximated in the following way.

THEOREM 10.17. *Choose* $n_0 \in \mathbb{N}$ *such that* $M \leq \sqrt{\frac{T}{2}}\left(2\sqrt{n} - \frac{1}{\sqrt{n^3}}\right)$ *for all* $n \geq n_0$. *There exists a family* $(\mathcal{C}_{n,k})_{n\geq n_0, k=1,\dots,n}$ *of rooted graphs such that:*

(a) *For each* $n \geq n_0$, $(\mathcal{C}_{n,k})_{k=1,\dots,n}$ *can be considered as graphs with common vertex set* V_n *and common root* o_n. *Let* $A_{n,k}$ *be the adjacency matrix of* $\mathcal{C}_{n,k}$. *Then the increments* $(A_{n,1}, A_{n,2} - A_{n,1}, \dots, A_{n,n} - A_{n,n-1})$ *are monotonically independent with respect to the quantum probability space* $(B(l^2(V_n)), \langle \delta_{o_n}, \cdot \delta_{o_n}\rangle)$.

(b) *Denote by* $\mu_{n,k}$ *the distribution of* $A_{n,k}$. *Then*

$$\lim_{n\to\infty} \mu_{n,\lfloor tn/T\rfloor}(\sqrt{2n^3/T}\,\cdot) = \mu_t(\cdot)$$

with respect to weak convergence for all $t \in [0, T]$.

PROOF. Due to Lemma 8.26, there exists a sequence of continuous driving functions $U_m : [0, T] \to [-M, M]$ such that the corresponding solution $f_{m,t}$ to (8.2) converges locally uniformly to f_t for all $t \in [0, T]$ as $m \to \infty$. The proof shows that the driving functions can be chosen to be non-negative, i.e., $U_m : [0, T] \to [0, M]$. Define $\mu_{m,t}$ by $F_{\mu_{m,t}} = f_{m,t}$. Then Theorem 3.18 implies that $\lim_{m\to\infty} \mu_{m,t} = \mu_t$.

Let $\mathcal{C}_{n,k;m}$ be the graphs from Theorem 10.16 for the driving function U_m with distributions $\mu_{n,k;m}$. Note that $n \geq n_0$ and (10.3) together with the bound $U_m(t) \leq M$ imply that n is large enough to construct these graphs. Then

$$\lim_{n\to\infty} \mu_{n,\lfloor tn/T\rfloor;m}(\sqrt{2n^3/T}\,\cdot) = \mu_{m,t}(\cdot).$$

A diagonalization argument (note that there is a metric for probability measures on $\mathbb{R}$ that is compatible with weak convergence, e.g., the Lévy distance, see Remark 2.30) gives us a sequence $m(n)$ converging to ∞ such that

$$\lim_{n\to\infty} \mu_{n,\lfloor tn/T\rfloor;m(n)}(\sqrt{2n^3/T}\,\cdot) = \mu_t(\cdot).$$

Hence, the graphs $\mathcal{C}_{n,k} := \mathcal{C}_{n,k;m(n)}$ (where $\mathcal{C}_{n,k}$ is regarded as a subgraph of $\mathcal{C}_{n,n}$) satisfy all required conditions. □

REMARK 10.18. Each measure μ_t, generated by equation (10.6) with $\operatorname{supp}\nu_t \subset [0, M]$ for all $t \geq 0$, has a univalent F-transform, compact support, mean zero, and non-negative higher odd moments. However, not all probability measures with these properties arise in this way. To demonstrate this, we look at the Cauchy transforms of μ_t, which satisfy

$$\frac{\partial}{\partial t} G_{\mu_t}(z) = -\frac{\partial}{\partial z} G_{\mu_t}(z) \cdot G_{\nu_t}(z).$$

Denote by $m_n(t)$ and $k_n(t)$ the nth moment of μ_t and ν_t respectively. The power series expansion at ∞ yields

$$\frac{\dot{m}_1(t)}{z^2}+\frac{\dot{m}_2(t)}{z^3}+\frac{\dot{m}_3(t)}{z^4}+\cdots=$$
$$\left(\frac{1}{z^2}+\frac{2m_1(t)}{z^3}+\frac{3m_2(t)}{z^4}+\dots\right)\cdot\left(\frac{1}{z}+\frac{k_1(t)}{z^2}+\frac{k_2(t)}{z^3}+\dots\right).$$

We have $m_1(t)=0$, $m_2(t)=t$, and furthermore,

$$\dot{m}_3(t)=k_1(t)+2m_1(t)=k_1(t)\quad\text{for a.e. } t\geq 0.$$

As $k_1(t)\geq 0$ for a.e. $t\geq 0$, we have $m_3(T)=0$ for some $T>0$ only if $k_1(t)=0$ for a.e. $t\in[0,T]$, which implies $\nu_t=\delta_0$ for a.e. $t\in[0,T]$. So $m_3(T)=0$ implies that μ_T is the arcsine distribution $Arc(0,T)$.

Now consider a measure μ with compact support, symmetric with respect to 0, and having a univalent F-transform, e.g., the semicircle law $\mu=W(0,1)$. Then all odd moments are 0 and we see that such a measure can be generated in our setting only if μ is an arcsine distribution.

10.5. Exercises

E 10.5.1. Let $G = (V, A, o)$ be a finite rooted graph with n vertices $V = \{v_1, \ldots, v_n\}$. Denote by $\lambda_1 \leq \lambda_2 \leq \cdots \leq \lambda_n$ the eigenvalues of A.

(a) Prove that λ_n is bounded above by $\deg(G) = \max_{v \in V} \deg(v)$.

(b) Assume that G is connected and that $\lambda_n = \deg(G)$. Prove that G is regular, meaning that every vertex has the same degree.

E 10.5.2. For $n \in \mathbb{N}$, let $G_n = (V_n, A_n, o_n)$ be a rooted graph with $n = |V_n|$. Assume that G_n is complete, i.e., $x \sim y$ whenever $x \neq y$. Compute the distribution μ_n of G_n.

E 10.5.3. Compute the Cauchy transform of the distribution μ of the following rooted graph:

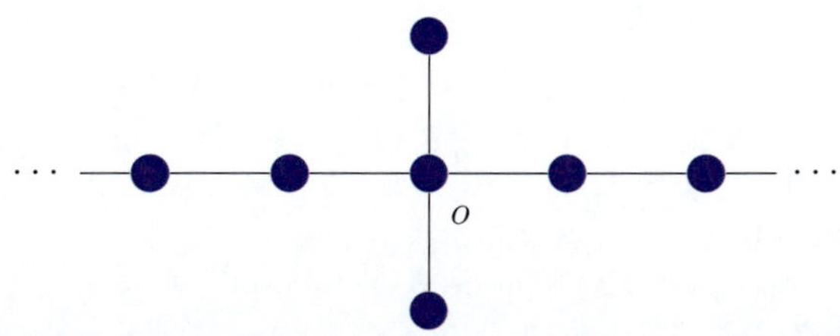

E 10.5.4. Consider the graph with vertex set $V = \mathbb{N}_0$, $o = 0$, and $j \sim k$ if and only if $|j - k| = 1$.

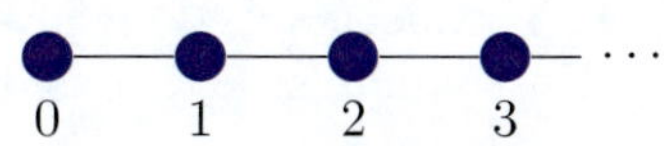

Show that its distribution is given by $W(0, 1)$.

10.6. Solutions

S 10.6.1.

(a) Let $w = (w_1, \ldots, w_n) \in \mathbb{R}^n$ be an eigenvector of A with eigenvalue λ_n and let $\max_{j=1,\ldots,n} |w_j| = |w_{j_{max}}|$. We may assume that $w_{j_{max}} > 0$. ($w_{j_{max}}$ cannot be 0. If it is negative, we consider $-w$ instead.)

We have $\lambda_n w_{j_{max}} = (Aw)_{j_{max}}$, where we denote by $(Aw)_{j_{max}}$ the j_{max}-entry of Aw. Therefore,

$$\begin{aligned}\lambda_n w_{j_{max}} &= (Aw)_{j_{max}} = \sum_{k=1}^{n} A_{j_{max},k} w_k \leq \sum_{k=1}^{n} A_{j_{max},k} w_{j_{max}} \\ &= \deg(v_{j_{max}}) w_{j_{max}} \leq \deg(G) w_{j_{max}}.\end{aligned} \tag{10.7}$$

(b) Now assume that G is connected and that $\lambda_n = \deg(G)$. Then we have equality in (10.7), which implies $\deg(v_{j_{max}}) = \deg(G)$ and $w_{j_{max}} = w_k$ for all k with $A_{j_{max},k} = 1$. For all such k, we have $\max_{j=1,\ldots,n} |w_j| = w_k$, and we can repeat the argument from (a) to obtain $\lambda_n w_k \leq \deg(v_k) w_k$. Thus, $\deg(v_k) = \lambda_n = \deg(G)$. As G is connected, we obtain by induction that $\deg(G) = \deg(v)$ for all $v \in V$.

S 10.6.2. We have $A_{n;x,x} = 0$ and $A_{n;x,y} = 1$ for all $x, y \in V$ with $x \neq y$. First, the vector $w = (1, 1, \ldots, 1)/\sqrt{n}$ is a normalized eigenvector of A with eigenvalue $n - 1$. Let $I_n \in \mathbb{R}^{n \times n}$ be the identity matrix and consider the matrix

$$I_n - (-1)A_n = I_n + A_n.$$

It is a rank-one matrix, and thus, -1 is an eigenvalue of A_n with multiplicity $n-1$. We conclude that μ_n is given by

$$\mu_n = \langle \delta_o, w \rangle^2 \delta_{n-1} + (1 - \langle \delta_o, w \rangle^2)\delta_{-1} = \frac{1}{n}\delta_{n-1} + \frac{n-1}{n}\delta_{-1}.$$

S 10.6.3. The graph is the star product of the graphs from Example 10.3 and Example 10.2:

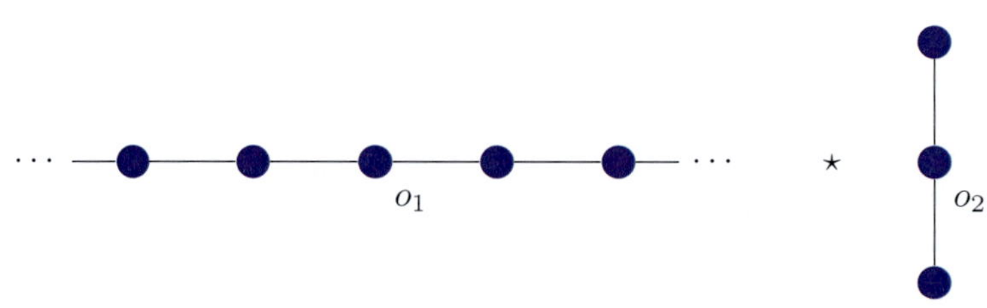

Denote by α and β the distribution of the left and right graph respectively. Then

$$B_\alpha(z) = z - \sqrt{z^2 - 4}, \quad B_\beta(z) = z - \frac{2}{\frac{1}{z-\sqrt{2}} + \frac{1}{z+\sqrt{2}}}.$$

Thus, $B_\mu(z) = B_{\alpha \uplus \beta}(z) = B_\alpha(z) + B_\beta(z)$, which leads to

$$G_\mu(z) = \frac{1}{\sqrt{z^2 - 4} - \frac{2}{z}}.$$

S 10.6.4. We think of the graph as being embedded in $\mathbb{Z}$ (as in Example 10.3).

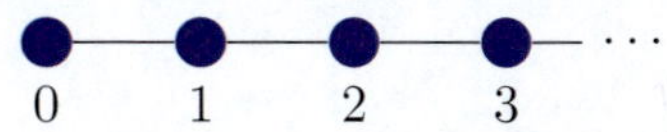

The number of n-step walks from o to itself is 0 for odd n. For $n = 2k$, we choose k positive steps and k negative steps such that we stay within $\mathbb{N}_0$. Let K_{2k} be the set of such walks and let M_{2k} be the set of all $2k$-step walks on $\mathbb{Z}$ from o to itself.

Then a walk $(0, v_1, \ldots, v_{2k-1}, 0)$ from M_{2k} belongs to $M_{2k} \setminus K_{2k}$ if and only if it visits -1, i.e., at least one of the vertices $v_1, \ldots, v_{2k-1}$ is equal to -1. Let p be the smallest index such that $v_p = -1$. Now consider the reflection of this walk with respect to -1 (i.e., $v \mapsto -v-2$) up to v_p, that is, the walk $(-2, -v_1-2, \ldots, -v_p-2 = v_p, v_{p+1}, \ldots, 0)$. The walks obtained in this way correspond to the set of all $2k$-step walks from -2 to 0. The size of this set is $\binom{2k}{k+1}$, as we choose $k+1$ positive steps. Thus

$$\begin{aligned}|K_{2k}| = |M_{2k}| - |M_{2k} \setminus K_{2k}| &= \binom{2k}{k} - \binom{2k}{k+1} \\ &= \binom{2k}{k} - \frac{k}{k+1}\binom{2k}{k} = \frac{1}{k+1}\binom{2k}{k}.\end{aligned}$$

This number is also called the k-th *Catalan number.* In Exercise 4.6.8, we have seen that it is equal to the $2k$-th moment of the semicircle distribution $W(0, 1)$.

CHAPTER 11

Outlook on Higher Dimensional Generalizations

Many notions that we have studied can be extended to more general settings. In this expository chapter, we look at two important generalizations: Loewner chains in higher dimensions and abstract probability spaces with operator-valued expectations. Both fields are under active research and questions regarding how results or even notions from the one-dimensional case can be generalized to these two settings, quickly lead to open problems.

11.1. Loewner theory in complex analysis of several variables

Let $D \subset \mathbb{C}^n$ be a domain. A function $f : D \to \mathbb{C}$ is called holomorphic if, for every $p = (p_1, \ldots, p_n) \in D$, f can be represented by a convergent power series

$$f(z_1, \ldots, z_n) = \sum a_{k_1,\ldots,k_n}(z_1 - p_1)^{k_1} \cdot \cdots \cdot (z_n - p_n)^{k_n}$$

in a neighborhood of p. A function $f = (f_1, \ldots, f_n) : D \to E$, where $E \subset \mathbb{C}^n$ is another domain, is called holomorphic if all $f_1, \ldots, f_n$ are holomorphic and it is called biholomorphic if f is bijective and f^{-1} is holomorphic as well.

These definitions are straightforward and seemingly harmless, but they have rather surprising implications when $n \geq 2$. One surprise concerns the extension of holomorphic functions. Consider the following domain H, called *Hartogs' figure*:

$$H := \{(z, w) \in \mathbb{D}^2 \mid |z| < 1/2 \text{ or } |w| > 1/2\} \subset \mathbb{D}^2.$$

By using Cauchy's integral formula, one can show that every holomorphic $f : H \to \mathbb{C}$ extends to a holomorphic function on $\mathbb{D}^2$. We cite another result of this form, see [**99**, Theorem 1.9.1].

THEOREM 11.1 (Hartogs' extension theorem). *Let $n > 1$, $D \subset \mathbb{C}^n$ be a domain, and $K \subset D$ a compact subset such that $D \setminus K$ is connected. Then every holomorphic $f : D \setminus K \to \mathbb{C}$ extends to a holomorphic function on D.*

In these two examples, the boundaries of the domains seem to be partially invisible to holomorphic functions. Such phenomena are entirely absent in one-dimensional complex analysis. In higher dimensions, it thus becomes essential to define and characterize "proper" domains for holomorphic functions. To prevent the extension of all holomorphic functions, one defines a domain of holomorphy as follows.

DEFINITION 11.2. A domain $D \subset \mathbb{C}^n$ is called a *domain of holomorphy* if the following holds: if $V \subset U \subset \mathbb{C}^n$ are domains with the properties

(a) $V \subset D$,
(b) for every holomorphic $f : D \to \mathbb{C}$ there is a holomorphic $F : U \to \mathbb{C}$ such that $f|_V = F|_V$,

then $U \subset D$.

In one dimension, every domain $D \subset \mathbb{C}$ is a domain of holomorphy: This is clear if $D = \mathbb{C}$, so let $D \subsetneq \mathbb{C}$ and choose $V \subset U \subset \mathbb{C}$ with the properties (a) and (b) and $U \not\subset D$. There exists a point $p \in \partial D \cap U$ and the function $f(z) = \frac{1}{z-p}$ is holomorphic in D. Because of $f|_V = F|_V$ and the identity principle, we have $F(z) = \frac{1}{z-p}$ on U as U is a domain. But F is not holomorphic in U.

Domains of holomorphy can be characterized by a property called pseudoconvexity. In particular, every convex domain is a domain of holomorphy. Therefore, for any norm in $\mathbb{C}^n$, the corresponding unit ball is a domain of holomorphy. For introductions to complex analysis in several variables we refer to [**94**, **99**, **110**].

If we want to study (decreasing) Loewner chains $(f_t)_{t\geq 0}$ on a domain $D \subset \mathbb{C}^n$, it is reasonable to ensure that D is a domain of holomorphy. But then, each $f_t(D)$ is also a domain of holomorphy, and for $n \geq 2$ we cannot expect $(f_t(D))_{t\geq 0}$ to describe deformations of D as arbitrary as in one dimension.

This is further strengthened by the "failure" of the Riemann mapping theorem in higher dimensions. For $n = 1$, if $D \subsetneq \mathbb{C}$ is a simply connected domain, then the Riemann mapping theorem guarantees that any simply connected subdomain $E \subseteq D$ can be mapped biholomorphically onto D. We also used it for proving our embedding results, Theorems 6.38 and 7.8. In higher dimensions, the homeomorphic equivalence of two domains is far less related to biholomorphic equivalence. Poincaré [**149**] already observed that the Euclidean unit ball

$$B_n = \{(z_1, \ldots, z_n) \in \mathbb{C}^n \,|\, |z_1|^2 + \cdots + |z_n|^2 < 1\}$$

cannot be mapped biholomorphically onto the polydisk $\mathbb{D}^n$ (the unit ball with respect to the maximum norm) for $n \geq 2$.

On a positive note, the correspondence between Loewner chains and Herglotz vector fields can indeed be established on complete Kobayashi hyperbolic manifolds, see [**46**]. Special cases mostly treat unit balls in $\mathbb{C}^n$ (or in general Banach spaces) with respect to a norm and their corresponding normalized radial Loewner chains, see the book [**80**].

The infinitesimal generators associated with normalized radial Loewner chains on $\mathbb{D}$ are given by $G(\mathbb{D}) := \{z \mapsto -zp(z) \,|\, p \text{ holomorphic}, \mathrm{Re}(p) > 0, p(0) = 1\}$. For the Euclidean unit ball, the corresponding generators are given by

$$\begin{aligned} G(B_n) :=\{&h : B_n \to \mathbb{C}^n \,|\, h \text{ holomorphic}, h(0) = 0, h'(0) = -I_n, \\ &\mathrm{Re}\,\langle h(z), z\rangle < 0 \text{ for all } z \in B_n \setminus \{0\}\}. \end{aligned}$$

Here, I_n denotes the identity matrix. A decreasing Loewner chain $(f_t)_{t\geq 0}$ on B_n, normalized with $f_t(0) = 0$, $f_t'(0) = e^{-t}I_n$, now satisfies Loewner's partial differential equation

$$\frac{\partial}{\partial t} f_t(z) = f_t'(z)G(z,t) \quad \text{for a.e. } t \geq 0 \text{ and all } z \in B_n,\ f_0(z) = z,$$

and $G(\cdot, t) \in G(B_n)$ for a.e. $t \geq 0$. In case of the polydisk, the corresponding infinitesimal generators are given by

$$G(\mathbb{D}^n) := \{h \in \mathbb{D}^n \to \mathbb{C}^n \mid h \text{ holomorphic}, h(0) = 0, h'(0) = -I_n,$$
$$\operatorname{Re}\left(\frac{h_j(z)}{z_j}\right) < 0 \text{ when } \|z\|_\infty = |z_j| > 0\}.$$

We do not have slit mappings in higher dimensions, but we may try to generalize Loewner's slit equation as follows. The sets $G(B_n)$ and $G(\mathbb{D}^n)$ are convex sets. For a convex subset M, a point $p \in M$ is called an *extreme point* if it cannot be expressed as $p = tx + (1-t)y$ for $x, y \in M$ with $x \neq y$ and $t \in (0,1)$. The extreme points of $G(\mathbb{D})$ are all functions $-zp(z)$ where p is a function with a point measure in its Herglotz representation, i.e., $-zp(z) = -z\frac{u+z}{u-z}$ for some $u \in \partial\mathbb{D}$. These functions correspond exactly to the infinitesimal generators appearing in Loewner's radial slit equation.

QUESTION 11.3. How can the extreme points of $G(B_n)$ and $G(\mathbb{D}^n)$ be characterized? Some partial results can be found in [**188**].

Loewner's parametric representation can also be extended to B_n [**80**]: One can solve the ordinary Loewner equation $\frac{d}{dt}\varphi_t(z) = G(\varphi_t(z), t)$, where $t \mapsto G(z,t)$ is measurable for all $z \in B_n$ and $z \mapsto G(z,t)$ belongs to $G(B_n)$ for all $t \geq 0$. Then the limit $\lim_{t\to\infty} e^t\varphi_t$ exists and is a univalent function on B_n. Denote the set of all functions obtained in this way by $S^0(B_n)$. This class can be shown to be compact with respect to locally uniform convergence. It is a subclass of

$$S(B_n) := \{f : B_n \to \mathbb{C}^n \mid f \text{ univalent}, f(0) = 0, f'(0) = I_n\}.$$

For $n = 1$, the classes $S(B_n)$ and $S^0(B_n)$ coincide, but not for $n \geq 2$: As the map $(z_1, \ldots, z_{n-1}, z_n + h(z_1))$ belongs to $S(B_n)$ for any entire function $h : \mathbb{C} \to \mathbb{C}$ with $h(0) = h'(0) = 0$, $S(B_n)$ is not compact when $n \geq 2$.

QUESTION 11.4. Let $N \geq 2$ and $f = (f_1, \ldots, f_N) \in S^0(B_N)$, and consider the power series expansion $f_1(z_1, 0, \ldots, 0) = z_1 + \sum_{n=2}^\infty a_n z_1^n$. It is known that $|a_2| \leq 2$. Does the Bieberbach conjecture $|a_n| \leq n$ hold for all $n \geq 2$?

See [**49**] for some coefficient bounds for $S^0(B_N)$. This procedure also yields a compact class, $S^0(\mathbb{D}^n)$, of univalent functions on the polydisk, see [**154, 155**], and [**76, 169**] for coefficient estimates.

11.2. Operator-valued expectations

A *-algebra $\mathcal{A}$ over $\mathbb{C}$ is a vector space over $\mathbb{C}$ together with a $\mathbb{C}$-bilinear associative product $(a,b) \mapsto a \cdot b$ and a map $a \mapsto a^*$ with

$$(a^*)^* = a, \quad (ab)^* = b^*a^*, \quad (za + wb)^* = \overline{z}a^* + \overline{w}b^*$$

for all $a, b \in \mathcal{A}$ and $z, w \in \mathbb{C}$. We assume further that $\mathcal{A}$ is unital, i.e., there exists $1 \in \mathcal{A}$ with $1 \cdot a = a \cdot 1 = a$ for all $a \in \mathcal{A}$. We can now call $a \in \mathcal{A}$

- self-adjoint if $a = a^*$;
- unitary if $aa^* = a^*a = 1$;
- non-negative if $a = x^*x$ for some $x \in \mathcal{A}$.

An expectation on $\mathcal{A}$ is defined as a $\mathbb{C}$-linear functional $\varphi : \mathcal{A} \to \mathbb{C}$ with $\varphi(1) = 1$ and $\varphi(x^*x) \geq 0$ for all $x \in \mathcal{A}$.

More generally, let $\mathcal{B} \subset \mathcal{A}$ be a unital *-subalgebra. Then we can define a $\mathcal{B}$-valued expectation on $\mathcal{A}$ as a $\mathbb{C}$-linear functional $\varphi : \mathcal{A} \to \mathcal{B}$ such that $\varphi(x^*x)$ is non-negative for all $x \in \mathcal{A}$, $\varphi(b_1) = b_1$ and $\varphi(b_1 a b_2) = b_1 \varphi(a) b_2$ for all $a \in \mathcal{A}$, $b_1, b_2 \in \mathcal{B}$.

The triple $(\mathcal{A}, \mathcal{B}, \varphi)$ defines a non-commutative *-probability space. Note that we can regard φ as a conditional expectation on $\mathcal{A}$ (see the properties of $\mathbb{E}[\cdot|\mathcal{G}]$ from Theorem 2.36).

EXAMPLE 11.5. Let $\mathcal{A} = \mathbb{C}^{n\times n}$ with X^* being the adjoint, i.e., the conjugate transpose of $X \in \mathcal{A}$. We have seen that the mapping $X \mapsto \langle e_1, Xe_1 \rangle$, $e_1 = (1, 0, \ldots, 0)^T$, defines a $\mathbb{C}$-valued expectation. It simply maps $[X_{j,k}]_{j,k=1,\ldots,n}$ to the entry $X_{1,1}$. Now let $m \in \mathbb{N}$ and $n > m$. Then $\mathcal{B} = \mathbb{C}^{m\times m}$ is a unital *-subalgebra of $\mathcal{A}$, embedded by mapping each $b \in \mathcal{B}$ to the block matrix $\begin{pmatrix} b & 0 \\ 0 & 0 \end{pmatrix} \in \mathbb{C}^{n\times n}$. The function $[X_{j,k}]_{j,k=1,\ldots,n} \mapsto [X_{j,k}]_{j,k=1,\ldots,m}$ defines a $\mathcal{B}$-valued expectation on $\mathcal{A}$.

In general, the multiplication in $\mathcal{B}$ is not commutative. Thus, the distribution of $X \in \mathcal{A}$ will be encoded by the set of all $\varphi(b_1 X b_2 X \cdots X b_{n+1})$, where $b_1, \ldots, b_{n+1} \in \mathcal{B}$, $n \in \mathbb{N}$. In this purely algebraic setting, we can now define the five types of independence. Furthermore, we can use formal power series, instead of holomorphic functions, to define transforms such as the Cauchy transform. See, e.g., [**90**] for monotone independence. This general setting also has applications in random matrix theory, such as to band and block random matrices; see [**176**], which considers Gaussian random band matrices, and [**116, 117, 159, 181**].

It is natural to endow $\mathcal{A}$ with a norm that turns it into a Banach algebra. One further condition on the norm leads to the following notion, which is a standard concept in functional analysis.

A C^*-*algebra* is a *-algebra $\mathcal{A}$ which is a Banach algebra (i.e., $\mathcal{A}$ is a complete normed vector space and $\|ab\| \leq \|a\|\|b\|$) with

$$\|a^*a\| = \|a\|\|a^*\| \text{ for all } a \in \mathcal{A} \ (C^*\text{-property}).$$

EXAMPLE 11.6.

(1) The space of all bounded operators on a Hilbert space, as in Definition 4.9, yields an example of a C^*-probability space. One can verify that $\|a\| = \|a^*\|$ and then

$$\|a\|\|a^*\| = \|a\|^2 = \sup_{\|x\|=1} \|ax\|^2 = \sup_{\|x\|=1} \langle ax, ax \rangle =$$
$$\sup_{\|x\|=1} \langle x, a^*ax \rangle \leq \sup_{\|x\|=1} \|a^*ax\|\|x\| = \|a^*a\|.$$

This proves the C^*-property as we also have $\|a^*a\| \leq \|a^*\|\|a\|$.

(2) An example for a commutative C^*-algebra is given by the set of all complex-valued continuous functions on a compact Hausdorff space (with addition, multiplication, and conjugation defined pointwise, and the supremum norm).
(3) The set of all complex diagonal $n \times n$-matrices is a C^*-subalgebra of $\mathbb{C}^{n\times n}$, which is commutative and of the form (2).
(4) Let $\mathcal{A}$ be the set of all bounded holomorphic functions $f : \mathbb{D} \to \mathbb{C}$ with involution $f^*(z) = \overline{f(\overline{z})}$. Endowed with the supremum norm, pointwise addition and multiplication, this space is a Banach algebra. However, it does not satisfy the C^*-property. For $f(z) = e^{iz}$ we have $\|f^*f\| = \|e^{-iz}e^{iz}\| = \|1\| = 1$, but $\|f^*\|\|f\| = \|f\|^2 = (\sup_{z\in\mathbb{D}} |e^{iz}|)^2 = (\sup_{z\in\mathbb{D}} |e^{-\operatorname{Im}(z)}|)^2 = e^2$.

REMARK 11.7. The Gelfand-Naimark theorem ([**43**, Section II.6.4]) states that every C^*-algebra is isomorphic to a C^*-algebra consisting of bounded operators on a Hilbert space. Furthermore, commutative C^*-algebras can always be represented as complex-valued continuous functions on a locally compact Hausdorff space X that vanish at infinity ([**43**, II.2.2.4]). This space is unital if and only if X is compact, and then we obtain all continuous functions on X. (The term C^*-algebra was introduced by I. E. Segal in 1947, where "C" stands for "closed in the norm topology"; see [**174**, p. 75].)

Now let $(\mathcal{A}, \mathcal{B}, \varphi)$ be a *-probability space, where $\mathcal{A}$ is a unital C^*-algebra and $\mathcal{B} \subset \mathcal{A}$ a unital sub-C^*-algebra. Let $X \in \mathcal{A}$ be self-adjoint.

In this setting, the Cauchy transform $\varphi((z - X)^{-1})$ can now be considered as an analytic function defined on the upper half-plane

$$\bigcup_{\varepsilon>0} \{b \in \mathcal{B} \mid \operatorname{Im}(b) = (b - b^*)/(2i) > \varepsilon\}.$$

For $\|z\|$ large enough we have $\varphi((z - X)^{-1}) = \sum_{n=0}^{\infty} \varphi((z^{-1}X)^n z^{-1})$, and we see that this transform does not encode all mixed moments $\varphi(b_1 X b_2 X \cdots X b_{n+1})$.

This issue can be handled by using matrices $\mathcal{A}^{n\times n}$. For $A = (a_{j,k}) \in \mathcal{A}^{n\times n}$, we let $(a_{j,k})^* = (a^*_{k,j})$ and we define $\varphi(A)$ by entrywise extension. An element in $\mathcal{A}^{n\times n}$ is non-negative if it can be written as A^*A for some $A \in \mathcal{A}^{n\times n}$. (We note that $\mathcal{A}^{n\times n}$ is also a C^*-algebra, consisting of operators on the Hilbert C^*-module $\mathcal{A}^n$, see [**114**, p. 1-8].) Let us assume that φ is not only positive, but *completely positive*, i.e.,

$$\varphi(A^*A) \geq 0 \quad \text{for all } A \in \mathcal{A}^{n\times n} \text{ and all } n \in \mathbb{N}.$$

EXAMPLE 11.8. Let $\mathcal{A}$ be the space of all bounded operators on a Hilbert space H, $\mathcal{B} = \mathbb{C}$, and $\varphi(a) = \langle \psi, a\psi \rangle$ for some $\psi \in H$ with $\|\psi\| = 1$ and all $a \in \mathcal{A}$. Let $A = (a_{j,k}) \in \mathcal{A}^{n\times n}$, $n \in \mathbb{N}$. The (j,k)-entry of A^*A is given by $\sum_{l=1}^{n} a^*_{l,j} a_{l,k}$. Then $\varphi(A^*A) \geq 0$ if and only if

$$\sum_{j,k} \overline{v_j} \varphi\left(\sum_{l=1}^{n} a^*_{l,j} a_{l,k}\right) v_k \geq 0 \quad \text{for every } v = (v_1, \ldots, v_n)^T \in \mathbb{C}^n.$$

We have

$$\sum_{j,k} \overline{v_j}\varphi\left(\sum_{l=1}^{n} a_{l,j}^* a_{l,k}\right) v_k = \sum_{j,k,l} \overline{v_j}\left\langle \psi, a_{l,j}^* a_{l,k}\psi \right\rangle v_k = \sum_{j,k,l} \left\langle v_j\psi, a_{l,j}^* a_{l,k} v_k\psi \right\rangle$$

$$= \sum_{j,k,l} \left\langle a_{l,j} v_j\psi, a_{l,k} v_k\psi \right\rangle = \sum_{l} \left\langle \sum_{j} a_{l,j} v_j\psi, \sum_{k} a_{l,k} v_k\psi \right\rangle = \langle A\psi v, A\psi v\rangle \geq 0.$$

In the last expression, we view ψv as an element from the Hilbert space H^n and $A\psi v$ is defined by matrix multiplication. Thus φ is completely positive.

Let $I_n \in \mathcal{A}^{n\times n}$ be the identity. We now obtain a Cauchy transform that encodes the distribution of X and lives on a union of matricial upper half-planes:

$$G_X(z) = \varphi((z - XI_n)^{-1}),\, z \in \mathbb{H}(\mathcal{B}) = \bigcup_{n\in\mathbb{N}} \bigcup_{\varepsilon>0} \{b \in \mathcal{B}^{n\times n} \mid \operatorname{Im}(b) = (b - b^*)/(2i) > \varepsilon\}.$$

The F-transform is simply defined as $F_X = G_X^{-1}$ and as φ is completely positive, it maps $\mathbb{H}(\mathcal{B})$ into itself.

These types of transforms lead to the theory of "non-commutative holomorphic functions", which is discussed in [**102**], [**196**]. One can now classify those analytic functions that are Cauchy transforms and F-transforms (for bounded distributions, see [**196**]). For a Herglotz representation in this context, see [**148**]. For the case of monotone independence, we refer to the works [**10, 29, 90, 153, 177**]. In [**100**], the author studies (bounded) additive processes with monotonically independent increments and establishes a one-to-one correspondence between their distributions and Loewner chains consisting of F-transforms.

Higher-dimensional scenarios are relevant in multivariate settings, i.e., when considering tuples of random variables, see [**27**], and [**73, 81, 88, 97**] for monotone independence. We conclude this section by looking at monotonically independent pairs as discussed in [**81**].

EXAMPLE 11.9. In the setting from above, we take $\mathcal{B} = \mathbb{C}$ and we choose self-adjoint $a, b \in \mathcal{A}$ with $ab = ba$.

The distribution of (a, b) is given by the set of all mixed moments $\varphi(a^m b^n)$, where $m, n \geq 0$. These moments can be identified with a compactly supported probability measure μ on $\mathbb{R}^2$, which can be encoded via the transform

$$G_{(a,b)}(u, w) := \varphi\left((u - a)^{-1}(w - b)^{-1}\right)$$
$$= \int_{\mathbb{R}^2} \frac{1}{(u - x)(w - y)}\mu(dxdy), \quad (u, w) \in (\mathbb{C} \setminus \mathbb{R})^2.$$

(Several properties of this function can be found in [**97**].)

We now treat the pair (a, b) as the self-adjoint random variable $X = \begin{pmatrix} a & 0 \\ 0 & b \end{pmatrix}$ within $\mathcal{A}^{2\times 2}$, where the map $\varphi : \mathcal{A}^{2\times 2} \to \mathbb{C}^{2\times 2}$ is a $\mathbb{C}^{2\times 2}$-valued expectation. It suffices now to consider $\varphi((z - X)^{-1})$ only for $z \in \mathbb{C}^{2\times 2}$. As observed in [**27**], we can even restrict our considerations to triangular matrices. Let $z = \begin{pmatrix} u & v \\ 0 & w \end{pmatrix} \in \mathbb{C}^{2\times 2}$

such that z has no real eigenvalues, i.e., $(u,w) \in (\mathbb{C} \setminus \mathbb{R})^2$, $v \in \mathbb{C}$. Then $z - X$ is invertible and

$$G_X(z) = \varphi((z-X)^{-1}) = \varphi\left(\begin{pmatrix} \frac{1}{u-a} & -\frac{v}{(u-a)(w-b)} \\ 0 & \frac{1}{w-b} \end{pmatrix}\right) = \begin{pmatrix} G_a(u) & -vG_{a,b}(u,w) \\ 0 & G_b(w) \end{pmatrix},$$

$$F_X(z) = G_X(z)^{-1} = \begin{pmatrix} F_a(u) & vG_{a,b}(u,w)F_a(u)F_b(w) \\ 0 & F_b(w) \end{pmatrix}.$$

The function F_X maps the upper half-plane in $\mathbb{C}^{2\times 2}$ into itself. Let D be the set of all

$$z = \begin{pmatrix} u & v \\ 0 & w \end{pmatrix} \quad \text{with} \quad \operatorname{Im}(z) = \begin{pmatrix} \operatorname{Im}(u) & v/(2i) \\ \overline{v}/(-2i) & \operatorname{Im}(w) \end{pmatrix} > 0.$$

This is equivalent to $\operatorname{Im}(u) > 0, \operatorname{Im}(w) > 0$, $|v|^2 < 4\operatorname{Im}(u)\operatorname{Im}(w)$, and we can identify D with $D = \{(u,v,w) \in \mathbb{C}^3 \mid u \in \mathbb{H}, w \in \mathbb{H}, |v|^2 < 4\operatorname{Im}(u)\operatorname{Im}(w)\}$.

Now assume that $a_1, b_1, a_2, b_2 \in \mathcal{A}$ are self-adjoint with $a_1b_1 = b_1a_1$ and $a_2b_2 = b_2a_2$. Let X and Y be the corresponding diagonal matrices. Assume that (X,Y) is monotonically independent. Then $F_{X+Y} = F_X \circ F_Y$. We have

$$F_{X+Y}(z) = \begin{pmatrix} F_{a_1+a_2}(u) & vG_{a_1+a_2,b_1+b_2}(u,w)F_{a_1+a_2}(u)F_{b_1+b_2}(w) \\ 0 & F_{b_1+b_2}(w) \end{pmatrix},$$

$$(F_X \circ F_Y)(z) = \begin{pmatrix} F_{a_1}(F_{a_2}(u)) & H \\ 0 & F_{b_1}(F_{b_2}(w)) \end{pmatrix}$$

with

$$H = vG_{a_2,b_2}(u,w)F_{a_2}(u)F_{b_2}(w)G_{a_1,b_1}(F_{a_2}(u),F_{b_2}(w))F_{a_1}(F_{a_2}(u))F_{b_1}(F_{b_2}(w)).$$

This yields the relations

$$F_{a_1+a_2} = F_{a_1} \circ F_{a_2}, \quad F_{b_1+b_2} = F_{b_1} \circ F_{b_2}, \quad \text{and}$$
$$G_{a_1+a_2,b_1+b_2}(u,w) = F_{a_2}(u)F_{b_2}(w)G_{a_2,b_2}(u,w)G_{a_1,b_1}(F_{a_2}(u),F_{b_2}(w)).$$

One could now further investigate semigroups and Loewner chains on D induced by F-transforms of the above form.

APPENDIX A

Compactness of the Class S

Bieberbach's theorem readily implies the following result.

THEOREM A.1. *Let $f \in S$. Then we have for all $z \in \mathbb{D}$:*

$$\left|\frac{zf''(z)}{f'(z)} - \frac{2|z|^2}{1-|z|^2}\right| \le \frac{4|z|}{1-|z|^2}.$$

PROOF. Let $w \in \mathbb{D}$. Then

$$g(z) = \frac{f(\frac{z+w}{\overline{w}z+1}) - f(w)}{(1-|w|^2)f'(w)} = z + A_2 z^2 + \cdots \in S.$$

We obtain

$$A_2 = \frac{1}{2}\left((1-|w|^2)\frac{f''(w)}{f'(w)} - 2\overline{w}\right).$$

Bieberbach's theorem implies $|A_2| \le 2$. Now we replace w by z and obtain

$$\left|\frac{zf''(z)}{f'(z)} - \frac{2|z|^2}{1-|z|^2}\right| \le \frac{4|z|}{1-|z|^2}. \qquad \square$$

THEOREM A.2 (Distortion theorem). *Let $f \in S$. Then we have for all $z \in \mathbb{D}$:*

$$\frac{1-|z|}{(1+|z|)^3} \le |f'(z)| \le \frac{1+|z|}{(1-|z|)^3}.$$

PROOF. From $|w| \le C$, we obtain $-C \le \operatorname{Re} w \le C$. Thus, Theorem A.1 implies that

$$\frac{2|z|^2 - 4|z|}{1-|z|^2} \le \operatorname{Re}\left(\frac{zf''(z)}{f'(z)}\right) \le \frac{2|z|^2 + 4|z|}{1-|z|^2}.$$

As $f'(z) \ne 0$ and $f'(0) = 1$, there exists a branch of the logarithm $\log f'$ of f' with $\log f'(0) = 0$. By writing $z = re^{i\varphi}$, we obtain

$$r\frac{\partial}{\partial r}\operatorname{Re}(\log f'(re^{i\varphi})) = r\operatorname{Re}\left(\frac{e^{i\varphi}f''(re^{i\varphi})}{f'(re^{i\varphi})}\right) = \operatorname{Re}\left(\frac{zf''(z)}{f'(z)}\right).$$

Hence,

$$\frac{2r-4}{1-r^2} \le \frac{\partial}{\partial r}\operatorname{Re}(\log f'(re^{i\varphi})) \le \frac{2r+4}{1-r^2}, \quad \text{or}$$
$$\frac{2r-4}{1-r^2} \le \frac{\partial}{\partial r}(\log|f'(re^{i\varphi})|) \le \frac{2r+4}{1-r^2}.$$

Integration from 0 to R gives

$$\log\frac{1-R}{(1+R)^3} \le \log|f'(Re^{i\varphi})| - \log|f'(0)| \le \log\frac{1+R}{(1-R)^3}.$$

As $\log|f'(0)| = 0$, we have for all $z \in \mathbb{D}$:

$$\frac{1-|z|}{(1+|z|)^3} \leq |f'(z)| \leq \frac{1+|z|}{(1-|z|)^3}. \qquad \square$$

THEOREM A.3 (Growth theorem). *Let $f \in S$. Then we have for all $z \in \mathbb{D}$:*

$$\frac{|z|}{(1+|z|)^2} \leq |f(z)| \leq \frac{|z|}{(1-|z|)^2}.$$

PROOF. Let $z = re^{i\varphi} \in \mathbb{D}$. As $f(0) = 0$, we have

$$f(z) = \int_0^r f'(\rho e^{i\varphi})e^{i\varphi}d\rho.$$

Consequently,

$$|f(z)| \leq \int_0^r |f'(\rho e^{i\varphi})|d\rho \underset{\text{Theorem A.2}}{\leq} \int_0^r \frac{1+\rho}{(1-\rho)^3}d\rho = \frac{r}{(1-r)^2}.$$

Next, we prove the lower bound: If $|f(z)| \geq 1/4$, then the inequality holds trivially. So let $|f(z)| < 1/4$. Because of Koebe's $1/4$–theorem, the straight line segment G between 0 and $f(z)$ lies in the range of f. Let $C = f^{-1}(G)$. From $f(z) = \int_C f'(w)dw$, and since $f(C)$ is a straight line segment ($f'(w)dw$ has constant angle), we obtain

$$|f(z)| = \int_G |du| \underset{u=f(w)}{=} \int_C |f'(w)||dw| \geq \int_C |f'(w)|\, d|w|$$
$$\underset{\text{Theorem A.2}}{\geq} \int_C \frac{1-|w|}{(1+|w|)^3}d|w| = \frac{|z|}{(1+|z|)^2}. \qquad \square$$

THEOREM A.4. *The class S is sequentially compact, i.e., every sequence in S has a locally uniformly converging subsequence whose limit belongs to S.*

PROOF. Because of Theorem A.3, the class S is locally uniformly bounded. Montel's theorem implies that S is a normal family, i.e., every sequence $(f_n)_n \subset S$ has a locally uniformly convergent subsequence $(f_{n_k})_k$. The limit $f = \lim_{k\to\infty} f_{n_k}$ is either constant or injective according to Hurwitz's theorem. Also, $f(0) = 0$ and $f'(0) = 1$, so f is not constant. Hence, $f \in S$. $\square$

COROLLARY A.5. *For every $n \geq 2$, there exists an $M_n > 0$ such that*

$$|a_n| \leq M_n$$

for every $f(z) = z + \sum_{n=2}^{\infty} a_n z^n \in S$.

PROOF. Because of

$$a_n = \frac{1}{2\pi i}\int_{\partial(\frac{1}{2}\mathbb{D})} \frac{f(w)}{w^{n+1}}dw, \tag{A.1}$$

the functional $F_n : f \mapsto a_n(f)$ is continuous. As S is sequentially compact, the set $F_n(S) \subset \mathbb{C}$ is compact. $\square$

APPENDIX B

Continuous Extension of Univalent Functions

In this section, we regard the question of under what conditions a conformal mapping in $\mathbb{D}$ can be extended continuously to the boundary. We follow [**152**], where much more on the boundary behavior of conformal mappings can be found.

We denote by $B_r(z)$, $z \in \mathbb{C}$, $r > 0$, the Euclidean disk with center z and radius r. If γ is a continuously differentiable curve in $\mathbb{C}$, then $L(\gamma)$ denotes the Euclidean length of γ.

Let D and E be two conformally equivalent subdomains of $\mathbb{C}$. Then the boundaries ∂D and ∂E can be quite different from each other. For example, $\mathbb{D}$ is conformally equivalent to the domain

$$E = \mathbb{H} \setminus \left([0, i] \cup \bigcup_{n\in\mathbb{N}} [1/n, 1/n + i] \cup \bigcup_{n\in\mathbb{N}} [-1/n, -1/n + i] \right).$$

Let $f : \mathbb{D} \to E$ be conformal. One can show that there exists a point $q \in \partial\mathbb{D}$, such that every sequence $(z_n)_n \subset E$ with $z_n \to p \in [0, i]$ satisfies $f^{-1}(z_n) \to q$. This implies that f cannot be extended continuously to q.

We recall: A set $X \subset \mathbb{C}$ is called *connected* if X cannot be written as $X = A \cup B$, where $A, B \neq \emptyset$ are disjoint and open in X.

A compact set $X \subset \mathbb{C}$ is called *locally connected* if for every $x \in X$ and $r > 0$ we can find a set $U \subset X$ which is connected and open in X and satisfies $x \in U \subset B_r(x)$. This is equivalent to: for every $\varepsilon > 0$ there exists $\delta > 0$ such that for all points $a, b \in X$ with $|a - b| < \delta$ there exists a compact and connected set $B \subset X$ with $a, b \in B$ and $\sup_{z,w\in B} |z - w| < \varepsilon$.

THEOREM B.1. *Let $f : \mathbb{D} \to D$ be conformal and assume D is bounded. Then the following statements are equivalent:*

(a) *f can be extended continuously to $\overline{\mathbb{D}}$.*
(b) *∂D is locally connected.*

In the example above, ∂E is not locally connected at $z = i$ (and the unboundedness of E can be neglected as the same statements also hold for the domain $B_2(0) \cap E$).

LEMMA B.2. *Let $U \subset \mathbb{C}$ be open and let $f : U \to B_R(0)$ be univalent. Furthermore, let $c \in \mathbb{C}$ and $C(r) = U \cap \partial B_r(c)$. Then, for $0 < \rho < 1$, we have:*

$$\inf_{\rho<r<\sqrt{\rho}} L(f(C(r))) \le \frac{2\pi R}{\sqrt{\log 1/\rho}}.$$

In particular, there is a sequence $(r_n)_n$, $r_n \in (0,1)$, converging to 0 with

$$L(f(C(r_n))) \to 0 \quad \text{for } n \to \infty.$$

PROOF. We have

$$L(f(C(r)))^2 = \left(\int_{C(r)} |f'(z)||dz|\right)^2 \underset{\text{Cauchy–Schwarz inequality}}{\le}$$

$$\int_{C(r)} |dz| \int_{C(r)} |f'(z)|^2|dz| \le 2\pi r \int_{c+re^{i\varphi}\in U} |f'(c+re^{i\varphi})|^2 r\,d\varphi.$$

Integration with respect to r yields

$$\int_0^\infty L(f(C(r)))^2 r^{-1} dr = 2\pi \int_0^\infty \int_{c+re^{i\varphi}\in U} |f'(c+re^{i\varphi})|^2 r\,d\varphi\,dr \le 2\pi^2 R^2,$$

as the last integral yields the area of $f(U)$, which is bounded by πR^2. Hence,

$$2\pi^2 R^2 \ge \int_\rho^{\sqrt{\rho}} L(f(C(r)))^2 r^{-1} dr$$

$$\ge \inf_{\rho<r<\sqrt{\rho}} L(f(C(r)))^2 \cdot \int_\rho^{\sqrt{\rho}} r^{-1} dr = \inf_{\rho<r<\sqrt{\rho}} L(f(C(r)))^2 \cdot \frac{1}{2}\log(1/\rho). \qquad \square$$

For the proof of Theorem B.1 we need the following topological statements:

(i) Continuous extension theorem: If $f : \mathbb{D} \to \mathbb{C}$ is uniformly continuous (for every $\varepsilon > 0$ there is $\delta > 0$ such that for every $z, w \in \mathbb{D}$ the inequality $|z - w| < \delta$ implies $|f(z) - f(w)| < \varepsilon$), then f has a continuous extension to $\overline{\mathbb{D}}$.
(ii) Janiszewski's theorem: Let $A, B \subset \mathbb{C}$ be closed sets, such that $A \cap B$ is connected. If $a, b \in \mathbb{C}$ are neither disconnected by A nor by B (they lie in the same connected component of the complements $\mathbb{C} \setminus A$ and $\mathbb{C} \setminus B$ respectively), then they are not disconnected by $A \cup B$ either.

PROOF OF THEOREM B.1. Assume (a), then $f(\partial\mathbb{D})$ is a compact, continuous curve, and thus, it is locally connected.

Next, assume (b), i.e., ∂D is locally connected. We assume that $f(0) = 0$. So there exist $R_0, R > 0$ with $R_0 < R$ and

$$B_{R_0}(0) \subset D \subset B_R(0).$$

Let $0 < \varepsilon < R_0$ and $0 < \delta < \varepsilon$, such that, for $a, b \in \partial D$ with $|a - b| < \delta$, there exists a compact, connected set $B \subset \partial D$ with $a, b \in B$ and $\sup_{u,v \in B} |u - v| < \varepsilon$.

Next, we choose $0 < \rho < 1/4$ such that $2\pi R/(\sqrt{\log(1/\rho)}) < \delta$. Let $z, z' \in \mathbb{D}$ with $1/2 < |z|, 1/2 < |z'|$ and $|z - z'| < \rho$. Due to Lemma B.2 with $U = \mathbb{D}$ and $c = z$, there exists a $\rho < r < 1/2$ such that

$$(*) \qquad L(f(C)) < \delta < \varepsilon, \quad C = \mathbb{D} \cap \partial B_r(z).$$

We show that this implies

$$|f(z) - f(z')| < 2\varepsilon,$$

i.e., f is uniformly continuous and, consequently, has a continuous extension to $\overline{\mathbb{D}}$.

Assume that $|f(z) - f(z')| \geq 2\varepsilon$ and $\overline{C} \cap \partial\mathbb{D} \neq \emptyset$. (If $\overline{C} \cap \partial\mathbb{D} = \emptyset$, then $f(z)$ and $f(z')$ lie inside a Jordan domain whose boundary has length $< \delta < \varepsilon$.)

The curve $\overline{f(C)}$ has end points $a, b \in \partial D$; otherwise, the Euclidean length of $f(C)$ would be ∞. Because of $(*)$, we have $|a - b| < \delta$. Consequently, there exists a compact, connected set $B \subset \partial D$ with $a, b \in B$ and $\sup_{u,v \in B} |u - v| < \varepsilon$. Hence,

$$B \cup f(C) \subset B_\varepsilon(a), \quad 0 \notin B_\varepsilon(a).$$

Because of $|f(z) - f(z')| \geq 2\varepsilon$, the points 0 and $f(z')$ (or 0 and $f(z)$) are not disconnected by $B \cup f(C)$. As 0 and $f(z')$ are not disconnected by ∂D either, and because $(B \cup f(C)) \cap \partial D = B$ is connected, Janiszewski's theorem implies that $f(0) = 0$ and $f(z')$ are not disconnected by the set $(B \cup f(C)) \cup \partial D = f(C) \cup \partial D$. Then 0 and z' are not disconnected by $C \cup \partial\mathbb{D}$, a contradiction, because $|z - z'| < \rho < r$. □

Let $E \subset \mathbb{C}$ be connected. Then $p \in E$ is called a cut point of E if $E \setminus \{p\}$ is not connected.

THEOREM B.3. *Let $f : \mathbb{D} \to D$ be conformal, D bounded and ∂D locally connected. Let $a \in \partial D$ and*

$$A = f^{-1}(\{a\}), \quad m = |A| \leq \infty.$$

Then a is a cut point of ∂D if and only if $m > 1$. The connected components of $\partial D \setminus \{a\}$ have the form $f(I_k), k = 1, \ldots, m$, where $I_k \subset \partial\mathbb{D}$ are open circular arcs.

PROOF. f can be extended continuously to $\overline{\mathbb{D}}$ by Theorem B.1. The set $\partial\mathbb{D} \setminus A$ is open in $\partial\mathbb{D}$ and consists of countably many pairwise disjoint open circular arcs I_k, so

$$\partial D \setminus \{a\} = f(\partial\mathbb{D} \setminus A) = \bigcup_{k=1}^{m} f(I_k).$$

As I_k is connected, also $f(I_k)$ is connected.

Let $m > 1$. Then $f(I_j), f(I_k), j \neq k$, are not connected within $\partial D \setminus \{a\}$: Let C be a curve in $\mathbb{D}$, that connects the end points of I_k. Then $f(C) \cup \{a\}$ is a closed Jordan curve in $D \cup \{a\}$. Let C' be a curve in $\mathbb{D}$ with end points in I_j and I_k such that C and C' intersect exactly once. Then $f(C)$ and $f(C')$ intersect exactly once. So $f(I_j)$ and $f(I_k)$ lie in different components of $\mathbb{C} \setminus (f(C) \cup \{a\})$, and, consequently, they are not connected in $\partial D \setminus \{a\}$.

If $m = 1$, then $\partial D \setminus \{a\} = f(\partial\mathbb{D} \setminus \{p\})$ for a $p \in \partial\mathbb{D}$. Hence, $\partial D \setminus \{a\}$ is connected and a is not a cut point of ∂D. □

THEOREM B.4. *Let $f : \mathbb{D} \to D$ be a conformal mapping onto a bounded domain D. Then the following statements are equivalent:*

(a) *f can be extended to a homeomorphism on $\overline{\mathbb{D}}$.*
(b) *∂D is a Jordan curve.*

PROOF. The direction $(a) \Longrightarrow (b)$ is obvious.

Let ∂D be a Jordan curve. Then f can be extended continuously to $\overline{\mathbb{D}}$ because of Theorem B.1. Each $a \in \partial D$ is not a cut point of ∂D, so $|f^{-1}(\{a\})| = 1$ due to Theorem B.3, i.e., f is injective on $\overline{\mathbb{D}}$. □

Bibliography

[1] Marco Abate, *Iteration theory of holomorphic maps on taut manifolds*, Research and Lecture Notes in Mathematics. Complex Analysis and Geometry, Mediterranean Press, Rende, 1989. MR1098711

[2] L. Accardi, *Quantum probability: an historical survey*, Probability on algebraic structures (Gainesville, FL, 1999), 2000, pp. 145–159, DOI 10.1090/conm/261/04138.

[3] Luigi Accardi, *Wilhelm von Waldenfels (2-3-1932–12-3-2021), a pioneer of quantum probability*, Infin. Dimens. Anal. Quantum Probab. Relat. Top. **25** (2022), no. 4, Paper No. 2250016, 14, DOI 10.1142/S0219025722500163. MR4534167

[4] Luigi Accardi, Anis Ben Ghorbal, and Nobuaki Obata, *Monotone independence, comb graphs and Bose-Einstein condensation*, Infin. Dimens. Anal. Quantum Probab. Relat. Top. **7** (2004), no. 3, 419–435, DOI 10.1142/S0219025704001645. MR2085641

[5] Luigi Accardi, Romuald Lenczewski, and Rafał Sałapata, *Decompositions of the free product of graphs*, Infin. Dimens. Anal. Quantum Probab. Relat. Top. **10** (2007), no. 3, 303–334, DOI 10.1142/S0219025707002750. MR2354364

[6] Luigi Accardi and Yun Gang Lu, *Emergence of quantum theories from classical probability: historical origins, developments, and open problems*, J. Stoch. Anal. **1** (2020), no. 4, Art. 9, 16. MR4188590

[7] Gernot Akemann, Jinho Baik, and Philippe Di Francesco (eds.), *The Oxford handbook of random matrix theory*, Oxford University Press, Oxford, 2011. MR2920518

[8] L. A. Aksent′ev, *An application of the principle of the argument to the study of univalence conditions. I*, Izv. Vysš. Učebn. Zaved. Matematika **1968** (1968), no. 12(79), 3–15.

[9] Greg W. Anderson, Alice Guionnet, and Ofer Zeitouni, *An introduction to random matrices*, Cambridge Studies in Advanced Mathematics, vol. 118, Cambridge University Press, Cambridge, 2010. MR2760897

[10] Michael Anshelevich and John D. Williams, *Operator-valued monotone convolution semigroups and an extension of the Bercovici-Pata bijection*, Doc. Math. **21** (2016), 841–871. MR3548135

[11] David Applebaum, B. V. Rajarama Bhat, Johan Kustermans, and J. Martin Lindsay, *Quantum independent increment processes. I: From classical probability to quantum stochastic calculus*, Lecture Notes in Mathematics, vol. 1865, Springer-Verlag, Berlin, 2005. Lectures from the School "Quantum Independent Increment Processes: Structure and Applications to Physics" held in Greifswald, March 9–22, 2003; Edited by Michael Schürmann and Uwe Franz. MR2132092

[12] Octavio Arizmendi and Takahiro Hasebe, *Classical and free infinite divisibility for Boolean stable laws*, Proc. Amer. Math. Soc. **142** (2014), no. 5, 1621–1632, DOI 10.1090/S0002-9939-2014-12111-3. MR3168468

[13] Octavio Arizmendi, Takahiro Hasebe, and Noriyoshi Sakuma, *On the law of free subordinators*, ALEA Lat. Am. J. Probab. Math. Stat. **10** (2013), no. 1, 271–291. MR3083927

[14] S. Attal, *Lectures in Quantum Noise Theory*, `http://math.univ-lyon1.fr/~attal/chapters.html`.

[15] S. Attal and I. Nechita, *Discrete approximation of the free Fock space*, Séminaire de Probabilités XLIII, 2011, pp. 379–394, DOI 10.1007/978-3-642-15217-7_16.

[16] F. G. Avkhadiev and L. A. Aksent′ev, *The main results on sufficient conditions for an analytic function to be schlicht*, Russian Math. Surveys **30** (1975), 1–64.

[17] Z. D. Bai and Y. Q. Yin, *Necessary and sufficient conditions for almost sure convergence of the largest eigenvalue of a Wigner matrix*, Ann. Probab. **16** (1988), no. 4, 1729–1741. MR958213

[18] Ole E. Barndorff-Nielsen, Uwe Franz, Rolf Gohm, Burkhard Kümmerer, and Steen Thorbjørnsen, *Quantum independent increment processes. II: Structure of quantum Lévy processes, classical probability, and physics*, Lecture Notes in Mathematics, vol. 1866, Springer-Verlag, Berlin, 2006. Lectures from the School "Quantum Independent Increment Processes: Structure and Applications to Physics" held in Greifswald, March 9–22, 2003; Edited by Michael Schüermann and Franz. MR2213451

[19] Ole E. Barndorff-Nielsen, Thomas Mikosch, and Sidney I. Resnick (eds.), *Lévy processes: Theory and applications*, Birkhäuser Boston, Inc., Boston, MA, 2001, DOI 10.1007/978-1-4612-0197-7. MR1833689

[20] Robert O. Bauer, *Discrete Löwner evolution* (English, with English and French summaries), Ann. Fac. Sci. Toulouse Math. (6) **12** (2003), no. 4, 433–451. MR2060594

[21] R. O. Bauer, *Löwner's equation from a noncommutative probability perspective*, J. Theoret. Probab. **17** (2004), no. 2, 435–456, DOI 10.1023/B:JOTP.0000020702.23996.8f. MR2053711

[22] Robert O. Bauer, *Chordal Loewner families and univalent Cauchy transforms*, J. Math. Anal. Appl. **302** (2005), no. 2, 484–501, DOI 10.1016/j.jmaa.2004.08.017. MR2107849

[23] Jochen Becker, *Über die Lösungsstruktur einer Differentialgleichung in der konformen Abbildung* (German), J. Reine Angew. Math. **285** (1976), 66–74, DOI 10.1515/crll.1976.285.66. MR414865

[24] S. T. Belinschi, *Complex analysis methods in noncommutative probability*, ProQuest LLC, Ann Arbor, MI, 2005. PhD thesis, Indiana University, `http://gateway.proquest.com/openurl?url_ver=Z39.88-2004&rft_val_fmt=info:ofi/fmt:kev:mtx:dissertation&res_dat=xri:pqdiss&rft_dat=xri:pqdiss:3182615`.

[25] S. T. Belinschi, *A note on regularity for free convolutions*, Ann. Inst. H. Poincaré Probab. Statist. **42** (2006), no. 5, 635–648, DOI 10.1016/j.anihpb.2005.05.004.

[26] S. T. Belinschi and H. Bercovici, *Atoms and regularity for measures in a partially defined free convolution semigroup*, Math. Z. **248** (2004), no. 4, 665–674, DOI 10.1007/s00209-004-0671-y. MR2103535

[27] S. T. Belinschi, H. Bercovici, Y. Gu, and P. Skoufranis, *Analytic subordination for bi-free convolution*, J. Funct. Anal. **275** (2018), no. 4, 926–966, DOI 10.1016/j.jfa.2018.03.003. MR3807781

[28] S. T. Belinschi, M. Bożejko, F. Lehner, and R. Speicher, *The normal distribution is $\boxplus$-infinitely divisible*, Adv. Math. **226** (2011), no. 4, 3677–3698, DOI 10.1016/j.aim.2010.10.025.

[29] S. T. Belinschi, M. Popa, and V. Vinnikov, *On the operator-valued analogues of the semicircle, arcsine and Bernoulli laws*, J. Operator Theory **70** (2013), no. 1, 239–258, DOI 10.7900/jot.2011jun24.1963. MR3085826

[30] A. Ben Ghorbal and M. Schürmann, *On the algebraic foundations of non-commutative probability theory* (1999). Preprint.

[31] H. Bercovici, *Multiplicative monotonic convolution*, Illinois J. Math. **49** (2005), no. 3, 929–951, `http://projecteuclid.org/euclid.ijm/1258138229`.

[32] Hari Bercovici and Vittorino Pata, *A free analogue of Hinčin's characterization of infinite divisibility*, Proc. Amer. Math. Soc. **128** (2000), no. 4, 1011–1015, DOI 10.1090/S0002-9939-99-05087-X. MR1636930

[33] H. Bercovici and D. Voiculescu, *Lévy-Hinčin type theorems for multiplicative and additive free convolution*, Pacific J. Math. **153** (1992), no. 2, 217–248, `http://projecteuclid.org/euclid.pjm/1102635830`.

[34] Hari Bercovici and Dan Voiculescu, *Free convolution of measures with unbounded support*, Indiana Univ. Math. J. **42** (1993), no. 3, 733–773, DOI 10.1512/iumj.1993.42.42033. MR1254116

[35] E. Berkson and H. Porta, *Semigroups of analytic functions and composition operators*, Michigan Math. J. **25** (1978), no. 1, 101–115, `http://projecteuclid.org/euclid.mmj/1029002009`.

[36] S. Bernstein, *Démonstration du théorème de Weierstrass fondée sur le calcul des probabilités*, Comm. Kharkov Math. Soc. **13** (1912).

[37] P. Biane, *Processes with free increments*, Math. Z. **227** (1998), no. 1, 143–174, DOI 10.1007/PL00004363.

[38] Philippe Biane, Jim Pitman, and Marc Yor, *Probability laws related to the Jacobi theta and Riemann zeta functions, and Brownian excursions*, Bull. Amer. Math. Soc. (N.S.) **38** (2001), no. 4, 435–465, DOI 10.1090/S0273-0979-01-00912-0. MR1848256

[39] L. Bieberbach, *Über die Koeffizienten derjenigen Potenzreihen, welche eine schlichte Abbildung des Einheitskreises vermitteln*, Sitzungsberichte der Königlich Preussischen Akademie der Wissenschaften zu Berlin (1916), 940–955.

[40] Patrick Billingsley, *Probability and measure*, 3rd ed., Wiley Series in Probability and Mathematical Statistics, John Wiley & Sons, Inc., New York, 1995. A Wiley-Interscience Publication. MR1324786

[41] Patrick Billingsley, *Convergence of probability measures*, 2nd ed., Wiley Series in Probability and Statistics: Probability and Statistics, John Wiley & Sons, Inc., New York, 1999. A Wiley-Interscience Publication, DOI 10.1002/9780470316962. MR1700749

[42] Torben Maack Bisgaard and Zoltán Sasvári, *Characteristic functions and moment sequences: Positive definiteness in probability*, Nova Science Publishers, Inc., Huntington, NY, 2000. MR2014167

[43] B. Blackadar, *Operator algebras: Theory of C^*-algebras and von Neumann algebras; Operator Algebras and Non-commutative Geometry, III*, Encyclopaedia of Mathematical Sciences, vol. 122, Springer-Verlag, Berlin, 2006, DOI 10.1007/3-540-28517-2. MR2188261

[44] Stéphane Boucheron, Gábor Lugosi, and Pascal Massart, *Concentration inequalities: A nonasymptotic theory of independence*, Oxford University Press, Oxford, 2013. With a foreword by Michel Ledoux, DOI 10.1093/acprof:oso/9780199535255.001.0001. MR3185193

[45] Marek Bożejko and Takahiro Hasebe, *On free infinite divisibility for classical Meixner distributions*, Probab. Math. Statist. **33** (2013), no. 2, 363–375. MR3158562

[46] Filippo Bracci, Manuel D. Contreras, and Santiago Díaz-Madrigal, *Evolution families and the Loewner equation. II. Complex hyperbolic manifolds*, Math. Ann. **344** (2009), no. 4, 947–962, DOI 10.1007/s00208-009-0340-x. MR2507634

[47] Filippo Bracci, Manuel D. Contreras, and Santiago Díaz-Madrigal, *Evolution families and the Loewner equation I: the unit disc*, J. Reine Angew. Math. **672** (2012), 1–37, DOI 10.1515/crelle.2011.167. MR2995431

[48] Filippo Bracci, Manuel D. Contreras, Santiago Díaz-Madrigal, and Pavel Gumenyuk, *Boundary regular fixed points in Loewner theory*, Ann. Mat. Pura Appl. (4) **194** (2015), no. 1, 221–245, DOI 10.1007/s10231-013-0372-4. MR3303013

[49] F. Bracci and O. Roth, *Support points and the Bieberbach conjecture in higher dimension*, Complex analysis and dynamical systems, 2018, pp. 67–79, DOI 10.1007/978-3-319-70154-7_4.

[50] Steve Brooks, Andrew Gelman, Galin L. Jones, and Xiao-Li Meng (eds.), *Handbook of Markov chain Monte Carlo*, Chapman & Hall/CRC Handbooks of Modern Statistical Methods, CRC Press, Boca Raton, FL, 2011, DOI 10.1201/b10905. MR2742422

[51] C. Carathéodory, *Untersuchungen über die konformen Abbildungen von festen und veränderlichen Gebieten* (German), Math. Ann. **72** (1912), no. 1, 107–144, DOI 10.1007/BF01456892. MR1511688

[52] Manuel D. Contreras, Santiago Díaz-Madrigal, and Pavel Gumenyuk, *Loewner chains in the unit disk*, Rev. Mat. Iberoam. **26** (2010), no. 3, 975–1012, DOI 10.4171/RMI/624. MR2789373

[53] Manuel D. Contreras, Santiago Díaz-Madrigal, and Pavel Gumenyuk, *Local duality in Loewner equations*, J. Nonlinear Convex Anal. **15** (2014), no. 2, 269–297. MR3184323

[54] M. D. Contreras, S. Díaz Madrigal, and Ch. Pommerenke, *On boundary critical points for semigroups of analytic functions*, Math. Scand. **98** (2006), no. 1, 125–142, DOI 10.7146/math.scand.a-14987. MR2221548

[55] John B. Conway, *A course in functional analysis*, 2nd ed., Graduate Texts in Mathematics, vol. 96, Springer-Verlag, New York, 1990. MR1070713

[56] Romain Couillet and Mérouane Debbah, *Random matrix methods for wireless communications*, Cambridge University Press, Cambridge, 2011, DOI 10.1017/CBO9780511994746. MR2884783

[57] R. Couillet and Z. Liao, *Random Matrix Methods for Machine Learning*, Cambridge University Press, 2022.

[58] L. de Branges, *A proof of the Bieberbach conjecture*, Acta Math. **154** (1985), no. 1-2, 137–152, DOI 10.1007/BF02392821.

[59] M. De. Giosa and Y. G. Lu, *The free creation and annihilation operators as the central limit of the quantum Bernoulli process*, Random Oper. Stochastic Equations **5** (1997), no. 3, 227–236, DOI 10.1515/rose.1997.5.3.227. MR1483010

[60] Andrea del Monaco and Pavel Gumenyuk, *Chordal Loewner equation*, Complex analysis and dynamical systems VI. Part 2, Contemp. Math., vol. 667, Amer. Math. Soc., Providence, RI, 2016, pp. 63–77, DOI 10.1090/conm/667/13531. MR3511252

[61] Jean Dieudonné, *Recherches sur quelques problèmes relatifs aux polynômes et aux fonctions bornées d'une variable complexe* (French), Ann. Sci. École Norm. Sup. (3) **48** (1931), 247–358. MR1509314

[62] Peter L. Duren, *Univalent functions*, Grundlehren der mathematischen Wissenschaften [Fundamental Principles of Mathematical Sciences], vol. 259, Springer-Verlag, New York, 1983. MR708494

[63] A. Edelman and N. R. Rao, *Free probability, sample covariance matrices, and signal processing*, 2006 IEEE International Conference on Acoustics Speech and Signal Processing Proceedings (2006).

[64] Mark Elin, David Shoikhet, and Toshiyuki Sugawa, *Geometric properties of the nonlinear resolvent of holomorphic generators*, J. Math. Anal. Appl. **483** (2020), no. 2, 123614, 18, DOI 10.1016/j.jmaa.2019.123614. MR4037570

[65] Klaus-Jochen Engel and Rainer Nagel, *One-parameter semigroups for linear evolution equations*, Graduate Texts in Mathematics, vol. 194, Springer-Verlag, New York, 2000. With contributions by S. Brendle, M. Campiti, T. Hahn, G. Metafune, G. Nickel, D. Pallara, C. Perazzoli, A. Rhandi, S. Romanelli and R. Schnaubelt. MR1721989

[66] A. R. Feier, *Methods of Proof in Random Matrix Theory*, Harvard University, 2012.

[67] P. J. Forrester, *Log-gases and random matrices*, London Mathematical Society Monographs Series, vol. 34, Princeton University Press, Princeton, NJ, 2010, DOI 10.1515/9781400835416. MR2641363

[68] Uwe Franz, *Boolean convolution of probability measures on the unit circle* (English, with English and French summaries), Analyse et probabilités, Sémin. Congr., vol. 16, Soc. Math. France, Paris, 2008, pp. 83–94. MR2599263

[69] Uwe Franz, *Monotone and Boolean convolutions for non-compactly supported probability measures*, Indiana Univ. Math. J. **58** (2009), no. 3, 1151–1185, DOI 10.1512/iumj.2009.58.3578. MR2541362

[70] Uwe Franz, Takahiro Hasebe, and Sebastian Schleißinger, *Monotone increment processes, classical Markov processes, and Loewner chains*, Dissertationes Math. **552** (2020), 119, DOI 10.4064/dm808-1-2020. MR4152669

[71] U. Franz and N. Muraki, *Markov property of monotone Lévy processes*, Infinite dimensional harmonic analysis III, 2005, pp. 37–57, DOI 10.1142/9789812701503_0003.

[72] Theodore W. Gamelin, *Complex analysis*, Undergraduate Texts in Mathematics, Springer-Verlag, New York, 2001, DOI 10.1007/978-0-387-21607-2. MR1830078

[73] Malte Gerhold, *Bi-monotone Brownian motion*, Infinite dimensional analysis, quantum probability and related topics, QP–PQ: Quantum Probab. White Noise Anal., vol. 32, World Sci. Publ., Hackensack, NJ, [2024] ©2024, pp. 59–76. MR4678074

[74] B. V. Gnedenko and A. N. Kolmogorov, *Limit distributions for sums of independent random variables*, Addison-Wesley Publishing Co., Inc., Cambridge, MA, 1954. Translated and annotated by K. L. Chung. With an Appendix by J. L. Doob. MR62975

[75] V. V. Goryaĭnov and I. Ba, *Semigroup of conformal mappings of the upper half-plane into itself with hydrodynamic normalization at infinity* (Russian, with Russian and Ukrainian summaries), Ukraïn. Mat. Zh. **44** (1992), no. 10, 1320–1329, DOI 10.1007/BF01057676; English transl., Ukrainian Math. J. **44** (1992), no. 10, 1209–1217 (1993). MR1201130

[76] I. Graham, H. Hamada, and G. Kohr, *Extremal problems for mappings with g-parametric representation on the unit polydisc in* $\mathbb{C}^n$, Complex analysis and dynamical systems, 2018, pp. 141–167, DOI 10.1007/978-3-319-70154-7_9.

[77] Ian Graham, Hidetaka Hamada, and Gabriela Kohr, *Loewner chains and nonlinear resolvents of the Carathéodory family on the unit ball in* $\mathbb{C}^n$, J. Math. Anal. Appl. **491** (2020), no. 1, 124289, 29, DOI 10.1016/j.jmaa.2020.124289. MR4109415

[78] Ian Graham, Hidetaka Hamada, Gabriela Kohr, and Mirela Kohr, *Spirallike mappings and univalent subordination chains in* $\mathbb{C}^n$, Ann. Sc. Norm. Super. Pisa Cl. Sci. (5) **7** (2008), no. 4, 717–740. MR2483641

[79] I. Graham, H. Hamada, G. Kohr, and M. Kohr, *Extremal properties associated with univalent subordination chains in* $\mathbb{C}^n$, Math. Ann. **359** (2014), no. 1-2, 61–99, DOI 10.1007/s00208-013-0998-y.

[80] I. Graham and G. Kohr, *Geometric Function Theory in One and Higher Dimensions*, Monographs and Textbooks in Pure and Applied Mathematics, vol. 255, Marcel Dekker, Inc., New York, 2003, DOI 10.1201/9780203911624.

[81] Y. Gu, T. Hasebe, and P. Skoufranis, *Bi-monotonic independence for pairs of algebras*, J. Theoret. Probab. **33** (2020), no. 1, 533–566, DOI 10.1007/s10959-019-00884-2.

[82] J. Guan, Y. Feng, and M. Ying, *Decomposition of quantum Markov chains and its applications*, J. Comput. System Sci. **95** (2018), 55–68, DOI 10.1016/j.jcss.2018.01.005.

[83] Jack K. Hale, *Ordinary differential equations*, 2nd ed., Robert E. Krieger Publishing Co., Inc., Huntington, NY, 1980. MR587488

[84] Tarek Hamdi, *Monotone and boolean unitary Brownian motions*, Infin. Dimens. Anal. Quantum Probab. Relat. Top. **18** (2015), no. 2, 1550012, 19, DOI 10.1142/S0219025715500125. MR3356250

[85] Takahiro Hasebe, *Monotone convolution and monotone infinite divisibility from complex analytic viewpoint*, Infin. Dimens. Anal. Quantum Probab. Relat. Top. **13** (2010), no. 1, 111–131, DOI 10.1142/S0219025710003973. MR2646794

[86] T. Hasebe, *Free infinite divisibility for beta distributions and related ones*, Electron. J. Probab. **19** (2014), no. 81, 33, DOI 10.1214/EJP.v19-3448.

[87] Takahiro Hasebe and Ikkei Hotta, *Additive processes on the unit circle and Loewner chains*, Int. Math. Res. Not. IMRN **22** (2022), 17797–17848, DOI 10.1093/imrn/rnab157. MR4514456

[88] T. Hasebe and H. Saigo, *Joint cumulants for natural independence*, Electron. Commun. Probab. **16** (2011), 491–506, DOI 10.1214/ECP.v16-1647.

[89] T. Hasebe and H. Saigo, *The monotone cumulants*, Ann. Inst. Henri Poincaré Probab. Stat. **47** (2011), no. 4, 1160–1170, DOI 10.1214/10-AIHP379.

[90] Takahiro Hasebe and Hayato Saigo, *On operator-valued monotone independence*, Nagoya Math. J. **215** (2014), 151–167, DOI 10.1215/00277630-2741151. MR3263527

[91] T. Hasebe and N. Sakuma, *Unimodality of Boolean and monotone stable distributions*, Demonstr. Math. **48** (2015), no. 3, 424–439, DOI 10.1515/dema-2015-0030.

[92] Edwin Hewitt and Karl Stromberg, *Real and abstract analysis: A modern treatment of the theory of functions of a real variable*, Graduate Texts in Mathematics, No. 25, Springer-Verlag, New York-Heidelberg, 1975. Third printing. MR367121

[93] Akihito Hora and Nobuaki Obata, *Quantum probability and spectral analysis of graphs*, Theoretical and Mathematical Physics, Springer, Berlin, 2007. With a foreword by Luigi Accardi. MR2316893

[94] Lars Hörmander, *An introduction to complex analysis in several variables*, 3rd ed., North-Holland Mathematical Library, vol. 7, North-Holland Publishing Co., Amsterdam, 1990. MR1045639

[95] Ikkei Hotta and Sebastian Schleißinger, *Problems related to conformal slit-mappings*, Complex Var. Elliptic Equ. **69** (2024), no. 2, 215–233, DOI 10.1080/17476933.2022.2121820. MR4699682

[96] I. Hotta, S. Schleißinger, and T. Sugawa, *Nonlinear resolvents and decreasing Loewner chains*, J. Geom. Anal. **34** (2024), no. 4, Paper No. 99, 28, DOI 10.1007/s12220-023-01544-y.

[97] H.-W. Huang and J.-C. Wang, *Analytic aspects of the bi-free partial R-transform*, J. Funct. Anal. **271** (2016), no. 4, 922–957, DOI 10.1016/j.jfa.2016.04.026.

[98] D. Igarashi and N. Obata, *Asymptotic spectral analysis of growing graphs: odd graphs and spidernets*, Quantum probability, 2006, pp. 245–265, DOI 10.4064/bc73-0-18.

[99] Marek Jarnicki and Peter Pflug, *First steps in several complex variables: Reinhardt domains*, EMS Textbooks in Mathematics, European Mathematical Society (EMS), Zürich, 2008, DOI 10.4171/049. MR2396710

[100] David Jekel, *Operator-valued chordal Loewner chains and non-commutative probability*, J. Funct. Anal. **278** (2020), no. 10, 108452, 100, DOI 10.1016/j.jfa.2019.108452. MR4067990

[101] T. Jiang, *Wigner's semicircle law for Gaussian random matrices* (2021). University of Chicago, Mathematics REU 2021.

[102] Dmitry S. Kaliuzhnyi-Verbovetskyi and Victor Vinnikov, *Foundations of free noncommutative function theory*, Mathematical Surveys and Monographs, vol. 199, American Mathematical Society, Providence, RI, 2014, DOI 10.1090/surv/199. MR3244229

[103] Olav Kallenberg, *Foundations of modern probability*, 2nd ed., Probability and its Applications (New York), Springer-Verlag, New York, 2002, DOI 10.1007/978-1-4757-4015-8. MR1876169

[104] J. B. Keiper, *Power series expansions of Riemann's ξ function*, Math. Comp. **58** (1992), no. 198, 765–773, DOI 10.2307/2153215.

[105] J. Kempe, *Quantum random walks: An introductory overview*, Contemporary Physics **44** (2003), 307–327.

[106] Tom Kennedy, *A fast algorithm for simulating the chordal Schramm-Loewner evolution*, J. Stat. Phys. **128** (2007), no. 5, 1125–1137, DOI 10.1007/s10955-007-9358-1. MR2348786

[107] Tom Kennedy, *Numerical computations for the Schramm-Loewner evolution*, J. Stat. Phys. **137** (2009), no. 5-6, 839–856, DOI 10.1007/s10955-009-9866-2. MR2570752

[108] Julia Koch and Sebastian Schleißinger, *Three value ranges for symmetric self-mappings of the unit disc*, Proc. Amer. Math. Soc. **145** (2017), no. 4, 1747–1761, DOI 10.1090/proc/13350. MR3601565

[109] Wolfram Koepf, *Bieberbach's conjecture, the de Branges and Weinstein functions and the Askey-Gasper inequality*, Ramanujan J. **13** (2007), no. 1-3, 103–129, DOI 10.1007/s11139-006-0244-2. MR2281158

[110] Steven G. Krantz, *Function theory of several complex variables*, AMS Chelsea Publishing, Providence, RI, 2001. Reprint of the 1992 edition, DOI 10.1090/chel/340. MR1846625

[111] P. P. Kufarev, *A remark on integrals of Löwner's equation* (Russian), Doklady Akad. Nauk SSSR (N.S.) **57** (1947), 655–656. MR23907

[112] P. P. Kufarev, V. V. Sobolev, and L. V. Sporyševa, *A certain method of investigation of extremal problems for functions that are univalent in the half-plane* (Russian), Trudy Tomsk. Gos. Univ. Ser. Meh.-Mat. **200** (1968), 142–164. MR257336

[113] S. Lalley, G. Lawler, and H. Narayanan, *Geometric interpretation of half-plane capacity*, Electron. Commun. Probab. **14** (2009), 566–571, DOI 10.1214/ECP.v14-1517.

[114] E. C. Lance, *Hilbert C^*-modules*, London Mathematical Society Lecture Note Series, vol. 210, Cambridge University Press, Cambridge, 1995. A toolkit for operator algebraists, DOI 10.1017/CBO9780511526206. MR1325694

[115] Gregory F. Lawler, *Conformally invariant processes in the plane*, Mathematical Surveys and Monographs, vol. 114, American Mathematical Society, Providence, RI, 2005, DOI 10.1090/surv/114. MR2129588

[116] Romuald Lenczewski, *Asymptotic properties of random matrices and pseudomatrices*, Adv. Math. **228** (2011), no. 4, 2403–2440, DOI 10.1016/j.aim.2011.07.011. MR2836126

[117] Romuald Lenczewski, *Limit distributions of Gaussian block ensembles*, Acta Phys. Polon. B **46** (2015), no. 9, 1833–1850, DOI 10.5506/APhysPolB.46.1833. MR3403843

[118] D. A. Levin, Y. Peres, and E. L. Wilmer, *Markov Chains and Mixing Times*, American Mathematical Society, Providence, RI, 2009. With a chapter by James G. Propp and David B. Wilson, DOI 10.1090/mbk/058.

[119] Xian-Jin Li, *The positivity of a sequence of numbers and the Riemann hypothesis*, J. Number Theory **65** (1997), no. 2, 325–333, DOI 10.1006/jnth.1997.2137. MR1462847

[120] Joan Lind, Donald E. Marshall, and Steffen Rohde, *Collisions and spirals of Loewner traces*, Duke Math. J. **154** (2010), no. 3, 527–573, DOI 10.1215/00127094-2010-045. MR2730577

[121] Joan Lind and Steffen Rohde, *Space-filling curves and phases of the Loewner equation*, Indiana Univ. Math. J. **61** (2012), no. 6, 2231–2249, DOI 10.1512/iumj.2012.61.4794. MR3129109

[122] Joan R. Lind, *A sharp condition for the Loewner equation to generate slits*, Ann. Acad. Sci. Fenn. Math. **30** (2005), no. 1, 143–158. MR2140303

[123] Karl Löwner, *Untersuchungen über schlichte konforme Abbildungen des Einheitskreises. I* (German), Math. Ann. **89** (1923), no. 1-2, 103–121, DOI 10.1007/BF01448091. MR1512136

[124] Y. G. Lu, *An interacting free Fock space and the arcsine law*, Probab. Math. Statist. **17** (1997), no. 1, Acta Univ. Wratislav. No. 1928, 149–166. MR1455615

[125] Eugene Lukacs, *Characteristic functions*, Griffin's Statistical Monographs & Courses, No. 5, Hafner Publishing Co., New York, 1960. MR124075

[126] E. Lukacs and O. Szász, *On analytic characteristic functions*, Pacific J. Math. **2** (1952), 615–625, `http://projecteuclid.org/euclid.pjm/1103051639`.

[127] Hans Maassen, *Addition of freely independent random variables*, J. Funct. Anal. **106** (1992), no. 2, 409–438, DOI 10.1016/0022-1236(92)90055-N. MR1165862

[128] Donald E. Marshall and Steffen Rohde, *The Loewner differential equation and slit mappings*, J. Amer. Math. Soc. **18** (2005), no. 4, 763–778, DOI 10.1090/S0894-0347-05-00492-3. MR2163382

[129] V. A. Marčenko and L. A. Pastur, *Distribution of eigenvalues in certain sets of random matrices* (Russian), Mat. Sb. (N.S.) **72(114)** (1967), 507–536. MR208649

[130] Madan Lal Mehta, *Random matrices*, 3rd ed., Pure and Applied Mathematics (Amsterdam), vol. 142, Elsevier/Academic Press, Amsterdam, 2004. MR2129906

[131] Paul-André Meyer, *Quantum probability for probabilists*, Lecture Notes in Mathematics, vol. 1538, Springer-Verlag, Berlin, 1993, DOI 10.1007/978-3-662-21558-6. MR1222649

[132] James A. Mingo and Roland Speicher, *Free probability and random matrices*, Fields Institute Monographs, vol. 35, Springer, New York; Fields Institute for Research in Mathematical Sciences, Toronto, ON, 2017, DOI 10.1007/978-1-4939-6942-5. MR3585560

[133] B. Mohar, *The spectrum of an infinite graph*, Linear Algebra Appl. **48** (1982), 245–256, DOI 10.1016/0024-3795(82)90111-2.

[134] B. Mohar and W. Woess, *A survey on spectra of infinite graphs*, Bull. London Math. Soc. **21** (1989), no. 3, 209–234, DOI 10.1112/blms/21.3.209.

[135] Peter Mörters and Yuval Peres, *Brownian motion*, Cambridge Series in Statistical and Probabilistic Mathematics, vol. 30, Cambridge University Press, Cambridge, 2010. With an appendix by Oded Schramm and Wendelin Werner, DOI 10.1017/CBO9780511750489. MR2604525

[136] P. Mukhopadhyay, *An Introduction to the Theory of Probability*, World Scientific Publishing, 2011.

[137] Naofumi Muraki, *A new example of noncommutative "de Moivre-Laplace theorem"*, Probability theory and mathematical statistics (Tokyo, 1995), World Sci. Publ., River Edge, NJ, 1996, pp. 353–362. MR1467953

[138] N. Muraki, *Noncommutative Brownian motion in monotone Fock space*, Comm. Math. Phys. **183** (1997), no. 3, 557–570, DOI 10.1007/s002200050043.

[139] N. Muraki, *Monotonic convolution and monotone Lévy-Hinčin formula* (2000). Preprint.

[140] Naofumi Muraki, *Monotonic independence, monotonic central limit theorem and monotonic law of small numbers*, Infin. Dimens. Anal. Quantum Probab. Relat. Top. **4** (2001), no. 1, 39–58, DOI 10.1142/S0219025701000339. MR1824472

[141] Naofumi Muraki, *Towards "monotonic probability": Topics in information sciences and applied functional analysis*, Sūrikaisekikenkyūsho Kōkyūroku **1186** (2001), 28–35. (Japanese) (Kyoto, 2000). MR1853184

[142] Naofumi Muraki, *The five independences as natural products*, Infin. Dimens. Anal. Quantum Probab. Relat. Top. **6** (2003), no. 3, 337–371, DOI 10.1142/S0219025703001365. MR2016316

[143] Zeev Nehari, *Conformal mapping*, Dover Publications, Inc., New York, 1975. Reprinting of the 1952 edition. MR377031

[144] Alexandru Nica and Roland Speicher, *Lectures on the combinatorics of free probability*, London Mathematical Society Lecture Note Series, vol. 335, Cambridge University Press, Cambridge, 2006, DOI 10.1017/CBO9780511735127. MR2266879

[145] Nobuaki Obata, *Spectral analysis of growing graphs: A quantum probability point of view*, SpringerBriefs in Mathematical Physics, vol. 20, Springer, Singapore, 2017, DOI 10.1007/978-981-10-3506-7. MR3617953

[146] B. Øksendal, *Stochastic Differential Equations*, Sixth, Universitext, Springer-Verlag, Berlin, 2003, DOI 10.1007/978-3-642-14394-6.

[147] Etienne Pardoux, *Markov processes and applications: Algorithms, networks, genome and finance*, Translated and revised from the 2007 French edition, Wiley Series in Probability and Statistics, John Wiley & Sons, Ltd., Chichester; Dunod, Paris, 2008, DOI 10.1002/9780470721872. MR2488952

[148] J. E. Pascoe, Benjamin Passer, and Ryan Tully-Doyle, *Representation of free Herglotz functions*, Indiana Univ. Math. J. **68** (2019), no. 4, 1199–1215, DOI 10.1512/iumj.2019.68.7730. MR4001470

[149] H. Poincaré, *Les fonctions analytiques de deux variables et la représentation conforme*, Rend. Circ. Mat. Palermo **23** (1907), 185–220.

[150] Ch. Pommerenke, *On the Loewner differential equation*, Michigan Math. J. **13** (1966), 435–443. MR206245

[151] Christian Pommerenke, *Univalent functions*, Studia Mathematica/Mathematische Lehrbücher [Studia Mathematica/Mathematical Textbooks], Band XXV, Vandenhoeck & Ruprecht, Göttingen, 1975. With a chapter on quadratic differentials by Gerd Jensen. MR507768

[152] Ch. Pommerenke, *Boundary behaviour of conformal maps*, Grundlehren der mathematischen Wissenschaften [Fundamental Principles of Mathematical Sciences], vol. 299, Springer-Verlag, Berlin, 1992, DOI 10.1007/978-3-662-02770-7. MR1217706

[153] M. Popa, *A combinatorial approach to monotonic independence over a C^*-algebra*, Pacific J. Math. **237** (2008), no. 2, 299–325, DOI 10.2140/pjm.2008.237.299.

[154] T. Poreda, *On the univalent holomorphic maps of the unit polydisc in $\mathbf{C}^n$ which have the parametric representation. I. The geometrical properties* (English, with Polish summary), Ann. Univ. Mariae Curie-Skłodowska Sect. A **41** (1987), 105–113 (1989). MR1049182

[155] T. Poreda, *On the univalent holomorphic maps of the unit polydisc in $\mathbf{C}^n$ which have the parametric representation. II. The necessary conditions and the sufficient conditions* (English, with Polish summary), Ann. Univ. Mariae Curie-Skłodowska Sect. A **41** (1987), 115–121 (1989). MR1049183

[156] Dmitri Prokhorov and Kristina Samsonova, *Value range of solutions to the chordal Loewner equation*, J. Math. Anal. Appl. **428** (2015), no. 2, 910–919, DOI 10.1016/j.jmaa.2015.03.065. MR3334955

[157] D. V. Prokhorov, *Value regions in classes of conformal mappings* (English, with English and Russian summaries), Izv. Sarat. Univ. (N.S.) Ser. Mat. Mekh. Inform. **19** (2019), no. 3, 258–279, DOI 10.18500/1816-9791-2019-19-3-258-279. MR4014262

[158] Martin L. Puterman, *Markov decision processes: discrete stochastic dynamic programming*, Wiley Series in Probability and Mathematical Statistics: Applied Probability and Statistics, John Wiley & Sons, Inc., New York, 1994. A Wiley-Interscience Publication. MR1270015

[159] Reza Rashidi Far, Tamer Oraby, Wlodek Bryc, and Roland Speicher, *On slow-fading MIMO systems with nonseparable correlation*, IEEE Trans. Inform. Theory **54** (2008), no. 2, 544–553, DOI 10.1109/TIT.2007.913239. MR2444540

[160] S. Reich and D. Shoikhet, *Generation theory for semigroups of holomorphic mappings in Banach spaces*, Abstr. Appl. Anal. **1** (1996), no. 1, 1–44, DOI 10.1155/S1085337596000012.

[161] Simeon Reich and David Shoikhet, *Semigroups and generators on convex domains with the hyperbolic metric* (English, with English and Italian summaries), Atti Accad. Naz. Lincei Cl. Sci. Fis. Mat. Natur. Rend. Lincei (9) Mat. Appl. **8** (1997), no. 4, 231–250. MR1631605

[162] M. Riesz, *Sur le problème des moments. troisième note.*, Arkiv för matematik, Astronomi och Fysik **17** (1923), 1–52.

[163] Oliver Roth and Sebastian Schleißinger, *Rogosinski's lemma for univalent functions, hyperbolic Archimedean spirals and the Loewner equation*, Bull. Lond. Math. Soc. **46** (2014), no. 5, 1099–1109, DOI 10.1112/blms/bdu054. MR3262210

[164] Walter Rudin, *Fourier analysis on groups*, Interscience Tracts in Pure and Applied Mathematics, No. 12, Interscience Publishers (a division of John Wiley & Sons, Inc.), New York-London, 1962. MR152834

[165] Walter Rudin, *Real and complex analysis*, 3rd ed., McGraw-Hill Book Co., New York, 1987. MR924157

[166] Stephan Ruscheweyh, *Convolutions in geometric function theory: Fundamental Theories of Physics*, Séminaire de Mathématiques Supérieures [Seminar on Higher Mathematics], vol. 83, Presses de l'Université de Montréal, Montreal, QC, 1982. MR674296

[167] St. Ruscheweyh and T. Sheil-Small, *Hadamard products of Schlicht functions and the Pólya-Schoenberg conjecture*, Comment. Math. Helv. **48** (1973), 119–135, DOI 10.1007/BF02566116. MR328051

[168] Ken-iti Sato, *Lévy processes and infinitely divisible distributions*, Cambridge Studies in Advanced Mathematics, vol. 68, Cambridge University Press, Cambridge, 1999. Translated from the 1990 Japanese original; Revised by the author. MR1739520

[169] S. Schleißinger, *On the parametric representation of univalent functions on the polydisc*, Rocky Mountain J. Math. **48** (2018), no. 3, 981–1001, DOI 10.1216/RMJ-2018-48-3-981.

[170] S. Schleißinger, *Loewner's differential equation and spidernets*, Complex Anal. Oper. Theory **13** (2019), no. 8, 3899–3921, DOI 10.1007/s11785-019-00939-6.

[171] Konrad Schmüdgen, *The moment problem*, Graduate Texts in Mathematics, vol. 277, Springer, Cham, 2017. MR3729411

[172] O. Schramm, *Scaling limits of loop-erased random walks and uniform spanning trees*, Israel J. Math. **118** (2000), 221–288, DOI 10.1007/BF02803524.

[173] O. Schramm and D. B. Wilson, *SLE coordinate changes*, New York J. Math. **11** (2005), 659–669, `http://nyjm.albany.edu:8000/j/2005/11_659.html`.

[174] I. E. Segal, *Irreducible representations of operator algebras*, Bull. Amer. Math. Soc. **53** (1947), 73–88, DOI 10.1090/S0002-9904-1947-08742-5. MR20217

[175] J. H. Shapiro, *Composition Operators and Classical Function Theory*, Universitext: Tracts in Mathematics, Springer-Verlag, New York, 1993, DOI 10.1007/978-1-4612-0887-7.

[176] Dimitri Shlyakhtenko, *Random Gaussian band matrices and freeness with amalgamation*, Internat. Math. Res. Notices **20** (1996), 1013–1025, DOI 10.1155/S1073792896000633. MR1422374

[177] Michael Skeide, *Independence and product systems*, Recent developments in stochastic analysis and related topics, World Sci. Publ., Hackensack, NJ, 2004, pp. 420–438. MR2200526

[178] S. Sodin, *The classical moment problem*, 2019. Lecture notes.

[179] R. Speicher, *A new example of "independence" and "white noise"*, Probab. Theory Related Fields **84** (1990), no. 2, 141–159, DOI 10.1007/BF01197843.

[180] R. Speicher, *On universal products*, Free probability theory (Waterloo, ON, 1995), 1997, pp. 257–266, DOI 10.1090/fic/012/12.

[181] Roland Speicher and Carlos Vargas, *Free deterministic equivalents, rectangular random matrix models, and operator-valued free probability theory*, Random Matrices Theory Appl. **1** (2012), no. 2, 1150008, 26, DOI 10.1142/S2010326311500080. With an appendix by Tobias Mai. MR2934714

[182] R. Speicher and R. Woroudi, *Boolean convolution*, Free probability theory (Waterloo, ON, 1995), 1997, pp. 267–279, DOI 10.1090/fic/012/13.

[183] Richard S. Sutton and Andrew G. Barto, *Reinforcement learning: an introduction*, 2nd ed., Adaptive Computation and Machine Learning, MIT Press, Cambridge, MA, 2018. MR3889951

[184] Leon A. Takhtajan, *Quantum mechanics for mathematicians*, Graduate Studies in Mathematics, vol. 95, American Mathematical Society, Providence, RI, 2008, DOI 10.1090/gsm/095. MR2433906

[185] A. Tamir, *Applications of Markov Chains in Chemical Engineering*, Elsevier Science, 1998.

[186] T. Tao, *Topics in Random Matrix Theory*, Graduate Studies in Mathematics, vol. 132, American Mathematical Society, Providence, RI, 2012, DOI 10.1090/gsm/132.

[187] C. A. Tracy and H. Widom, *Distribution functions for largest eigenvalues and their applications*, Proceedings of the International Congress of Mathematicians, Vol. I (Beijing, 2002), 2002, pp. 587–596.

[188] Mircea Iulian Voda, *Loewner Theory in Several Complex Variables and Related Problems*, ProQuest LLC, Ann Arbor, MI, 2011. Thesis (Ph.D.)–University of Toronto (Canada). MR3022485

[189] Dan Voiculescu, *Addition of certain noncommuting random variables*, J. Funct. Anal. **66** (1986), no. 3, 323–346, DOI 10.1016/0022-1236(86)90062-5. MR839105

[190] Dan Voiculescu, *Multiplication of certain noncommuting random variables*, J. Operator Theory **18** (1987), no. 2, 223–235. MR915507

[191] D. Voiculescu, *Limit laws for random matrices and free products*, Invent. Math. **104** (1991), no. 1, 201–220, DOI 10.1007/BF01245072.

[192] Dan Voiculescu, *Lectures on free probability theory*, Lectures on probability theory and statistics (Saint-Flour, 1998), Lecture Notes in Math., vol. 1738, Springer, Berlin, 2000, pp. 279–349, DOI 10.1007/BFb0106703. MR1775641

[193] Jiun-Chau Wang, *The central limit theorem for monotone convolution with applications to free Lévy processes and infinite ergodic theory*, Indiana Univ. Math. J. **63** (2014), no. 2, 303–327, DOI 10.1512/iumj.2014.63.5229. MR3233209

[194] E. P. Wigner, *Characteristic vectors of bordered matrices with infinite dimensions*, Ann. of Math. (2) **62** (1955), 548–564, DOI 10.2307/1970079.

[195] E. P. Wigner, *On the distribution of the roots of certain symmetric matrices*, Ann. of Math. (2) **67** (1958), 325–327, DOI 10.2307/1970008.
[196] J. D. Williams, *Analytic function theory for operator-valued free probability*, J. Reine Angew. Math. **729** (2017), 119–149, DOI 10.1515/crelle-2014-0106.
[197] A. Zygmund, *A remark on characteristic functions*, Ann. Math. Statistics **18** (1947), 272–276, DOI 10.1214/aoms/1177730443.

Index

Selected Published Titles in This Series

294 **Sebastian Schleissinger,** Univalent Functions in Quantum Probability Theory, 2025
293 **Bennett Chow, Brett Kotschwar, and Ovidiu Munteanu,** Ricci Solitons in Dimensions 4 and Higher, 2025
292 **Francisco Urbano,** Morse Index of Minimal Submanifolds, 2025
291 **Tadashi Ochiai,** Iwasawa Theory and Its Perspective, Volume 3, 2025
288 **Anna Geyer and Dmitry E. Pelinovsky,** Stability of Nonlinear Waves in Hamiltonian Dynamical Systems, 2025
287 **Jacques Sauloy,** Basic Modern Theory of Linear Complex Analytic q-Difference Equations, 2024
286 **Zalman Balanov, Wieslaw Krawcewicz, Dmitrii Rachinskii, Hao-Pin Wu, and Jianshe Yu,** Degree Theory and Symmetric Equations Assisted by GAP System, 2025
285 **Niles Johnson and Donald Yau,** Bimonoidal Categories, E_n-Monoidal Categories, and Algebraic K-Theory, 2024
284 **Donald Yau,** Bimonoidal Categories, E_n-Monoidal Categories, and Algebraic K-Theory, 2024
283 **Donald Yau,** Bimonoidal Categories, E_n-Monoidal Categories, and Algebraic K-Theory, 2024
282 **Michael Kapovich and Pranab Sardar,** Trees of Hyperbolic Spaces, 2024
281 **Horst R. Thieme,** Discrete-Time Dynamics of Structured Populations and Homogeneous Order-Preserving Operators, 2024
280 **Tadashi Ochiai,** Iwasawa Theory and Its Perspective, Volume 2, 2024
279 **Ellen Elizabeth Eischen, Wee Teck Gan, Aaron Pollack, and Zhiwei Yun,** Automorphic Forms Beyond GL_2, 2024
278 **Delaram Kahrobaei, Ramón Flores, Marialaura Noce, Maggie E. Habeeb, and Christopher Battarbee,** Applications of Group Theory in Cryptography, 2024
277 **George Metcalfe, Francesco Paoli, and Constantine Tsinakis,** Residuated Structures in Algebra and Logic, 2023
276 **Balázs Bárány, Károly Simon, and Boris Solomyak,** Self-similar and Self-affine Sets and Measures, 2023
275 **Alekos Vidras and Alain Yger,** Multidimensional Residue Theory and Applications, 2023
274 **Willi Freeden and M. Zuhair Nashed,** Recovery Methodologies: Regularization and Sampling, 2023
273 **Gennadiy Feldman,** Characterization of Probability Distributions on Locally Compact Abelian Groups, 2023
272 **Tadashi Ochiai,** Iwasawa Theory and Its Perspective, Volume 1, 2023
271 **Hideto Asashiba,** Categories and Representation Theory, 2022
270 **Leslie Hogben, Jephian C.-H. Lin, and Bryan L. Shader,** Inverse Problems and Zero Forcing for Graphs, 2022
269 **Ralph S. Freese, Ralph N. McKenzie, George F. McNulty, and Walter F. Taylor,** Algebras, Lattices, Varieties, 2022
268 **Ralph S. Freese, Ralph N. McKenzie, George F. McNulty, and Walter F. Taylor,** Algebras, Lattices, Varieties, 2022
267 **Dragana S. Cvetković Ilić,** Completion Problems on Operator Matrices, 2022
266 **Kate Juschenko,** Amenability of Discrete Groups by Examples, 2022
265 **Nabil H. Mustafa,** Sampling in Combinatorial and Geometric Set Systems, 2022
264 **J. Scott Carter and Seiichi Kamada,** Diagrammatic Algebra, 2021

For a complete list of titles in this series, visit the
AMS Bookstore at **www.ams.org/bookstore/survseries/**.

Kolibri 360 GmbH
Ferdinand-Jühlke-Straße 7
99095 Erfurt, Deutschland
produktsicherheit@kolibri360.de